Power and Distribution Transformers

First edition published 2021
by CRC Press
6000 Broken Sound Parkway NW, Suite 300, Boca Raton, FL 33487-2742

and by CRC Press
2 Park Square, Milton Park, Abingdon, Oxon, OX14 4RN

CRC Press is an imprint of Taylor & Francis Group, LLC

ISBN: 978-0-367-53593-3 (hbk)
ISBN: 978-0-367-54297-9 (pbk)
ISBN: 978-1-003-08857-8 (ebk)

Typeset in Times
by codeMantra

Power and Distribution Transformers
Practical Design Guide

K.R.M. Nair

CRC Press
Taylor & Francis Group
Boca Raton London New York

CRC Press is an imprint of the
Taylor & Francis Group, an **informa** business

Contents

Preface

This book is written based on the 50+ years of experience of the author in the transformer industry. The initial drafts of the manuscript were based on the notes prepared by the author during training fresh engineers joining the industry. Over a period of time, practising engineers also found these useful and suggested to add more topics to the existing notes and publish it as a book for the benefit of the engineers of the industry, utility, consultants etc.

This book is divided into 30 chapters and 6 annexures covering practical design procedures, including formulae and parameters pertaining to distribution and power transformers of different types. Although the examples and parameters used in this book are focused on transformers up to 36 kV, the design procedures and concepts can be extended for higher voltage classes of up to 132 kV for oil-filled transformers. Chapter 2 gives a brief history of the transformer technology development and reviews the emerging trends. Step-by-step design procedures including practical calculation methods of 3-phase oil-filled distribution and power transformers are given in Chapters 3–12. Design inputs and a basic flow chart of designing a 3-phase oil-filled transformer are covered in Chapter 3.

Chapters 4–6 explain the calculation of core area, dimensions of core laminations, the designing of the low-voltage and high-voltage windings, and calculation of I^2R loss and load loss. The various methods of calculating eddy current and stray losses are also covered.

Chapter 7 explains the calculation of reactances of various winding configurations such as two windings, extended winding, zig-zag connection and auto connection. The basis of using the finite element method for calculating the reactance of a transformer is also briefly explained in this chapter.

Methods of calculating core weight, core loss and excitation current are explained in Chapter 8. The watts per kilogram of typical commercially available Cold Rolled Grain Oriented (CRGO) steel, HiB steel and domain-refined materials are given here. For quick application, these characteristics are converted to convenient formulas by curve fitting and are included in this chapter.

Lightning and switching surge response of transformer winding are covered in Chapter 9. The selection guide of surge arresters is also explained.

Calculation of inrush current peak value and the methods of reducing the peak values are covered in Chapter 10.

Procedures to calculate the dimensions of the core and coil assembly, clamping beams, dimensions of the tank and conservator and oil quantity are given in Chapter 11.

Winding gradient calculation, oil temperature rise calculation and estimation of cooling surface area required are detailed in Chapter 12.

A major proportion of the oil-filled distribution transformers are sealed type with gas/air cushion inside. Chapter 13 covers the pressure calculation of sealed-type transformers and completely filled corrugated-type transformers. This chapter also covers the calculation of the strength of a rectangular tank when it is pressure-tested, including the design of stiffness of the tank. The design of spacing of the cover bolt, selection of gasket and recommended tightening torques are also given.

The transformer shall withstand the thermal and dynamic stresses from short circuit. Chapter 14 explains the calculation procedures of these stresses on the windings, clamping structures and the capability to withstand these stresses.

Design of rectifier transformers required for the drives and industrial application is given in Chapter 15. The designs of extended delta, extended star and polygon connections including sample calculations are covered in this chapter.

With the increase in urbanization, there is an increasing requirement for dry-type cast resin transformers for power supply to skyscrapers, malls, hospitals etc. The design aspects of cast resin transformers including cooling and ventilation designs are covered in Chapter 16.

Earthing transformers are intended to provide a neutral point for an ungrounded system to provide a return path for the fault current. The design procedures including typical examples are given in Chapter 17.

Amorphous alloy metals are used instead of conventional CRGO steel for the transformer core to reduce no-load loss. The practical design procedures are explained in Chapter 18.

The designs of air core and gapped core reactors including sample calculations are provided in Chapter 19.

Scott-connected transformers are useful for industries where single-phase supply is required from a 3-phase balanced supply line. Chapter 20 covers its design.

Chapter 21 covers the design principles of autotransformers with a sample design of a 50 MVA auto transformer. Transformers for special applications including furnace transformers, dual ratio, multi-winding, and traction transformer are covered in Chapter 22. The design requirements of transformers for use in explosive atmospheres (ATEX / IECEX certified) are also explained.

There is an increasing need for renewable energy generation to control global warming, and the design required for the transformers for renewable energy is detailed in Chapter 23.

The health of a transformer is checked by online and offline diagnostic and monitoring methods. A review of the various techniques for diagnosing an oil-filled transformer is covered in Chapter 24.

The life time CO_2 emission of a transformer is a crucial factor for deciding the eco-friendliness of the product and is gaining universal importance. The calculation of the carbon footprint of a typical transformer is explained with sample calculations in Chapter 25.

Transformers required for operation in such areas as those with seismic activity, nuclear power station require design features to withstand the vertical and horizontal seismic motion. The seismic response calculation of transformers is explained in Chapter 26.

The emerging smart grid requires transformers suitable for automatic voltage drop compensation, harmonic isolation, and smart communications. Chapter 27 covers the design features of solid-state transformers intended for smart grid applications.

Transformer design is a complex process where the designer tries to optimize the performance with the available resources ensuring compliance with the specified and mandated constraints and limitations. Chapter 28 covers design optimization of oil-filled transformers.

The tanks, enclosures and parts of almost all oil-filled transformers are made using steel and therefore require corrosion protection. Typical surface treatment and corrosion protection coatings suitable for transformers are explained in Chapter 29.

Chapter 30 gives the calculation of commonly required technical parameters of transformers and related systems. The subjects covered include overloading capacity, altitude correction factor, and effect of solar radiation on increasing temperature.

This book contains six annexures, including design calculations of some distribution and power transformers. A brief introduction to the application of finite element methods applicable for transformer design, procedure for calculating the total owning cost (TOC) of transformers, list of relevant IEC and ANSI/IEEE standards are included.

Acknowledgments

First of all, I thank God for giving me the divine help and motivation to complete this book.

I would like to take this opportunity to express my sincere appreciation to all my colleagues at Federal Transformers Co. LLC, Abu Dhabi, UAE; who contributed directly or indirectly to this project. A very special thanks to Ms. Remya Krishnan; Transformer Design Engineer at Federal Transformers, for her untiring and efficient assistance in editing, compiling and also for technically contributing to the successful completion of this book. I am highly indebted to the suggestions and comments from the reviewer.

There are no words to describe how grateful I am to my wife, Prasanna; daughter, Meera; and granddaughter, Reema; for the care, love and immense support.

I gratefully acknowledge the support given by Codemantra for manuscript editing.

K. R. M. Nair

Author

K. R. M. Nair is an accomplished transformer technology expert with more than 50 years of industry experience. He is presently working as an independent consultant.

Nair was the chief operating officer of Federal Transformers Co. LLC, UAE for more than 20 years. Since the company's modest beginning in 1999, he is responsible for leading the company in to fast track development and in transforming it to a multi-product, multi-location company. Prior to joining Federal Transformers, he was the head of Technology with Emirates Transformers, UAE. Previously, Nair was the technical director of Vijai Electricals, Hyderabad, India (now Toshiba T&D) for 10 years, where he developed cast resin transformers and amorphous core transformers for the first time in India. A notable achievement during this tenure was the introduction of a new annealing process for amorphous ribbon for which the company has a patent.

After a short stint as a lecturer in the College of Engineering, Trivandrum, India, Nair started his career in 1968 with M/s. Transformers and Electricals (TELK), Kerala, India. He worked for 13 years and received long-term technology exposure of M/s. Hitachi, Japan, including intensive training in their Japanese factory. His professional contributions during this period include dedicated software development for short circuit force calculation in the 1970s and design of 220 kV class cascade voltage transformers (also first time in India).

Nair's academic qualification include PhD in Electrical Engineering, MBA and an LLB. He is a life member of IEEE (USA), life member of Institution of Engineers (India) and life member of Indian Council of Arbitration.

1 Transformer Design

1.1 INTRODUCTION

Equipment employing electrical energy has been and would continue to be used increasingly for the improvement of human convenience, comfort and productivity. Transformer is a vital link in the power system, which brings the electrical energy to the user for the consumption. Several millions of transformers are in operation around the world supplying power to a multitude of applications. The power capacity of these units ranges from very large ultrahigh-voltage units weighing several hundred tons to tiny microtransformers in electronic circuits, which are produced by lithographic process and weighing milligrams.

The demand for transformers increases continuously consequent to urbanization, population growth, industrialization of emerging economies, replacement of the ageing units from the existing system, etc.

Though the transformer is a mature product existing for more than 130 years, its basic structure and principle remain unchanged. By using the improved materials and technology for design, processing, production, operation and maintenance, the efficiency and sophistication of the product are continuously improving.

1.2 DEFINITION OF TRANSFORMER

International Electrotechnical Commission (IEC) defines the transformer as:

> A static piece of apparatus with two or more windings which by electromagnetic induction transform a system of alternating voltage and current into another system of voltage and current, usually of different values and at the same frequency for the purpose of transmitting electric power.

The Institute of Electrical and Electronics Engineers (IEEE) defines the transformer as

> A static electric device consisting of two or more windings with or without a magnetic core for producing mutual coupling between electric circuits.

The transformer essentially consists of a conducting system to carry the current, an insulation system to isolate the turns and windings between them and with others, a magnetic system to boost up the flux and the associated components and parts to enable efficient, reliable and optimum power transfer. The conducting system and the magnetic system produce heat when the power transfer process takes place, and the heat generated needs to be transferred effectively so that the systems, materials and components, especially the insulation, function without loss of property for the entire lifetime of the equipment.

1.3 DESIGN OBJECTIVES

The transformer design is expected to meet the ever-evolving demands from the different stakeholders including the customers, regulating authorities, society and industry. The designer of the product needs to be equipped to select the optimum from among different and, at times, conflicting alternatives. These objectives include, but are not limited to, the following:

- Design the transformer by selecting the most suitable materials to reduce the cost of the product
- Optimize the lifetime owning cost of the product

- Design environmental-friendly and safe product
- Develop product suitable to meet the evolving requirement of the grid and customer
- Adapt new technologies and offer continuous improvements

The engineers working in the transformer industry need to be conversant with the design principles and fundamental concepts to equip themselves to produce efficient and reliable transformers and strive to develop new products and absorb new technologies. This book gives in the initial chapters structured, step-by-step design procedures of a conventional transformer, which is the essential foundation to work further for optimization, product improvement, automation, etc. The contents included guidelines, formula and procedures for designing different types of transformers with special reference to three-phase oil-filled power and distribution transformers up to 132 kV class. The designs of the following types of transformers are covered with worked-out examples in some cases.

- Amorphous core transformers
- Dry-type cast-resin transformers
- Current limiting reactors
- Earthing transformers
- Solid-state transformers
- Industrial transformers and special transformers including Scott connected transformers, rectifier transformer and multiwinding transformer.

The design parameters given in this book are based on the manufacturing practices expected to result in a good level of factor of safety to ensure continuous service for the designed lifetime. However, the parameters can be modified to suit the manufacturing processes available in the particular production set-up after establishing predictable accuracy and repeatability.

1.4   BASIC THEORY OF TRANSFORMER

The basic theories of the transformer are available in all textbooks, and, therefore, those are not repeated in this book. However, some concepts that are useful for the design engineer for practising and optimizing the design are given in this section.

1.4.1   ELECTROMAGNETIC TERMS AND CONCEPTS

a. The pressure of a magnetic field "H" in free space of area "A" produces a magnetic flux of magnitude ϕ

$$\phi = \mu_0 \ HA \tag{1.1}$$

where
ϕ = flux (webers)
A = area (m^2)
H = magnetic field (A/m)

$$\mu_0 = 4\pi \times 10^{-7}$$

(permeability of free space)
b. The magnetic flux density is a commonly used term, and it is defined by

$$B = \frac{\phi}{A} \tag{1.2}$$

where B = magnetic flux density (Tesla)

c. The above two relationships give the formula for B in terms of H as

$$B = \mu_0 H \tag{1.3}$$

d. In order to represent the response of a magnetic material, the term relative permeability is used, which is defined by

$$\mu = \frac{B}{H} \tag{1.4}$$

In practice, the term relative permeability is used to distinguish the magnetic properties of the material by defining

$$\mu_r = \frac{\mu}{\mu_0} \tag{1.5}$$

e. The magnetic resistance is called "reluctance" which is analogous to resistance in electric circuit.

$$R = \frac{\text{MMF}}{\phi} = \frac{\text{NI}}{\phi} \tag{1.6}$$

where
MMF = magnetomotive force
NI = ampere-turns
The reluctance of a uniform magnetic loop can be calculated by

$$R = \frac{l_m}{\mu_0 \mu_r A} \tag{1.7}$$

where
l_m = mean length of magnetic path (m)
A = area of cross section of the magnetic loop (m^2)

1.4.2 Charges, Electric Field and Magnetic Field

Magnetism was one of the oldest physical phenomena observed and investigated. The first scientific study of magnetism was made by William Gilbert who studied the earth's magnetic field and its effect on a magnetic needle. In the year 1820, Oersted found that an electric current produces a magnetic field that encircles a wire; from this, Ampere made the hypothesis that electric current is the source of every magnetic field. He made one of the fundamental laws in 1826, which is Ampere's circuital law.

$$\Phi_c \ H. \ dl = \text{NI} \tag{1.8}$$

The theory states that the magnetic field strength in a closed magnetic circuit is generated by the number of conductors, N, in the circuit carrying current I. In 1831, Faraday discovered another fundamental law of magnetism, which states that if the magnetic flux ϕ linking an electric circuit changes, it induces an electromotive force (emf) V proportional to the change in the flux ϕ

$$\text{i.e. } V = -\frac{d\phi}{dt} \tag{1.9}$$

The minus sign is as per Lenz's law, which states that the induced voltage is always in a direction opposing the flux change.

Maxwell consolidated the equations relating to electricity and magnetism and published 20 equations in 1873. Later on, these equations were further reduced to four, which are now acting as fundamentals to the analysis of electricity and magnetism.

The law relating to the attraction or repulsion of charges is Coulomb's law, which states that "Like charges repel and opposite charges attract with a force proportional to the charges and inversely proportional to the square of the distance between them."

$$F = K \frac{Q_1 \ Q_2}{r^2} \tag{1.10}$$

where
F = force (N)
Q_1 = charges (C)
r = distance between charges (m)

$$k = \frac{1}{4\pi\varepsilon_0} \cong 8.98755 \times 10^9 \ \text{Nm}^2/\text{C}^2$$

ε_o = electric permittivity of vacuum

1.4.3 MAXWELL'S EQUATIONS

The concepts of the equations are given here:

a. Gauss's Law
   The number of electric field lines entering the surface is equal to the electric field lines leaving the surface

$$\phi_s \ \bar{F}. \ dA = \delta_v \ \text{div} \ \bar{F} \ dv \tag{1.11}$$

$$\text{i.e.} \ \nabla. \ E = 0$$

It shows that an electric field can be expressed as the gradient of a potential function.

$$\bar{E} = - \bar{\nabla} \ V \tag{1.12}$$

E is given in V/m
b. No monopoles for a magnet

As per this law, the magnetic field lines must form a continuous loop

$$\nabla. \ B = 0 \tag{1.13}$$

It states that the divergence of the magnetic field is zero.
   B is the magnetic flux density (Tesla)
c. Faraday's Law

Faraday's law deals with the induced electric current. In a loop of wire, a current can flow by moving a magnet inside the loop or by moving the loop.

$$V = -N \frac{d\phi}{dt} \tag{1.14}$$

$$\nabla \times E = -\frac{\delta B}{\delta t} \tag{1.15}$$

d. Ampere's Law

Ampere's law states that a changing electric field produces a changing magnetic field. This can further result in a changing electric field and continue as oscillations. The oscillations create waves that travel at the speed of light in vacuum.

$$\nabla \times B = \mu_0 \varepsilon_0 \frac{\delta \epsilon}{\delta t} \tag{1.16}$$

$$\mu_0 = 4\pi \times 10^{-7} \tag{1.17}$$

$$\epsilon_0 = 8.554187582 \times 10^{-12} \tag{1.18}$$

1.5   POWER TRANSFER CAPACITY OF TRANSFORMER

The power transfer capacity for a specific transformer per unit area can be expressed by the Poynting vector, which is the cross product of the electric field with the magnetic field, i.e. $E \times H$, which is expressed as $\bar{S} = \frac{1}{\mu_0} \left(\bar{E} \times \bar{H} \right)$. This concept simulates the transformer to a transmission line.

Alternatively, the power transfer capacity can be expressed in terms of the design parameters of flux density, current density and the quantities of magnetic and conducting materials. The relationships are explained in the following paragraph:

$$Q = VA \tag{1.19}$$

where
   Q = capacity (VA)
   V = voltage (V)
   A = current (I)
   By expressing V and A in terms of the related parameters,

$$V = 4.44\ f\ B_m\ A_{\text{core}}\ T \tag{1.20}$$

where
   B_m = working flux density (Tesla)
   A_{core} = area of cross section of core (m²)
   T = number of turns
   Also, $A = \delta . A_{\text{cond}}$
   δ = current density (A/m²)
   A_{cond} = area of cross section of conductor (m²)

$$Q = 4.44\ f\ B_m\ A_{\text{core}}\ T\ \delta A_{\text{cond}} \tag{1.21}$$

This shows that the power transfer capacity for a specific quantity of material can be increased by increasing the flux density, increasing the current density and increasing the frequency. The possible solutions include the usage of magnetic material which can work at higher flux density, adopting conducting materials capable of carrying high currents without overheating and increasing the frequency of operation of the magnetic circuit. The design of the product with increased power transfer capacity necessarily shall not in any case compromise the optimum lifetime cost, improved safety and reliability.

2 Brief History of Transformers and the Emerging Trends

2.1 BRIEF HISTORY OF THE TRANSFORMER

The theory of electromagnetic induction was discovered in the 1830s. The first transformer was made by Ottó Bláthy, Miksa Déri and Károly Zipernowsky at the Ganz factory at Hungary in 1878. The first reliable transformer was built by William Stanley working in Westinghouse, USA, in 1885/1886. The first three-phase transformer was designed and made by Mikhail Dobrovolsky in 1891. GE (General Electric, USA) made the first oil-filled transformer in 1892. By 1900, 50 kV class transformers up to 2000 kVA were installed in several substations. The progress of the further development was very fast, and 400 kV class transformers were built by the 1950s. Today, 1200 kV class transformers are built and the maximum rating is limited by the constraints of transportation only. Several transformers of ratings above 1000 MVA (1 million kVA) are in operation today.

2.2 DEVELOPMENT OF MATERIALS FOR TRANSFORMER PRODUCTION

There was continuous improvement in the quality and performance characteristics of the active materials for producing transformer including conducting material (copper and aluminium), magnetic material (electrical steel) and insulation material (transformer oil, synthetic fluids, silicon liquids, vegetable-based oils, etc.). Some of the significant developments are given in this section.

2.2.1 CONDUCTING MATERIAL

Copper and aluminium are the major conducting materials used for transformer winding. Because of the higher mechanical strength and conductivity, copper is extensively used. However, consideration of cost reduction and higher availability favour the use of aluminium for transformers of smaller ratings, especially distribution transformers.

Various types of winding conductors such as round enamelled or paper covered wire, rectangular enamelled/paper covered strip, aluminium/copper foil, continuously transposed conductor (CTC), etc., are used. Conductor/winding insulation using higher thermal classes (Class F, Class H, Class C, etc.) with mylar, glass fibre, nomex and similar materials are employed for special applications of dry-type transformers, railways, industries and so on.

2.2.2 COOLING MEDIUM

Air was the cooling medium for the first generation of transformers. When the sizes became bigger, mineral oil was used for better heat dissipation. The need to avoid flammability/explosion has brought in synthetic liquids like Clophen (polychlorinated biphenyls) in the 1930. These were banned in 1979 due to severe environmental hazards.

Less flammable liquids like silicone liquid (polydimethylsiloxane), synthetic ester and vegetable-based oils were introduced later. SF6 gas-filled transformers were produced by some manufacturers in the last few decades. The majority of transformers are produced using mineral oil today.

2.2.3 Magnetic Material (Electrical Steel)

The purpose of the core is to provide a low reluctance path for the flux linking the primary and secondary of the transformer.

The flux flowing through the core is alternating, and it produces energy loss. Thus, the ideal core material shall have very high permeability (i.e. very low reluctance) and very low losses. Continuous development efforts are carried out to reduce the losses of the core steel and improve the performance. The various stages of the progress in the past are given in Table 2.1.

2.2.4 New and Emerging Technologies

The modern power system demands have triggered several developments and innovations. The present challenges arise from

- Continuous increase in ratings and voltages
- Complexities of the rapid change due to distributed power generation demand for DC power, integration of storage, renewable energy, electric vehicle charging, etc.
- Increased safety and reliability requirements
- Development of smart grid
- Continuous monitoring and accurate diagnosis of the performance
- Environmental demands like reduction of CO_2 emission
- Optimization of resources including reduction of total owning cost

The present status of the technologies and some of the emerging developments are discussed in the following sections.

TABLE 2.1
Improvement of Specific Loss of Core Materials

Time Period	Core Material	Specific Loss w/kg at 50 HZ	Remarks
1880s	High-grade wrought iron	-	Early stage of transformer
1900	Steel with addition of small amount of silicon	-	-
1906	Hot-rolled silicon steel	7 w/kg at 1.5 T	0.35 mm thick steel sheets were used
1920	Anisotropic property of silicon steel crystals was established	-	-
1934	Concept of orienting grain with magnetically favourable direction was invented	-	Goss texture
1939	First commercial Cold Rolled Grain Oriented (CRGO) silicon steel was produced	1.5 w/kg at 1.5 T	0.32 mm thick lamination
1968	High-permeability (Hi-B) CRGO was commercially produced	1.20 w/kg at 1.7 T	Nippon steel, Japan produced the first Hi-B core steel
1980	Domain-refined CRGO steel	0.9 w/kg at 1.7 T	Thickness of lamination 0.23 mm
1978	Amorphous steel for transformer application was developed	0.15 w/kg at 1.3 T	Thickness of metal 0.025 mm
1990	Laser scribed/domain-refined CRGO steel	0.85 w/kg at 1.7 T	Thickness 0.23 mm
2012	Improved laser-scribed/domain-refined CRGO steel	0.8 w/kg at 1.7 T	Thickness 0.23 mm
2016	Low-loss domain-refined CRGO steel	0.75 w/kg at 1.7 T	Thickness 0.23 mm
2018	Low-loss domain-refined CRGO steel	0.70 w/kg at 1.7 T	Thickness 0.23 mm

2.2.5 NEW TECHNOLOGIES IN COMMERCIAL USE

At present, there are transformers in commercial use wherein new concepts and materials are employed.

2.2.5.1 Amorphous Core Transformers

The conventional core material is CRGO steel which has a crystalline structure, whereas amorphous steel is non-crystalline. The amorphous steel is manufactured by a rapid quenching process. This material has very low specific losses and therefore reduces the no-load loss of the transformer. A comparison of the major parameters of CRGO steel and amorphous metal for application to transformer is given in Table 2.2.

Amorphous core transformers have about 25%–30% of the core losses of a CRGO core transformer when designed by keeping all parameters like load loss, impedance and temperature rise unchanged. The reduction of no-load loss by 70% gives substantial savings of energy losses, total owning cost and greenhouse emission. A comparison of no-load loss of typical ratings of oil-filled transformers is given in Table 2.3.

2.2.5.2 Symmetrical Wound Core (3D Core) Transformer

Conventional design of three-phase transformer is by using planar core, i.e. stacked or wound core where the limbs and yokes are coplanar. These designs have an inherent magnetic asymmetry because the outer phases and the centre phases show different magnetic reluctance and the flux pattern is asymmetric.

TABLE 2.2
Comparison of CRGO Steel and Amorphous Metal

Sl. No.	Properties	CRGO Steel	Amorphous Metal
1	Density (g/cm³)	7.65	7.15
2	Resistivity (μΩ)	45	130
3	Thickness (mm)	0.3/0.27/0.23	0.025
4	Space factor	0.97	~0.86
5	Saturation flux density (Tesla)	2.03	1.56
6	Typical core loss at working flux density at 50 Hz (w/kg)	0.75–1.22	0.20

TABLE 2.3
No-Load Losses of CRGO Transformer and Amorphous Core Transformer

kVA	Number of Phases	No-Load Loss (Watts) (50 Hz Design)	
		CRGO Core Transformer	Amorphous Core Transformer
10	1	40	10
15	1	60	15
25	3	100	25
63	3	180	45
100	3	260	60
1000	3	1700	380

The 3D core construction uses three identical wound core rings, which are made by continuously winding diagonally slit core material to a mandrel. The three wound rings are arranged in an equilateral triangular form at the base with two adjacent rings forming the core limb. This construction gives complete magnetic symmetry and reduces the core losses by 15%–20% when compared to the planar core design and also offers material saving.

2.2.5.3   Biodegradable Oil-Filled Transformer

The demand for environmental protection has prompted the development of biodegradable liquids in place of mineral oil. Vegetable-based oils suitable for use in transformer having high flash point are available commercially. This oil has higher viscosity when compared to mineral oil, and 10%–12% extra cooling area is required to get cooling performance comparable to mineral oil.

2.2.5.4   Gas-Insulated Transformers

Inert gases can be used for filling the transformer instead of mineral oil or fluids. SF6 (sulphur hexafluoride) gas is the most commonly used for filling such transformers till now. SF6 is an inert gas, and its dielectric strength increases with pressure.

For distribution transformers, SF6 gas is filled at a positive pressure only (typical pressure is +0.14 to +.40 MPa gauge pressure).

Comparative features of gas-insulated transformer with oil-immersed transformers are given in Table 2.4.

SF6gas, if released to the atmosphere, has 23,900 times the global warming potential (GWP) when compared to CO_2. Consequently, gases to replace SF6 are now developed. C5 perfluroketone ($C_5F_{10}O$) is a good substitute which has a dielectric strength of 1.79 times SF6 gas at the same pressure. The GWP of $C_5F_{10}O$ is 1.

Other similar gases are C_4PFN (iso-C_4 fluronitrate).

Gas-insulated transformers are compatible with gas-insulated substations.

2.2.6   APPLICATION OF EMERGING TECHNOLOGIES TO TRANSFORMER DESIGN AND MANUFACTURE

This section discusses the recent developments and use of emerging technologies, including materials used in design and production of transformer.

2.2.6.1   Transformers Using High-Temperature Superconductors

Superconductivity was first observed at a temperature of 4 K by Heike Onnes in 1911. This is low-temperature superconductivity as the phenomenon is at very low temperature. In the 1980s, superconducting metal alloys were developed which become superconductors at 77K achievable by

---

TABLE 2.4

Comparative Features of Oil-Immersed Transformers and Gas-Insulated Transformers

S. No.	Feature	Oil-Immersed Transformer	Gas-Insulated Transformer
1	Applicable standard/insulating medium	IEC 60076-1 Insulating oil/fluid	IEC 60076-15 SF6
2	Solid insulation	Paper, press board	PET film, nomex
3	Leakage of cooling/insulating medium	Visual detection is easy	Detection by pressure gauge
4	Flammability	High with mineral oil Less with other fluids	Non-flammable

liquid nitrogen. The first generation of these materials was bismuth based, and these were called the first-generation superconductors. The next (second) generation of superconductors developed by 2005 such as yttrium barium copper oxide (YBCO) superconducting wires is used to manufacture transformers. This material has critical temperature of above 90 K now.

Several transformer manufacturers have manufactured prototype units using YBCO wires. Three types of construction can be used for superconducting transformers, and the conceptual arrangements are given in Table 2.5 (Figures 2.1–2.3).

TABLE 2.5

Different Types of Superconducting Transformer

Warm Iron Core	Cold Iron Core	Conduction Cooled
• Iron core is at room temperature • Complete winding & insulation is inside a cryostat	• Total iron core and winding assembly are inside a cryostat	• Iron core at room temperature • Windings are cooled by a cryocondenser arrangement

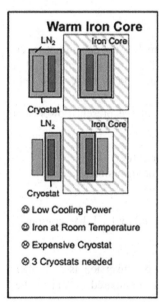

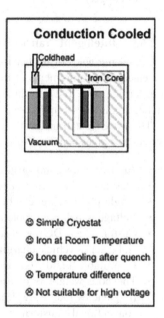

FIGURE 2.1 Warm iron core.	FIGURE 2.2 Cold iron core.	FIGURE 2.3 Conduction cooled.

Advantages	Disadvantages	Advantages	Disadvantages	Advantages	Disadvantages
• Low colling power • Iron at Room Temperature (Core)	• 3 Cryostats Required for 3 phase Transformer • Expensive cryostat	• Simple Cryostat • Simple cooling Interface	• High Cooling power	• Simple Cryostat • Iron at Room Temperature	• Long Recooling • Temperature Difference • Suitable for Low Voltage only

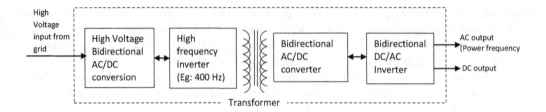

FIGURE 2.4 Conceptual design of "intelligent transformer".

Most of the prototypes developed now use the warm iron core-type design.
Some of the advantages of superconducting transformers are as follows:

- High overload capacity with low global warming potential (CO_2 emission)
- Withstands overloads without loss of life
- Very low load loss (this is due to the very high current carrying capacity of 10–50 times of copper)

2.2.6.2 Intelligent Transformer for Smart Grid

The conventional transformer is a static link in the power system which is not capable of meeting the emerging requirements of the power system and the consumer. The concept of an "intelligent transformer" was first introduced by Electric Power Research Institute (EPRI), USA. The conceptual design is shown in Figure 2.4.

The intelligent transformer offers solution to the shortcomings of the present power supply system and gives benefits to the customer including the following:

- Supply voltage, and variation is eliminated in the solid-state circuitry.
- Presence of harmonics, and spikes are removed in the conversion inversion process.
- Single-phase power can be taken.
- Voltage unbalance problem is eliminated.
- Customer gets DC power so that AC/DC converters are not required for loads like computer and electric vehicle charging.
- Flammable liquids like oil are not used as the total system in dry type.
- The core weight is considerably reduced because the voltage conversion is at higher frequencies like a few kilohertz which reduce the core weight compared to that of the conventional transformer.
- Reactive support to the power system can be given.

2.3 REPLACEMENT OF COPPER/ALUMINIUM BY CARBON NANOTUBE

Carbon nanotubes (CNTs) are isotopes of carbon, which have a cylindrical nanostructure. CNTs show very high conductivity, high mechanical strength and very low thermal coefficient of resistance. Rice University, USA, has developed a process for producing high conductivity CNT suitable for winding wires, cables, etc. Further, Teijin Aramid (Japanese – Dutch Company) in collaboration with Rice University has developed a spinning technology to produce multifibre yarn for production of motors, cables, transformers, etc.

A comparison of the performance parameters of the CNT yarn with copper and aluminium is given in Table 2.6.

TABLE 2.6

Comparison of Copper, Aluminium and CNT

Material	Density (kg/m³)	Electrical conductivity at 750°C (S/m/m)	Electrical resistivity at 200°C (Ω/m)	Temperature coefficient of resistivity (per°C)	Thermal conductivity at 273k (W/mK)
Copper	8960	50×10^8	1.68×10^{-8}	0.00386	400
Aluminium	2760	30×10^8	2.65×10^{-8}	0.00429	235
CNT	1500	100×10^8	1×10^{-8}	−0.001 to 0.002	635

The above comparison is based on the present day development of CNT. However, the conductivity of the material is expected to increase several times in the near future as the theoretical level of conductivity is several 100 times more than copper. Also, the temperature coefficient of resistivity is very low; and thus, the transformer can work at a higher temperature without increase of resistivity and losses. The density of the material is nearly one-sixth of copper, and therefore, the overall weight of the transformer is reduced.

2.4 USE OF ARTIFICIAL INTELLIGENCE (AI) FOR TRANSFORMER DESIGN AND DIAGNOSTICS

Several optimization programmes are used by the transformer industry today, which aim to reduce the overall cost of the transformer. The objective function is formulated based on the given constraints, and then, it is optimized to get the overall initial cost or overall owning cost, using established mathematical techniques. These include linear programming, non-linear programming and geometric programming.

AI-based techniques are now being used, and some of the techniques reported are as below:

- Stochastic methods
- Artificial neural networks (ANNs)
- Genetic algorithm (GA)
- Swarm intelligence
- Decision tree, etc.

In order to monitor the performance of the transformer and to identify incipient faults before culminating into a breakdown, several monitoring and diagnostic methods are employed now. More and more expert systems and application of AI are now being used to arrive at optimum conclusions of the complex monitoring and test data.

2.5 USE OF ADDITIVE MANUFACTURING (3D PRINTING) FOR TRANSFORMER PRODUCTION

The rapid growth of 3D printing technology offers scope for its applications to the production of transformer components and parts of the transformer. Several complex parts and components can be produced with high precision without the loss of materials.

The most useful developments expected in this direction include the following:

- Magnetic circuit (core) with perfect symmetry and uniform reluctance path

- Winding construction by additive manufacture to get optimum impulse distribution, mechanical strength and heat dissipation
- Tailor-made components and interfaces to adapt to the site conditions and performance requirements

The magnetic, conduction, insulation and heat dissipation requirements of the complex system can be analysed using the finite element method to get the optimum elemental size for layer-by-layer deposition of the material mixture using additive manufacturing system.

3 Design Procedures

3.1 DESIGN INPUT DATA

Preparation of the design input data and document is the first step of the design. The data shall include, but not limited to, the following:

- Customer specifications including drawings
- Amendments (if any) to the specifications
- Offer given to the customer including technical datasheets, drawings and correspondences
- Purchase order from customer
- Applicable standards
- Mandatory regulations
- Existing designs and test results of similar transformers, if any
- Salient features of the customer specifications including
  - kVA
  - Number of phases
  - Frequency
  - Type of cooling
  - Application – indoor, outdoor, excitation, furnace, mining, traction, etc.

• Service conditions	-	• Maximum ambient air temperature, • Daily average ambient temperature, • Yearly weighted average ambient temperature, • Minimum ambient temperature, • Altitude: Make special note, if altitude is in excess of 1000 m from mean sea level
• Seismic level	-	• If specified.
• Unusual service	-	• Frequency variation in excess allowed by the standard • Voltage variation in excess allowed by especially short-time overvoltage requirement for station auxiliary, unit auxiliary and generator transformer • Overfluxing requirement, if any • Special overload conditions including emergency overloading

---

- Rated voltage of HV at no load
- Connection of HV
- Rated voltage of LV at no load
- Connection of LV
- Vector group reference number
- Tappings and voltage variation
- Whether constant flux voltage variation or variable flux voltage variation or mixed variation
- Whether taps are required on HV or LV
- Tap range
- Tap steps

- Whether on-load taps, off-load taps or off-circuit taps (to get specific requirement of customer, if off-load taps are specified).
- Guaranteed temperature rise of winding: °C.
- Guaranteed oil temperature rise: °C.
- Percent impedance and allowable tolerance.
- Whether no-load losses and load losses are capitalized or not? If capitalization is used, what are the rates of capitalization for no-load loss and load loss?
- Whether maximum values of losses are specified.

Whether losses are subject to tolerance and if so, tolerance limits. Special care shall be taken to this aspect while quoting losses as per applicable standards.

- Allowable flux density – check with customer specification and overfluxing requirement
- Allowable current density – if specified by customer
- Allowable noise level
- Limitation for overall size and dimensions, if any
- Radiators/cooling fins – whether type or size is specified
- Colour of paint and painting scheme
- Fittings and accessories required
- External terminal arrangement – whether outdoor bushings, cable box, etc.
- Data for parallel operation, if specified.
- Requirement of special tests

Temperature rise test	-	Including test requirement, e.g. whether losses at lowest tap (maximum losses) are to be fed and whether overloading is to be tested
Impulse test	-	
Other type/special tests	-	• Short-circuit test • Noise level test • Zero sequence impedance • Capacitance and tan delta measurement • Magnetic balance test • Harmonic analysis • Pressure test on tank/transformer • Vacuum test on tank/transformer • Dissolved gas analysis • Frequency response analysis, etc.

3.2   DESIGN FLOW CHART [FOR DESIGN WITH LV PARALLEL CONDUCTORS AND HV LAYER WINDINGS]

Note: At any stage of the above design if the calculated result is found to be not meeting the guaranteed value or the permissible figure as per the standard or safety factor, the design is revised by changing the parameters.

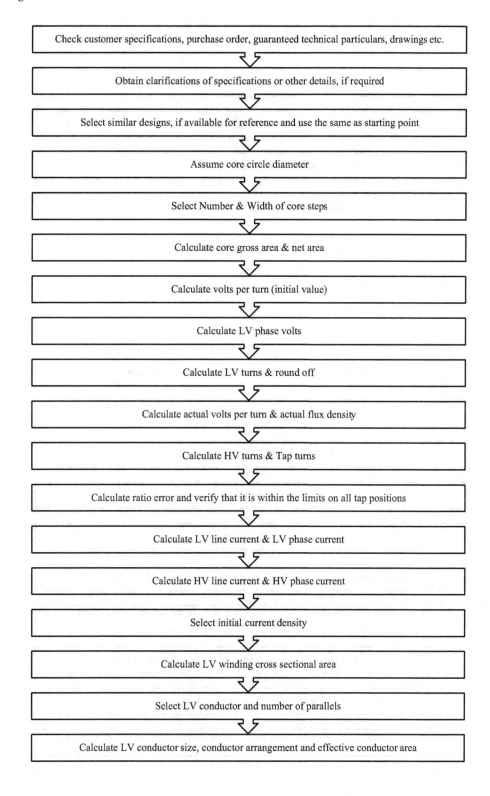

Check customer specifications, purchase order, guaranteed technical particulars, drawings etc.

Obtain clarifications of specifications or other details, if required

Select similar designs, if available for reference and use the same as starting point

Assume core circle diameter

Select Number & Width of core steps

Calculate core gross area & net area

Calculate volts per turn (initial value)

Calculate LV phase volts

Calculate LV turns & round off

Calculate actual volts per turn & actual flux density

Calculate HV turns & Tap turns

Calculate ratio error and verify that it is within the limits on all tap positions

Calculate LV line current & LV phase current

Calculate HV line current & HV phase current

Select initial current density

Calculate LV winding cross sectional area

Select LV conductor and number of parallels

Calculate LV conductor size, conductor arrangement and effective conductor area

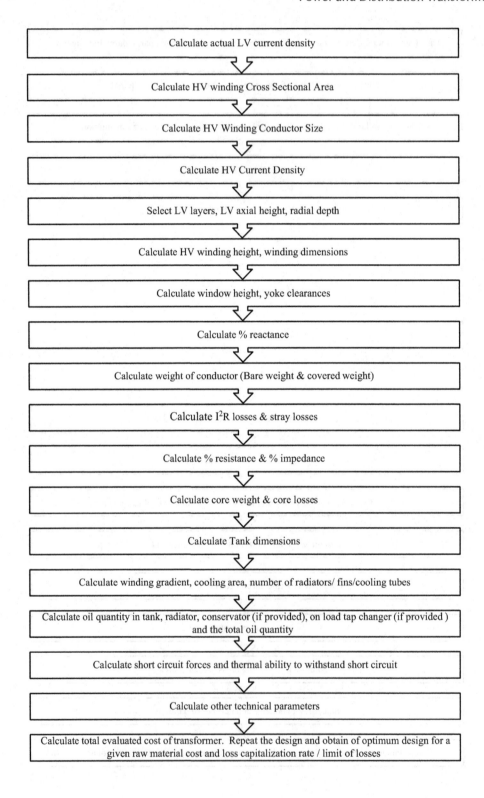

Calculate actual LV current density

Calculate HV winding Cross Sectional Area

Calculate HV Winding Conductor Size

Calculate HV Current Density

Select LV layers, LV axial height, radial depth

Calculate HV winding height, winding dimensions

Calculate window height, yoke clearances

Calculate % reactance

Calculate weight of conductor (Bare weight & covered weight)

Calculate I^2R losses & stray losses

Calculate % resistance & % impedance

Calculate core weight & core losses

Calculate Tank dimensions

Calculate winding gradient, cooling area, number of radiators/ fins/cooling tubes

Calculate oil quantity in tank, radiator, conservator (if provided), on load tap changer (if provided) and the total oil quantity

Calculate short circuit forces and thermal ability to withstand short circuit

Calculate other technical parameters

Calculate total evaluated cost of transformer. Repeat the design and obtain of optimum design for a given raw material cost and loss capitalization rate / limit of losses

4 Core Design
Core Area Calculation

4.1 CALCULATION OF CORE DIAMETER AND CORE AREA

4.1.1 SELECTION OF CORE CIRCLE

A design can be started by making an assumption of a core circle. This can be done based on previous designs of similar transformers. Typical range of core circle diameters is given in Table 4.1 for guidance.

The above ranges may change when the transformer specifications, capitalization rate and cost of core/copper change substantially. Alternatively, an initial value of volt/turn can be calculated from the formula:

$$\frac{V}{T} \propto K \sqrt{\text{kVA}} \qquad (4.1)$$

TABLE 4.1
Typical Core Circle Range for Various kVA Ratings

Three-Phase kVA	Core Circle Range (mm)
25	80–92
63	95–105
100	112–122
125	125–130
160	130–135
200	138–145
250	148–157
315	155–165
400	167–174
500	178–182
630	185–192
1000	212–222
1600	230–243
3150	278–290
5000	347–358
8000	360–375
10,000	375–400
16,000	420–465
20,000	500–560
31,500	580–680
50,000	620–680
63,000	700–760
100,000	780–840

---

where $\dfrac{V}{T}$ = Volt/Turn

    kVA = power of the three-phase transformer
K = Costant $\cong 0.35 - 0.45$
From V/T, the core diameter is calculated as follows:

$$\frac{V}{T} = 4.44 \ f \ B_{\max} A \tag{4.2}$$

where
    f = frequency
    $B_{\max}$ = maximum flux density (Tesla)
    A = net area of cross section of core (m²).
    By substituting the values of f and $B_{\max}$ in the above equation, the net area is calculated.
    A_g = gross area of the core cross section

$$A_g = \frac{A}{K_s} \tag{4.3}$$

where
    K_s = core stacking factor
    = 0.97 for Cold Rolled Grain Oriented (CRGO) steel

$$A_g = \frac{\pi}{4} d^2 K_j \tag{4.4}$$

K_f = Ratio of core gross area to area of core circle.

4.1.2 Selection of Core Step Widths

For any particular core diameter, the first and foremost point to be decided is the maximum allowable width of laminations so that the maximum area of the core circle is filled. For building the circular cross section of core in N steps, N different widths of laminations will have to be selected.

4.1.2.1 Optimum Width of Core Steps to Get Maximum Core Area

This is applicable for core-type stacked core construction. The core area can be maximized for a given number of steps, by selecting optimum step widths (if the number of steps is infinity, the core cross section becomes a circle).

    One-fourth of the core circle is shown Figure 4.1 with parameters
    R = Radius of core circle
    $x_1, x_2, x_3 \ldots\ldots$ = core stack thickness on one half of the core circle
    $y_1, y_2, y_3 \ldots\ldots$ = half of the step width

$$\text{Total core area} = 4\left[x_1 y_1 + (x_2 - x_1) y_2 + \ldots \right] \tag{4.5}$$

$$R^2 = x_i^2 + y_i^2 \tag{4.6}$$

$$\text{Area of } n^{\text{th}} \text{ step} = 4\left[Y_i (x_i - x_{i-1}) \right] \tag{4.7}$$

$$\text{here } Y_i = \sqrt{R^2 - x_i^2} \tag{4.8}$$

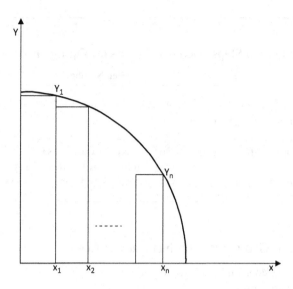

FIGURE 4.1 Core stack arrangement for 1/4th core circle.

Total core area is the sum of areas of all "n" steps.

$$A_n = 4 \sum_{n=1}^{n} (x_i - x_{i-1})\sqrt{R^2 - x_i^2} \qquad (4.9)$$

The independent variable is x_i
  To maximize the function,

$$\left.\begin{array}{l} \dfrac{\delta A_m}{\delta x_i} = 0 \\[3mm] \dfrac{\delta A_m}{\delta x_i} < 0 \end{array}\right\} \text{Condition for maxima} \qquad (4.10)$$

This gives a set of non-linear equations, and iterative methods are required to get the solutions.
  The equation to be solved is

$$\left(R^2 - x_1^2\right)\left[x_{i+1}^2 - x_i(3x_i - 2x_i - 1)\right] + x_i^2 (x_i - x_{i-1})^2 = 0 \qquad (4.11)$$

$$\left(n \text{ equations for values of } i = 1 \text{ to } i = n\right)$$

Normally, the above equation is normalized by dividing by R^4, so that the solution coordinates in terms of $\dfrac{x}{R}$ are obtained. For typical values of number of steps used, the optimum widths can be calculated by using the factors, given in Table 4.2.

  The step widths as calculated above will give different widths for different core diameters, and the number of widths to be handled becomes very large. Hence, widths are generally standardized in multiples of 5 or 10 mm and the standard widths are used for all practical purposes. Optimum combination of widths can be obtained by calculating widths using above factors and rounding off to the nearest 5 or 10 mm, or as per the desired optimization plan.

TABLE 4.2

Typical Values of Number of Steps Used and Factors for Optimum Widths

No. of Steps	% of Circle Filled (Gross Area)	Step Numbers										
		1	2	3	4	5	6	7	8	9	10	11
3	85.1	0.906	0.707	0.424	-	-	-	-	-	-	-	-
5	90.8	0.949	0.846	0.707	0.534	0.314	-	-	-	-	-	-
7	93.4	0.967	0.901	0.814	0.707	0.581	0.434	0.254	-	-	-	-
9	94.8	0.976	0.929	0.868	0.762	0.707	0.605	0.497	0.370	0.216	-	-
11	95.8	0.982	0.943	0.893	0.832	0.762	0.707	0.648	0.555	0.450	0.333	0.190

4.1.3   CALCULATION OF GROSS AREA AND NET AREA OF CORE

Figure 4.2 shows the core steps and stacks of a core

D = core circle diameter (mm)

$W_1, W_2, \ldots W_n$ = core step width (mm)

T_1, T_2, T_3 = stack thickness (mm)

$$\text{Gross area} = W_1 T_1 + W_2 T_2 + \ldots W_n T_n \qquad (4.12)$$

$$T_1 = \sqrt{D^2 - W_1^2}$$

$$T_2 = \sqrt{D^2 - W_2^2} \; - T_1$$

$$T_3 = \sqrt{D^2 - W_3^2} \; -(T_1 + T_2)$$

$$T_n = \sqrt{\left(D^2 - W_n^2\right)} \; -(T_1 + T_2 + \ldots T_{n-1})$$

$$\text{Net area} = \text{Gross area} \times \text{Stacking factor}$$

Stacking factor of 0.97 can be used for CRGO steel.

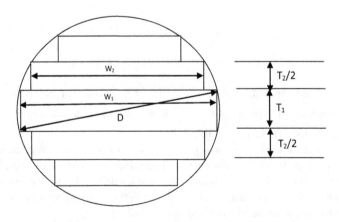

FIGURE 4.2   Core steps and stacks.

Calculation of Stack thickness of steps:
The stack thickness of each step is calculated after considering the following factors:
- Thickness of lamination
- Type of core design, i.e. step lap or normal
- Thickness of lamination
  The standard thicknesses used are 0.30, 0.27 and 0.23 mm. Therefore, the stack thicknesses $T_1\ T_2\ ...\ T_n$ will have to be rounded off to the integral multiples of 0.3, 0.27 or 0.23. The number of laminations shall be rounded off to the lowest integer.

Example – Non-step-lap constructions
Thickness of lamination = 0.27 mm
Calculated stack thickness = 40 mm
Number of laminations $= \dfrac{40}{0.27} = 148.15$
This is to be rounded off to 148
∴ Actual stack thickness = 148×0.27 = 39.96
For calculation purposes, this can be taken as 40.

Example – Step-lap construction
When step-lap construction is used, a defined number of laminations are taken together as a "book". The typical arrangements are as follows.
- Single sheet assembly
- Double sheet assembly
  In single sheet assembly, the number of laminations of one width will be only one in a "book". A book can consist of three, five or seven such single sheets. This is generally called 3 step lap, 5 step lap, 7 step lap, etc., with single sheet assembly.
  Calculations of stack thickness of steps with step-lap cutting
  Five step-lap single sheet assembly
  The method is explained by an example
  Core circle – 200 mm
  Step widths used – 180, 160, 140, … etc.
  CRGO thickness – 0.27 mm

The stack thickness of each step is initially calculated as in Table 4.3.

The maximum step width of 180 mm is in the middle, and its stack thickness is 87.17. The other steps 160, 140, etc., will be split equally below and above the middle step, as shown in Figure 4.3.

Single sheet assembly gives better building factor and therefore reduced losses.

In double sheet assembly, the number of laminations of one width will be two in a "book". Therefore, the number of laminations in a book of 3 step lap, 5 step lap, etc., is

3 step lap – 6 laminations
5 step lap – 10 laminations
7 step lap – 14 laminations

TABLE 4.3
Stack Thickness Calculation

Step Width	Stack Thickness	Remarks
180	87.17	These are theoretical maximum
160	32.83	thicknesses of stacks
140	22.83	

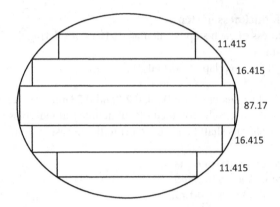

FIGURE 4.3   Core stack thicknesses.

Seven step-lap arrangement and double sheet assembly can be used when the lamination width and length are comparatively small. The consideration is that if the lamination sizes are big, one book with 14 laminations will be difficult to handle.

The no load loss will be higher for double sheet assembly when compared to that of single sheet assembly. Similarly, the lowest loss is obtained with 7 step-lap arrangement and 5 step lap gives more loss than 7 step lap and 3 step lap gives more loss when compared to 5 step lap, etc.

In the above example, 5 step-lap single sheet assembly is used. Hence, the "books" will consist of five laminations of 0.27 mm each (1.35 mm is the thickness of one book). Now, the stack thickness will have to be adjusted so that an integer number of books only are used. Fractional number of books is not used. For middle step, any integer number of books can be used. For other steps, even number is required since books are to be equally divided between top and bottom.

The number of books of each step is given in Table 4.4.

The number of books is calculated using the nominal thickness of the lamination (i.e. 0.3, 0.27, 0.23, etc.). However, if the laminations have consistently negative tolerance on thickness, the calculated weight and stack thickness will not be obtained when laminations are assembled. Additional filling up is required in such cases.

The number of books for middle step can be odd or even. The stack thicknesses will have to be recalculated on this basis, and continuing with the example, the calculations are given in Table 4.5.

With the above results of books and stacks, the calculations will have to be repeated to check whether the stack thickness or the number of books needs revision.

Five step-lap double sheet assembly

In this case, $5 \times 2 = 10$ laminations will be used in one book (two laminations of equal width $\times 5$ steps). The number of books will be half of the books calculated for single sheet assembly. The total stack and number of laminations in a stack remain unchanged, as given in Table 4.6.

TABLE 4.4
Number of Books of Each Step

Width	Calculated Stack (Total)	Calculated Stack (Each Side)	Number of Books (Total)	Number of Books (Each Side)
180	87.17	-	64.57	-
160	32.83	16.415	24.318	12.159
140	22.83	11.415	16.91	8.455
-	-	-	-	-

TABLE 4.5
Stack Thickness Recalculation

Width	Calculated Stack	Number of Books (Odd or Even for Middle) (Even for Other)	Actual Stack	New Feasible Stack	Number of Books
180	87.17	64	86.4	-	64
160	32.83	12 × 2	32.4	33.6	24 = 12 × 2
140	22.83	8 × 2	21.6	24.03	16 = 8 × 2
I	I	I	I	I	I
I	I	I	I	I	I
I	I	I	I	I	I
I	I	I	I		

TABLE 4.6
Final Stack Thickness

Width	Number of Books (Double Sheet Assembly)	Stack Thickness
180	32	86.4
160	6×2	16.2×2
140	4×2	10.8×2
I	I	I
I	I	I
I	I	I
I		I

5 Winding Design

5.1 CALCULATION OF VOLTS PER TURN

Volts per turn is calculated by the emf equation

$$V = 4.44 \ f \ B_{\text{max}} \ A \tag{5.1}$$

V = Volts/Turn
f = frequency (Hz)
B_{max} = Maximum Flux density (T)
A = Area of cross section (m^2)

For starting a calculation, the value of maximum flux density (B_m) is arrived at on the following considerations.

1. If customer specifies the maximum permissible flux density, use the specified value.
2. If limit of saturation flux density and overfluxing limit are specified by customer,

$$B_{\text{max}} = \frac{B_s}{\text{Per unit overfluxing limit}} \tag{5.2}$$

B_s = Saturation flux density
When overfluxing limit is not specified, use per unit overfluxing limit as 1.125.
3. When saturation flux density is not specified, use 1.9 T as saturation flux density.
4. When short-time overvoltage withstand capability is specified, check the overfluxing requirement prior to fixing the working flux density. If short-time overvoltage is limited to 40% above rated voltage for 5 seconds and 25% above rated voltage for 1 minute, the working flux density can be decided on the basis given above.
5. The working flux density as arrived at on the above basis may require reduction, if the no-load capitalization value is comparatively high or if sound level to be guaranteed is low.

5.2 CALCULATION OF PHASE VOLTS AND PHASE CURRENTS

1. Specifications give line-to-line voltages. Phase voltage and phase current of HV and LV shall be calculated as below.
2. Delta connected winding – (three-phase two-winding transformer);

$$\text{Phase voltge} = \text{Line voltage} \tag{5.3}$$

$$\text{Line current} = \frac{\text{kVA} \times 1000}{\sqrt{3} \ \text{Line voltage}} \tag{5.4}$$

$$\text{Phase current} = \frac{\text{Line current}}{\sqrt{3}} = \frac{I_L}{\sqrt{3}} \tag{5.5}$$

3. Wye (star) connected winding (three-phase two-winding transformer):

$$\text{Phase voltage} = \frac{\text{Line voltage}}{\sqrt{3}} = \frac{V_L}{\sqrt{3}} \tag{5.6}$$

$$\text{Line current} = \frac{\text{kVA} \times 1000}{\sqrt{3} \text{ Line voltage}} \tag{5.7}$$

$$\text{Phase current} = \text{Line current} \tag{5.8}$$

4. Zigzag winding (interstar winding):

$$\text{Line current} = \frac{\text{kVA} \times 1000}{\sqrt{3} \times \text{Line voltage}} \tag{5.9}$$

$$\text{Phase current} = \text{Line current} \tag{5.10}$$

Line voltage is the vector sum of two voltages which are 120° out of phase as shown in Figure 5.1.

$$V_L = \sqrt{3} \times V_{ph} \tag{5.11}$$

$$V_{ph} = \sqrt{3} \times \text{zig voltage} = \sqrt{3} \times \text{zag voltage} \tag{5.12}$$

$$\text{Zig winding Voltage} = \frac{V_L}{3}; \text{Zag winding Voltage} = \frac{V_L}{3} \tag{5.13}$$

Note:
1. When taps are provided, calculate the line currents and phase currents corresponding to the highest, normal (rated) and lowest tap positions.
2. Above calculation gives the voltages at no load. Take special note when the voltage specified by customer is at full load at specified power factor, in which case the no-load voltage of secondary (load side) winding shall be calculated by considering the regulation and the voltage at no load shall be used for the calculation of number of turns. However, this is done in special cases only and the guaranteed value remains as the voltage at no load only.

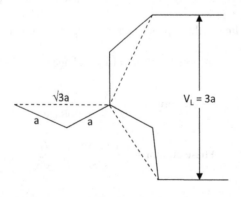

FIGURE 5.1 Voltage distribution in zigzag windings.

5.3 CALCULATION OF NUMBER OF TURNS

1. LV winding turn
   a. For Δ or γ winding

$$LV \text{ winding turns} = \frac{LV \text{ No load phase voltage}}{\text{Volts per turn}} \qquad (5.14)$$

   b. For zigzag winding
   a = zig winding volts = zag winding volts
   $3a$ = line voltage

$$\text{Zig winding turns} = \frac{\text{No load line voltage}}{3 \times \text{Volts per turn}} \qquad (5.15)$$

$$\text{Zag winding turns} = \frac{\text{No load line voltage}}{3 \times \text{Volts per turn}} \qquad (5.16)$$

When calculating turns as per above, the number of turns will not normally be an integer. The number of turns is rounded off to the nearest integer subject to the following.

– When actual flux density is calculated on the basis of rounded off turns, the flux density shall not exceed the values as per the calculation of volts per turn.
– It is preferable to select even number of turns for LV winding, when the number of LV layers is even. When foil winding or single-layer LV winding or disc winding is used, the number of turns can be odd or even number.
– For zigzag winding, the number of LV turns shall be even and preferably a multiple of four when two-layer winding is used.

2. HV winding turns
   a. For Δ or γ winding

$$\text{Number of HV turns at rated tap} = LV \text{ turns} \left(\frac{\text{HV No load phase Voltage (Rated Tap)}}{LV \text{ No Load Phase Voltage}} \right) \qquad (5.17)$$

   b. Zigzag winding on HV

$$\text{zig winding turns} = \frac{\text{HV Line Voltage}}{3 \times \dfrac{\text{Volt}}{\text{turn}}} \qquad (5.18)$$

$$\text{zag winding turns} = \frac{\text{HV Line Voltage}}{3 \times \dfrac{\text{Volt}}{\text{turn}}} \qquad (5.19)$$

Round off the turns to the nearest whole number.

5.3.1 CALCULATION OF VOLTAGE AND TURNS OF EXTENDED DELTA WINDING

Extended delta connection is used to get desired phase shifts for rectifier/converter transformers or phase shifting transformer.

A six-pulse rectifier requires a conventional three-phase supply with full-wave rectification. However, this system will have harmonics of 5, 7, 11, etc.

In order to reduce the harmonics, rectifiers of 12, 18, 24 pulse, etc., are used.

A 12-pulse power converter/rectifier is possible by connecting in parallel two six-pulse convert-ers and supplying the converters with 2 three-phase voltages shifted by 30°. This can be easily obtained by providing two secondary windings on the transformer, one with delta connection and the other with star connection so that the two secondaries have 30° phase shift.

However, the impedance between the primary and the 2 secondaries may not be equal in practi-cal designs due to the difference in turns, phase voltage and phase currents of the two secondaries.

When good impedance balance of the secondaries is required, design using two core coil assem-blies is used; one core coil assembly is designed with extended delta primary having +15° shift and normal delta secondary. The second core coil assembly is with extended delta primary having −15° shift and normal delta secondary. This design gives 30° phase shift between secondaries.

However, this design is expensive when compared to the design using a delta connected primary and 2 secondaries, one delta connected and the other star connected. The connection scheme of an extended delta-delta connected 12 pulse converted is shown in Figures 5.2 and 5.3.

The winding connection and vector diagram of extended delta connection with $\theta°$ lagging is shown in Figure 5.4. θ is the angle of lagging

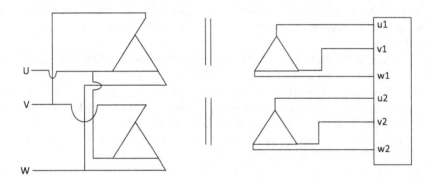

FIGURE 5.2 Phasor diagram of extended delta connection with $\theta°$ lagging.

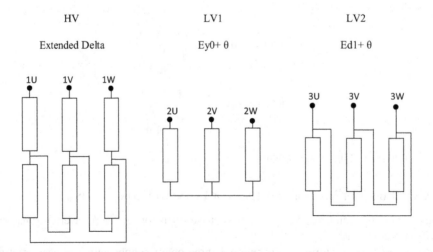

FIGURE 5.3 Winding connection and vector diagram of extended delta connection with $\theta°$ lagging.

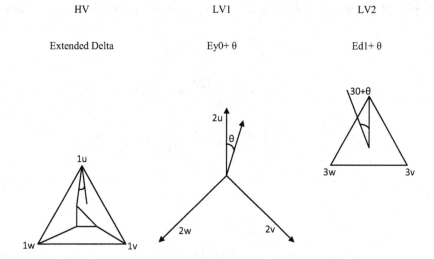

HV   LV1   LV2

Extended Delta   Ey0+ θ   Ed1+ θ

FIGURE 5.4   Vector diagram of extended delta connection with θ^0 lagging.

- The 2 LV windings will be axially paced (loosely coupled windings).
- The HV winding will have to be two axially placed windings connected in parallel for balancing

For any phase shift of "θ" required, the voltages of the delta and extended delta windings are related as per the following.

V = line-to-line voltage of the transformer
A = voltage of the extended delta winding
V' = secondary line-to-line voltage
B = voltage of the delta winding
θ = desired phase shift deg

$$\frac{A}{Sin\theta} = \frac{V}{Sin120} = \frac{B}{Sin\,(60-\theta)} \tag{5.20}$$

$$\therefore B = V\frac{Sin\,(60-\theta)}{Sin\,120} \tag{5.21}$$

For a 15° phase shift (i.e. θ = 15)
Assume

N_E = number of turns of extended delta winding
N_T = number of turns of delta (main) winding
N_2 = number of turns of the delta connected secondary winding

$$\frac{N_E}{N_E + N_T} = \frac{A}{B} = \frac{Sin\theta}{Sin\,(60-\theta)} \tag{5.22}$$

$$\frac{N_2}{(N_E + N_T)} = \frac{V'}{B} = \frac{V_1}{V}\frac{Sin\,(120)}{Sin\,(60-\theta)} \tag{5.23}$$

In order to generate a phase shift of 15°, the number of turns of windings is determined as below.

$$N_T = \left(\frac{\text{Sin } 45}{\text{Sin } 15} - 1\right)N_E = \sqrt{3}N_E \tag{5.24}$$

$$N_2 = \frac{V'\sin 120}{V\sin 15} N_E \tag{5.25}$$

$$N_2 = \frac{V'}{V} \frac{2\sqrt{3}}{\sqrt{3} \cdot \sqrt{2 + \sqrt{2}}} N_E = \frac{V'}{V} \frac{2}{\sqrt{3} \cdot \sqrt{2 + \sqrt{2}}} N_T \tag{5.26}$$

If $K = \dfrac{V}{V'}$

$$N_T = \sqrt{3} \, N_E$$

$$N_2 \approx \frac{3.34606}{K} N_E$$

The voltage magnitudes are

$$A = V \frac{\text{Sin } 15}{\text{Sin } 120} \tag{5.27}$$

$$B = V \frac{\text{Sin } 45}{\text{Sin } 120} \tag{5.28}$$

$$B' = B - A = V \frac{(\text{Sin } 45 - \text{Sin } 15)}{\text{Sin } 120} \tag{5.29}$$

While calculating the current density and conductor size, the following points are to be noted.

- The extended windings carry the line current
- The ampere-turn density of the main winding and extended winding needs to be made almost equal for good impedance control and better short-circuit strength
- Important Notes
  a. Any variation from the specified voltage ratio caused by the use of rounded off turns shall be calculated accurately at all tap positions, and the ratio error shall be within the limits specified by the standard or Purchase Order. Chapter 15 (Section 15.5 gives sample calculation of ratio error and phase angle error of a transformer with extended delta winding. Ratio error allowed is the lower of ± 0.5% of nominal voltage ratio or ± 10% of the percent impedance.)
  b. If a transformer is designed to operate in parallel with another transformer, the turns ratio of both transformers must be the same on each tap position.

5.3.2   CALCULATION OF TAP TURNS

Tappings are normally provided on transformers for the purpose of changing voltage ratio. The following essential information shall be checked before commencing design.

  i. Whether taps are no load/off circuit/on load
  ii. Tap range (upper and lower limits)

iii. Tap steps percent
iv. Total number of tap steps
v. Whether constant flux voltage variation (CFVV), variable flux voltage variation (VFVV) or combined voltage variation (CbVV).

5.3.2.1 Categories of Voltage Variation

Voltage variation when taps are provided can be classified into three.

i. Constant flux voltage variation (CFVV):
  Constant flux voltage variation occurs when:
  - The tapping voltage is equal to rated voltage for any untapped winding and rated voltage multiplied by the tapping factor for the tapped winding.
  - In other words, when tappings are provided in the winding where voltage variation occurs and the per turn voltage remains constant irrespective of the tap positions when the voltage corresponding to the tap position is applied to the tapped winding, the voltage variation is called constant flux voltage variation.
- Calculation of tap turns for CFVV:
  In CFVV, taps shall be provided on winding in which voltage variation is required. Thus, if HV voltage variation is required, the taps shall be provided on HV winding, and the number of turns is directly proportional to the voltage of the winding.

$$\text{Turns required at any tap position} = \frac{\text{Phase voltage corresponding to the tap position}}{\text{Volts per turn}}$$

(round off the turns to the nearest integer)
ii. Variable Flux Voltage Variation (VFVV):
  Voltage variation when tapping voltage is constant for the tapped winding and equal to its rated voltage.
  In other words, if taps are provided on one winding where the applied voltage is constant and voltage variation is required for untapped winding, the voltage variation is variable flux voltage variation. The flux density in core changes with every tap position.
- Calculation of tap turns for VFVV:
  In VFVV, taps are provided on the winding where voltage applied is constant. As a result, the volts per turn changes when taps are changed.
  In VFVV, with higher turns on tapped winding, output voltage will be lower, and conversely, with lower turns on tapped winding, output voltage will be higher.
iii. Combined Voltage Variation (CbVV):
  Voltage variation combines CFVV and VFVV.
  Below a certain tapping factor, the tapping voltage of the untapped winding is constant and above this factor, the tapping voltages of the tapped windings are constant.
  The difference between turns of adjacent taps is called the turns per tap step. In the case of transformers with tappings provided on disc winding, it is convenient if the turns per tap step are exactly an integer. This can be achieved by selecting the volts per turn, such that:

$$\text{Volts per turn} = \frac{\text{Step Voltage}}{\text{Turns per tap step}}$$

- Calculation of tap turns for zigzag winding
  If taps are required on zigzag winding, these shall be equally given on zig winding and zag winding; that is, there will be taps on both zig winding and zag winding with 50% of the step voltage. If taps are provided on one of the windings only or if taps are distributed

with 50% taps on zig winding and 50% taps on zag winding, the output voltage will have a phase shift.

Important Note:

- Calculate ratio error on all tap positions, and if ratio error exceeds the limits at any tap position, readjustment of turns is necessary.

5.4 CALCULATION OF THE CROSS SECTION AREA OF CONDUCTOR

1. Effective area of round conductor $= \dfrac{\pi}{4} d^2$

$$d = \text{Dia of bare conductor (mm)}$$

2.

$$\text{Effective area of rectangular condutor} = \begin{cases} (bt - 0.85) \text{ for } t \geq 3.5 \\ (bt - 0.55) \text{ for } t \leq 3.4 \geq 2.25 \\ \\ (bt - 0.35) \text{for } t \leq 2.24 \geq 1.6 \\ (bt - 0.20) \text{ for } t < 1.6 \end{cases} \qquad (5.30)$$

$$b = \text{Width of conductor}$$

$$t = \text{Thickness of conductor}$$

The effective area of rectangular conductor is less than the product of width and thickness due to rounding off of the corners. The corners are rounded off for avoiding sharp edges which will cut the insulation and to reduce the electrical stress concentration at the corners.

3.

$$\text{Effective area of foil} = bt$$

$$\text{where } b = \text{width of the foil in (mm)}$$

$$t = \text{thickness of the foil (mm)}$$

4. The effective current density is obtained by dividing the current by the effective area.

5.5 SELECTION OF CURRENT DENSITY

Selection of current density depends on the following factors:

a. Capitalization value of load losses and cost of conductor material or guaranteed value of load losses
b. Winding gradient
c. Short-circuit strength requirement

Generally, current density of various windings is selected in such a way that the winding gradients are equal. Whenever capitalization value of load loss is higher, the current density chosen shall be lower.

When impedance of transformer is low, the current density may be limited by the short-circuit strength.

Typical values of current densities, when capitalization is not used, are given below.

Aluminium – 1.5–2.0 A/mm²

Copper – 2.0–3.6 A/mm²

Higher current densities can be used if the losses, gradient and short-circuit stresses are within acceptable limit.

When capitalization of load loss is used, the lower limit of current density can be calculated by equating the cost of load loss (load loss capitalization) produced by using unit quantity of conductor material to the cost of unit quantity of conductor plus the apportioned value of manufacturing cost.

When load loss capitalization figure is specified, the following procedure can be used for determining the initial current density to commence a design.

a. Design with Copper Winding:

A = load loss capitalization ($/W)

C = cost of copper conductor ($/kg)

K = value addition factor

δ = current density (A/mm²)

$$\delta = \sqrt{\frac{K\ C}{2.4\ A}}$$

The value of K changes from manufacturer to manufacturer

b. Design with Aluminium Winding:

A = load loss capitalization ($/W)

C = cost of aluminium conductor ($/kg)

K = value addition factor

δ = current density (A/mm²)

$$\delta = \sqrt{\frac{K\ C}{13\ A}}$$

5.6   SELECTION OF CONDUCTOR SIZES

1. Bare Size of Conductor

The conductor cross-sectional area is obtained by dividing the phase current by the current density.

$$A = \frac{I_{ph}}{\delta}$$

A = cross-sectional area of conductor (mm²)

I_{ph} = phase current (A)

δ = current density (A/mm²)

From the above area, determine the nearest standard size of conductor, calculate the effective area and then calculate actual current density. For round conductors, use the standard wire gauge sizes. For rectangular conductors, select strip sizes based on the following.

a.

$$6 \leq \frac{\text{Bare width}}{\text{Bare thickness}} \geq 2$$

b. Widths less than 5 mm and thickness less than 1.2 mm are normally not preferred.
c. Widths greater than 16 mm and thickness greater than 4.5 mm are not preferred. The eddy current loss in winding increases in proportion to the square of the conductor thickness, and hence, thin conductors are preferred.
d. Use conductor thicknesses in increments of 0.05 mm and widths in increments of 0.1 mm.

2. Covered Size of Conductor

Paper covering or enamelling is used as conductor insulation. The type of covering shall be selected on the following basis.

a. Customer specification
b. Overall economy. If customer has not specified the type of covering, the selection is based on overall economy. The overall size of enamel covered conductor is less than that of paper covered conductor, and hence, enamel covering can be tried as an alternative to evaluate the overall economy.
c. It is uneconomical to use paper covered round wire for diameters less than 1.0 mm.

Typical covering of conductors is given in Table 5.1.

Standard Wire Gauges

Bare sizes of conductors as per standard wire gauge are given Tables 5.2 and 5.3

5.7 SELECTION OF TYPES OF WINDINGS

The following types of windings are generally used.

a. Spiral winding – for low-voltage application (single layer and multilayer)
b. Helical winding – high current, low voltage (normally single or two layers)

TABLE 5.1
Typical Covering of Conductors

Voltage Class of Winding (kV)	Type of Conductor	Type of Covering	Increase in Overall Size Due to Covering	Remarks
11, 22 and 33	Round	Enamel – medium covering	Approx. 0.09 mm (use maximum values as given in Table 5.3)	For crossover coils and layer windings
11, 22 and 33	Round	Paper covering	0.25 mm	For crossover coils and layer windings
11	Rectangular	Paper covering	0.4 mm	Layer winding/disc winding
LV, e.g. 0.433	Rectangular	Enamel – medium covering	Approx. 0.11 mm	For spiral winding
LV, e.g. 0.433	Rectangular	Paper covering	0.3 mm up to 100 kVA 0.4 mm above 100 kVA	
22	Rectangular	Paper covering	- 0.5 mm	Layer winding/disc winding
33	Rectangular	Paper covering	0.6 mm	Layer winding/disc winding

---

TABLE 5.2
Bare Sizes of Conductors as per Standard Wire Gauge

Gauge No.	Diameter		Area	
	Inches	Mm	Inches²	mm²
7/0	0.500	12.700	0.1964	126.709
6/0	0.464	11.786	0.1691	109.096
5/0	0.432	10.973	0.1466	94.580
4/0	0.400	10.160	0.1257	81.096
3/0	0.372	9.449	0.1087	70.128
2/0	0.348	8.839	0.09512	61.419
1/0	0.324	8.230	0.08245	53.193
1	0.300	7.620	0.07069	45.593
2	0.276	7.010	0.05983	38.599
3	0.252	6.401	0.04988	32.180
4	0.232	5.893	0.04227	27.270
5	0.212	5.385	0.03530	22.774
6	0.192	4.877	0.02895	18.677
7	0.176	4.470	0.02433	15.696
8	0.160	4.064	0.02011	12.974
9	0.144	3.658	0.01629	10.509
10	0.128	3.251	0.01287	8.303
11	0.116	2.946	0.01057	6.780
12	0.104	2.642	0.00850	5.483
13	0.092	2.337	0.00665	4.290
14	0.080	2.032	0.00503	3.245
15	0.072	1.829	0.00407	2.525
16	0.064	1.626	0.00322	2.066
17	0.056	1.422	0.00246	1.587
18	0.048	1.219	0.001810	1.168
19	0.040	1.016	0.001257	0.810
20	0.036	0.914	0.001018	0.656
21	0.032	0.813	0.000804	0.518
22	0.028	0.711	0.000616	0.397
23	0.024	0.610	0.000452	0.291
24	0.022	0.559	0.000380	0.245
25	0.020	0.508	0.000314	0.202
26	0.018	0.457	0.0002545	0.1541
27	0.0164	0.4166	0.0002112	0.1262
28	0.0148	0.3759	0.0001720	0.1100
29	0.0136	0.3454	0.0001453	0.0935
30	0.0124	0.3150	0.0001208	0.0779
31	0.0116	0.2946	0.0001057	0.0681
32	0.0108	0.2743	0.0000916	0.0589
33	0.0100	0.2540	0.0000785	0.0507
34	0.0092	0.2337	0.0000665	0.0429
35	0.0084	0.2134	0.0000554	0.0357

Note:
1. Select round conductors from Table 5.2.
2. Use conductors of bare diameter of 4.064 (SWG-8) or above in special cases.

TABLE 5.3
Limits of Overall Diameters and Increase in Diameters of Enamelled Round Wires

Medium Covering (All Dimensions in Millimetres)

Nominal Conductor Diameter	Overall Diameter Min	Increased in Diameter Min	Overall Diameter Max	Increase in Diameter Max
(1)	(2)	(3)	(4)	(5)
0.020	0.027	0.003	0.029	0.005
0.025	0.035	0.004	0.038	0.007
0.032	0.044	0.006	0.047	0.009
0.040	0.055	0.008	0.059	0.013
0.050	0.069	0.010	0.075	0.017
0.063	0.86	0.013	0.093	0.020
0.071	0.096	0.015	0.103	0.022
0.080	0.106	0.016	0.113	0.024
0.090	0.118	0.017	0.125	0.026
0.100	0.131	0.019	0.139	0.028
0.112	0.145	0.020	0.154	0.029
0.125	0.161	0.022	0.171	0.032
0.140	0.178	0.024	0.188	0.034
0.160	0.201	0.025	0.213	0.037
0.180	0.225	0.027	0.238	0.040
0.200	0.248	0.028	0.263	0.043
0.224	0.275	0.030	0.291	0.046
0.250	0.304	0.031	0.321	0.048
0.280	0.338	0.032	0.357	0.051
0.315	0.375	0.034	0.394	0.053
0.355	0.418	0.036	0.437	0.056
0.400	0.446	0.038	0.486	0.058
0.450	0.520	0.041	0.541	0.062
0.500	0.573	0.044	0.594	0.065
0.560	0.636	0.046	0.657	0.067
0.630	0.710	0.049	0.732	0.071
0.710	0.795	0.051	0.818	0.074
0.750	0.837	0.052	0.860	0.075
0.800	0.890	0.054	0.914	0.078
0.850	0.942	0.055	0.966	0.079
0.900	0.995	0.057	1.020	0.081
0.950	1.047	0.058	1.071	0.082
1.000	1.098	0.059	1.123	0.084
1.060	1.160	0.060	1.185	0.085
1.120	1.222	0.061	1.247	0.087
1.180	1.281	0.062	1.309	0.088
1.250	1.356	0.063	1.362	0.089
1.320	1.428	0.064	1.454	0.091
1.400	1.511	0.066	1.538	0.093
1.500	1.613	0.067	1.640	0.094
1.600	1.717	0.068	1.745	0.096
1.700	1.819	0.069	1.847	0.097
1.800	1.922	0.071	1.950	0.099
1.900	2.024	0.072	2.052	0.100
2.000	2.126	0.073	2.154	0.101

TABLE 5.4

General Guideline for Selection of the Type of Winding for a Particular Type of Transformer

kVA (3 Phase)	HV Winding Type	LV Winding Type
16	Crossover coil/layer	Spiral
25	Crossover coil/layer	Spiral
63	Crossover coil/layer	Spiral
100	Crossover coil/layer	Spiral
200	Crossover coil/layer	Spiral
315	Crossover coil/layer	Spiral/foil
500	Layer	Spiral/foil
630	Layer	Spiral/foil
800	Layer	Spiral/foil
1000	Layer	Spiral/foil
1600	Layer	Spiral/foil
3150	Layer/disc	Spiral/helical/foil if LV is less than 2.4 kV
5000	Layer/disc	Spiral/helical/foil if LV is less than 2.4 kV
6300	Layer/disc	Spiral/helical/disc
8000	Layer/disc	Spiral/helical/disc
10,000	Layer/disc	Spiral/helical/disc

c. Foil winding – high current, low voltage
d. Crossover coils – for HV windings of distribution transformers (11, 22, 33 kV, etc.)
e. Disc windings – for HV windings
f. Layer-type winding – for HV windings

The selection of a type of winding for a particular voltage class depends on the current density selected (consequently the area of cross section) and the number of turns to be accommodated and the test voltages. The Table 5.4 gives a general guideline for the selection of the type of winding for a particular type of transformer.

5.8 TAP CHANGERS AND TAP CHANGER CONNECTIONS

Tap changer is used for adding or cutting out turns of primary or secondary winding of the transformer. The following types of tap changers are used:

5.8.1 Off-Circuit Tap Changer

The tap changer is operated when transformer is switched off, and static voltages if any are grounded. Basic transformer winding circuit arrangements using off-circuit tap changers are linear, single bridging and double bridging as shown in Figures 5.5–5.7, respectively.

5.8.2 Off-Load Tap Changer

Off-load tap changer is different from off-circuit tap changer, in that off-load tap changer shall be capable of operation with the load disconnected and the winding still energized.

It should be noted that in many specifications, the terms off-circuit tap changer, off-load tap changer and no-load tap changer are used with the same meaning. Whenever off-load tap changer or no-load tap changer is specified, it is essential to obtain the clarification regarding the requirement

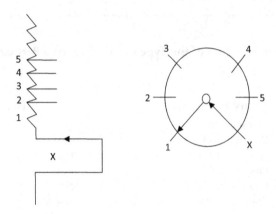

FIGURE 5.5   Linear tap switch.

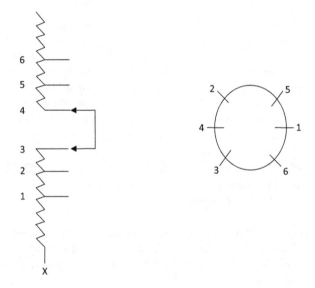

FIGURE 5.6   Single bridge tap switch.

All winding arrangements, i.e. linear, single bridging and double bridging explained in Section 5.8.1, can be used here.

5.8.3   ON-LOAD TAP CHANGER

On-load tap changers are provided to change turns ratio of transformer to regulate the voltage when the transformer is delivering normal load. Resistor transition-type on-load tap changers are most commonly used now. The common type of winding arrangement and tap changer connections are shown in Figures 5.8–5.10.

While selecting a tap changer, the following points shall be checked:

- Rated voltage of winding
- Insulation class of winding including test voltages (induced overvoltage, separate source and impulse)
- Total range of voltage variation required

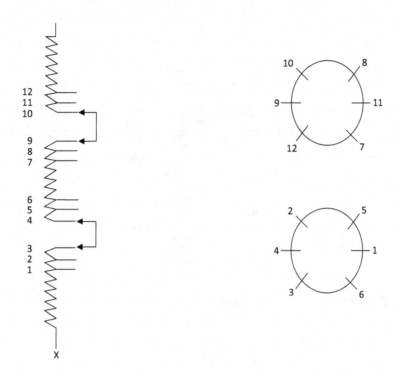

FIGURE 5.7 Double bridging tap switch.

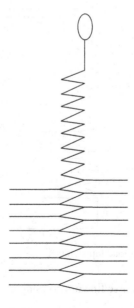

FIGURE 5.8 Linear-type on-load tap switch.

- Step voltage
- Rated current and maximum current
- Insulation level to ground at the point where tap changer is connected
- Fault level of the system
- Short-circuit requirements

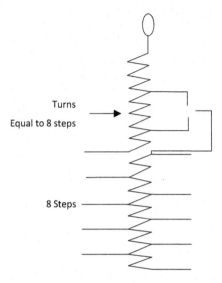

Turns
Equal to 8 steps

8 Steps

FIGURE 5.9   Coarse fine-type on-load tap changer.

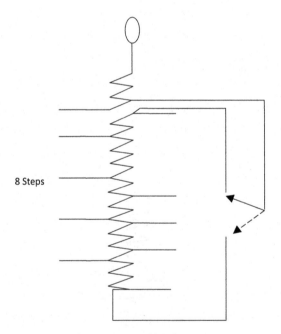

8 Steps

FIGURE 5.10   Reversing switch-type on-load tap changer.

- Tapping arrangement (linear, coarse fine or reversing)
- Tap changer protection required
- Whether parallel operation is required, and if so, details of existing transformer and tap changer control schematic.

5.9   CALCULATION OF AXIAL HEIGHT OF WINDING

Axial heights of different types of windings are calculated as below. All the calculated heights are the final heights of winding after drying and clamping

5.9.1 SPIRAL WINDING

$$\text{Axial height}\,(\text{Physical}) = (N+1)W\,N_F + 1.15W\,N_t \tag{5.31}$$

where
  N = turns per layer
  W = width of covered conductor
  N_t = number of transpositions in one layer
  N_F = number of parallel conductors arranged in flat direction

The axial height to be used for the calculation of reactance is the electrical height, which is calculated by the following formula:

$$\text{Axial height}\,(\text{Electrical}) = N\,W\,N_F + 1.15W\,N_t \tag{5.32}$$

5.9.2 FOIL WINDING

Axial height (electrical) = foil width
Axial height (wound) = foil width + 2 × edge strip

5.9.3 CROSSOVER COILS

Axial height of each crossover coil and the total height of the winding shall be calculated.
  Axial height of crossover coil = $(n + 1)\,d$
  N = number of turns per layer
  d = covered diameter of conductor

$$\text{Axial height of winding} = M(n+1)d + (m-1)T \tag{5.33}$$

where
  M = number of crossover coils
  T = distance between crossover coils (thickness of radial spacer between one set of crossover coils)

Above formula is used only when all crossover coils have equal height and the clearance (axial distance) between crossover coils is equal. When axial heights of crossover coils are different and the clearances are different, the axial height is the sum of the axial heights of crossover coils and the sum of the clearances between coils as shown in Figure 5.11.

$$W = W_1 + W_2 + W_3 + W_4 + T_1 + T_2 + T_3$$

5.9.4 DISC WINDING

The axial height is the sum of the widths of discs plus the sum of the ducts between discs.
  The axial height shall be calculated after preparing a winding diagram similar to Figure 5.12, indicating the type of discs, size of discs and the sizes of ducts between discs. The total axial height so calculated shall be the final dimension after drying and clamping. Hence, the spacer sizes to be used during winding shall be calculated after considering the compressibility of the press board used in the design.
  Use compressibility factor of 3% for pre-compressed grade K press board and 5% for pre-compressed grade three press board.

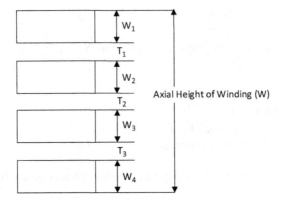

FIGURE 5.11   Crossover coil winding arrangement.

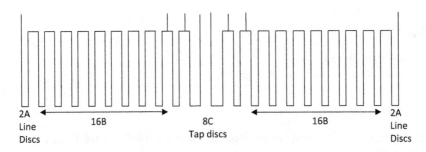

FIGURE 5.12   Disc winding representation.

A disc winding is represented in Figure 5.12. The conductor sizes and insulation thicknesses of the winding are shown in Table 5.5. Calculation of axial heights of conductors and that of spacers are shown in Tables 5.6 and 5.7.

The coil axial height is as wound condition = 450.8 + 199.6 = 650.4 (rounded off to 650).

- Calculation of winding axial height after drying and compression:
    This is calculated on the basic of the compressibility of press board spacer and paper covering of the conductor. Assuming that the compressibility of grade K pre-compressed press board is 3% and that of conductor insulation is 10%, the compressed final height of winding is calculated.
    Total axial height of bare conductor = 424
    Compressed height of conductor insulation – 0.9 × 26.8 = 24.12
    Compressed height of press board spacer – 0.97 × 199.6 = 193.6
    Total axial height of winding after compression = 641.7 (rounded off to 642 mm)

TABLE 5.5
Conductor Size and Insulation Thickness

Disc Designation	Quantity	Bare size of Conductors	Covered Size of Conductor
A	4	2.8 × 10	3.5 × 10.7
B	32	2.8 × 10	3.4 × 10.6
C	8	3 × 8	3.6 × 10.6

TABLE 5.6
Axial Height of Winding

Disc Designation	Quantity	Bare Conductor Width	Covered Width of Conductor	Axial Height of Bare Conductor	Axial Height of Covered Conductor	Total Axial Height of Paper
A	4	10	10.7	40	42.8	2.8
B	32	10	10.6	320	339.2	19.2
C	8	8	8.6	64	68.8	4.8
Total	44			424	450.8	26.8

TABLE 5.7
Spacer Size as Wound (Before Compression)

Spacer Designation	Grade of Spacer	Thickness of Spacer	Quantity	Total Height of Spacer (as Wound)
O	K-pre-compressed	4	4	16
V	K-pre-compressed	5	30	150
X	K-pre-compressed	3.2	8	25.6
D	K-pre-compressed	8	1	8
Total				199.6

5.10 ELECTRICAL CLEARANCES OF OIL-FILLED THREE-PHASE TRANSFORMERS

The typical clearances required for oil-filled three-phase transformers are given in Tables 5.9–5.17. If customer specification requires higher clearance, the same shall be used.

The HV–LV clearances have been calculated on the basis that the average AC stress during induced over voltage test is 5.5 kV/mm or less.

The maximum stress in oil can be calculated using the following procedure, when the HV–LV gap insulation is built up with oil and press board gap.

Concentric windings with test voltage V

$$e_i = \frac{V}{r_i E_i \left[\dfrac{1}{E_1} \ln \dfrac{r_2}{r_1} + \dfrac{1}{E_2} \ln \dfrac{r_3}{r_2} + \ldots \right]} \tag{5.34}$$

Where,

V = Total voltage across the gap (kV)

e_i = stress at the dielectric under consideration (kV/mm)

r_i = radius of dielectric under consideration (mm)

$E_1, E_2, E_3 \ldots$ = Dielectric permittivity of medium

E_i = Dielectric permittivity of the medium under consideration

Example

HV–LV gap of transformer

a. Assuming that the gap consists of oil only
$E = 2.2$ (for oil) H–L gap = 10 mm
Test voltage = 28 kV AC rms
Table 5.8 gives the maximum stress and average stress in the oil gap corresponding to different radii of electrodes. It is evident that when the radius is reduced the maximum stress increases.

b. HV–LV gap with press board cylinder
Test voltage = 28 kV AC rms
HV–LV gap = 10 mm
Radius of winding and press board cylinder are as Figure 5.13

$$\text{Stress in gap 1}\left(\text{gap between LV \& cylinder}\right) = \frac{28}{100 \times 2.2 \left[\dfrac{1}{2.2}\log_e \dfrac{104}{100} + \dfrac{1}{4.5}\log_e \dfrac{106}{104} + \dfrac{1}{2.2}\log_e \dfrac{110}{106} \right]}$$

$$= 3.272 \text{ kV/mm}$$

∴ Voltage drop in gap 1 = 3.272 × 4 = 13.08 kV

TABLE 5.8

Maximum Stress and Average Stress of Oil Gap

r_1 (mm)	r_2 (mm)	Maximum Stress (kV/mm)	Average Stress (kV/mm)
10	20	4.03	2.8
100	110	2.93	2.8
200	210	2.87	2.8
500	510	2.83	2.8
1000	1010	≈2.8	2.8

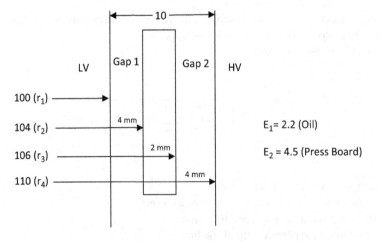

FIGURE 5.13 Typical radius of winding and pressboard cylinder.

TABLE 5.9
11 kV Class with 28 kV Test Voltage, 75 kV BIL

Description	Ratings (kVA)					
	25/63/100	160/200/250	315/400	500	630/1000	1600
Core–LV (radial dimension)	3.5	3.5	4	6	6	8
LV–HV (radial dimension)	9.5 / 10	9.5 / 10	9.5 / 10	10	10	10
HV–HV (radial dimension)	9 / 10	9 / 10	10	10	10	10
HV–Yoke: Top	25	25	25	25	30 / 30	30
Bottom	25	25	25	25	25 / 25	25
LV–Yoke: Top	9	9	17	19	23 / 40	20
Bottom	9	9	8	8	8 / 10	
Bet. crossover coils	8	8	8	8	8 / 8	8
Bet. tap coils	12	12	12	12	12	12
HV side– tank long side	55 (with tap) 50 (Without Tap)	65 (with tap)	65 (with tap)	65 (with tap)	75 (with tap)	80 (with tap)
LV side– tank long side	30	50	65	65	75	100
Winding– tank short side	30	35	40	40	40	40
Delta– winding (centre)	30 (with tap) 20 (without tap)	30 (with tap)	30 (with tap)	30 (with tap)	35 (with tap)	35 (with tap)
Delta–tank (centre)	35 (with tap) 30 (without tap)	35 (with tap)	35 (with tap)	35 (with tap)	40 (with tap)	45 (with tap)
No. of crossover coils	4	6	8	8	8	8

$$\text{Stress on press board} = \frac{28}{104 \times 4.5 \left[\dfrac{1}{2.2} \log_e \dfrac{104}{100} + \dfrac{1}{4.5} \log_e \dfrac{106}{104} + \dfrac{1}{2.2} \log_e \dfrac{110}{106} \right]}$$

$$= 1.53 \text{ kV/mm}$$

Voltage drop in press board $= 1.53 \times 2 = 3.06$ kV

$$\text{Stress in gap2} = \frac{28}{2.2 \times 106 \left[\dfrac{1}{2.2} \log_e \dfrac{104}{100} + \dfrac{1}{4.5} \log_e \dfrac{106}{104} + \dfrac{1}{2.2} \log_e \dfrac{110}{106} \right]}$$

$$= 3.0868 \text{ kV/mm}$$

TABLE 5.10
22 kV Class with 50 kV Test Voltage, 125 kV BIL, 433 V Class on LV

Description	Ratings (kVA)						
	25 / 63 / 100	160 / 200 / 250	315 / 400	500	630	800 / 1000	1600
LV–core (radial dimension)	3/3.5	3.5	4	5	6	6	8
LV–HV (radial dimension)	13	13	13	13	13	13	13
HV–HV (radial dimension)	13	13	13	13	13	13	13
HV–Yoke: Top	40	40	50	50	50	50	50
Bottom	40	40	40	40	40	40	40
LV–Yoke	------------MINIMUM 10 mm EDGE STRIP------------						
Bet. crossover coils	10	10	10	10	10	10	10
Bet. tap coils	12	12	12	12	12	12	15
HV side– tank long side	70 (without tap)	70 (with tap)	70 (with tap)	70 (with tap)	75 (with tap)	75 (with tap)	80 (with tap)
LV side– tank long side	40	55	65	65	75	90	100
Winding– tank short side	40	40	45	45	45	45	45
Delta– winding	30	30	30	30	30	30	35
Delta– tank	40	40	40	40	45	45	45
No. of crossover coils	6	8	10	10	10	10	10

TABLE 5.11
33 kV/433 V, 170 BIL, 70 kV AC

Description	Ratings (kVA)						
	50/100	160/200/250	315/400	500	630	800/1000	1600
LV–core (radial dimension)	3/3.5	3.5	4	5	6	6	8
LV–HV (radial dimension)	18	18	18	18	18	18	18
HV–HV (radial dimension)	18	18	18	18	18	18	18
HV–Yoke: Top	55	55	60	60	60	60	60
Bottom	55	55	55	55	55	55	55
LV–Yoke	MINIMUM 10 mm EDGE STRIP						
Bet. crossover coils	8	8	8	8	8	8	-
Bet. tap coils	12	12	12	12	12	12	
HV side– tank long side	140	140	140	140	140	140	140
LV side– tank long side	70	80	95	100	110	110	125
Winding– tank short side	70	70	70	70	70	70	70
Delta– winding	70	70	70	70	70	70	70
Delta– tank	70	70	70	70	70	70	70
No. of crossover coils	12	12	12	12	16	16	16

TABLE 5.12
33 kV/11 kV, 170 kV BIL, 70 kV AC for HV and 75 KV BIL 28 kV AC for 11 kV

Description	Ratings (kVA)			
	1600	3150	5000	8000 and 10,000
LV–Core (radial dimension)	18	18	20	20
LV–HV (radial dimension)	18	18	18	18
HV–HV (radial dimension)	18	18	18	18
HV-Yoke: Top	65	65	70	70
Bottom	55	55	60	60
LV–Yoke	55	55	55	55
HV side– tank long side	150	150	150	150
HV–tap lead	55	55	55	55
Tap lead–tank	95	95	95	95
LV side–tank long side	75	75	-	-
Winding–tank short side	75	75	75	75
	(non-tap switch side)	(non-tap switch side)	(non-tap switch side)	(non-tap switch side)
	150	150	150	150
	(tap switch side)	(tap switch side)	(tap switch side)	(tap switch side)

TABLE 5.13
Core–LV Insulation, HV–LV Insulation, Interphase Barrier, Insulation to Yoke, Number of Axial Spacers (Runners)

kVA	LV Voltage (Voltage Class of Winding Immediately after Core)	Core–LV Clearance (mm)	Number of Axial Runners	Size of Runner (Thickness × Width)	Core–LV Insulation Arrangement
25, 63, 100	433 V (voltage up to 1.1 kV class insulation)	3.5	-	-	Tape core with Webbing Tape 0.5 mm thick. Apply one press board wrap of 1.0 mm
160, 200, 250	433 V (voltage up to 1.1 kV class insulation)	3.5	-	-	Tape core with Webbing Tape 0.5 mm thick. Apply one press board wrap of 1.0 mm
315, 400	433 V (voltage up to 1.1 kV class insulation)	4.0	-	-	Tape core with Webbing Tape to build up 0.75 mm thickness and 1.0 mm press board wrap
500	433 V (voltage up to 1.1 kV class insulation)	6.0	-	-	Tape core with Webbing Tape to build up 1.0 mm thickness and 1.6 mm press board wrap
630, 800, 1000	433 V (voltage up to 1.1 kV class insulation)	6.0	10	-	Tape core with Webbing Tape to build up 1.0 mm thickness and 1.6 mm press board wrap
1600	433 V (voltage up to 1.1 kV class insulation)	8.0	12	-	Tape core with Webbing Tape to build up 2 mm thickness and 1.6 mm press board wrap
1000, 1600	3.3 kV	10	12	4 × 18	Core binding 2.0 mm thick + 3 mm cylinder + 4 mm runner

(Continued)

TABLE 5.13 (*Continued*)

Core–LV Insulation, HV–LV Insulation, Interphase Barrier, Insulation to Yoke, Number of Axial Spacers (Runners)

kVA	LV Voltage (Voltage Class of Winding Immediately after Core)	Core–LV Clearance (mm)	Number of Axial Runners	Size of Runner (Thickness × Width)	Core–LV Insulation Arrangement
	6.6 kV	10	12	4 × 18	Core binding 2.0 mm thick + 3 mm cylinder + 4 mm runner
	11 kV	11	12	5 × 18	Core binding 2.0 mm thick + 3 mm cylinder + 5 mm runner
3150, 5000	3.3 kV	10	12	4 × 18	Core binding 2.0 mm thick + 3 mm cylinder + 4 mm runner
	6.6 kV	10	12	4 × 18	Core binding 2.0 mm thick + 3 mm cylinder + 4 mm runner
	11 kV	12	12	6 × 18	Core binding 2.0 mm thick + 3 mm cylinder + 6 mm runner
8000, 10,000	6.6 kV	13	16	6 × 18	Core binding 0.25 mm thick + 50 mm wide resiglass tape × 7 turns + 3 mm thick cylinder + 6 mm runner
8000, 10,000	6.6 kV	13	16	6 × 18	Core binding 0.25 mm thick + 50 mm wide resiglass tape × 7 turns + 3 mm thick cylinder + 6 mm runner
	11 kV	13	16	6 × 18	Core binding 0.25 mm thick + 50 mm wide resiglass tape × 7 turns + 3 mm thick cylinder + 6 mm runner

TABLE 5.14

HV-LV Insulation Arrangement

Voltage Class of HV (kV)	HV – LV Clearance	HV – LV Gap Insulation Arrangement (Runner + Wrap/Cylinder)	Corrugated Board + Round Wood Rod
11	9	Runner – 3 mm on LV; Wrap – 1.0 mm × 2 Runner – 3.0 mm	2 mm Corrugated Board + 6 mm Wood Rod
	10	Runner – 4 mm on LV; Wrap – 1.0 mm × 2 Runner – 3.0 mm	2 mm Corrugated Board + 7 mm Wood Rod
	11	Runner – 4 mm on LV; Wrap – 1.0 mm× 2 Runner – 4.0 mm	2 mm Corrugated Board + 8 mm Wood Rod
	12	Runner – 5 mm on LV; Wrap – 1.0 mm × 2 Runner – 4.0 mm	2 mm Corrugated Board + 9 mm Wood Rod
22	13	Runner – 4 mm on LV; Wrap – 1.6 mm × 2 Runner – 5.0 mm	2 mm Corrugated Board + 10 mm Wood Rod
	14	Runner – 5 mm on LV; Wrap – 1.6 mm × 2 Runner – 5.0 mm	2 mm Corrugated Board + 11 mm Wood Rod
	15	Runner – 6 mm on LV; Wrap – 1.6 mm × 2 Runner – 5.0 mm	2 mm Corrugated Board + 12 mm Wood Rod
33	18	Runner – 4 mm on LV; Wrap – 3 mm × 2 Runner – 4.0 mm	3 mm Corrugated Board + 13 mm Wood Rod
	20	Runner – 5 mm on LV; Wrap – 3 mm × 2 Runner – 5.0 mm	3 mm Corrugated Board + 15 mm Wood Rod

TABLE 5.15
Interphase Barrier (HV-LV Insulation)

HV Test Voltage (kV$_{rms}$)	Arrangement of Interphase Barrier
28	2 × 1.6 mm Press Board
38	2 × 1.6 mm Press Board
50	2 × 3 mm Press Board
70	2 × 3 mm Press Board

TABLE 5.16
Winding to Yoke Insulation

kVA	Voltage class of HV and LV	Arrangement of Insulation to Yoke
25, 63, 100, 160	11 kV/433 V	HV-20 mm – Press Board Block (Top and Bottom). If customer specifies Angle Ring or Permawood Ring the same shall be provided and the remaining height to the filled by Press Board block.
200, 250	11 kV/433 V	HV – 20 mm – Top and Bottom 8 mm Press Board Spacer above HV Coil + 12 mm Permawood Ring
315, 400	11 kV/433 V	HV – Bottom – 25 mm 10 mm Press Board Spacer + 15 mm Permawood Ring HV – Top – 30 mm 10 mm Press Board Spacer + 20 mm Permawood common Ring LV – Bottom – 8 mm Edge Strip 8 mm Edge Strip + 20 mm Common Permawood ring with HV
500	11 kV/433 V	HV – Top – 30 mm 10 mm Press Board Spacer 20 mm Permawood Ring (Common with LV) HV – Bottom – 25 mm 10 mm Press Board Spacer 15 mm Permawood Ring LV – Bottom – 8 mm Edge Strip LV Top 8 mm Edge Strip + Common Permawood Ring with HV
630	11 kV/433 V	HV – Top – 35 mm 10 mm Press Board Spacer 20 mm Permawood Ring (Common with LV) HV – Bottom – 25 mm 10 mm Press Board Spacer 15 mm Permawood Ring LV – Bottom – 8 mm Edge Strip LV – Top 8 mm Edge Strip + Common Rings with HV

(Continued)

TABLE 5.16 (*Continued*)
Winding to Yoke Insulation

kVA	Voltage class of HV and LV	Arrangement of Insulation to Yoke
800, 1000	11 kV/433 V	HV – Top – 30 mm 8 mm Press Board Spacer + 12 mm Permawood HV Ring + 10 mm Permawood Ring (Common with LV) HV – Bottom – 25 mm 10 mm Press Board Spacer 15 mm Permawood HV Ring LV – Bottom – 10 mm Edge Strip LV – Top 30 mm including 10 mm common ring with HV
1600	11 kV/433 V	HV – Top – 40 mm 8 mm Press Board Spacer + 12 mm Permawood HV Ring + 20 mm Permawood Ring (Common with HV) + 4 mm free space HV – Bottom – 25 mm 10 mm Press Board Spacer + 15 mm ring LV – Bottom – 10 mm Edge Strip (minimum) LV – Top – 45 mm total including edge strip

Note: Common permawood rings and press board rings used for 315 to 1600 kVA slots shall be provided for cooling purposes at HV-LV Gap.

kVA	Voltage class of HV and LV	Arrangement of Insulation to Yoke
25, 63, 100, 160, 200 250	22 kV/433 V	HV – 40 mm each at Top and Bottom 12 mm Press Board Spacer + 12 mm Permawood Ring + 15 mm Common Ring made up by pasting 12 mm block on 3 mm Press Board Ring LV – Minimum 10 mm Edge Strip
315, 400	22 kV/433 V	HV – Top – 40 mm 12 mm Press Board Spacer + 12 mm Permawood HV Ring + 15 mm Common Ring with Petals made up by pasting 12 mm Permawood / Press Board on 3 mm press board ring HV – Bottom – 40 mm 12 mm Press Board Spacer 12 mm Permawood Ring LV – Minimum 10 mm Edge Strip
500	22 kV/433 V	HV – Top – 40 mm 12 mm Press Board Spacer + 12 mm Permawood HV Ring + 15 mm Common Ring with Petals made up by pasting 12 mm Permawood / Press Board Block on 3 mm Press Board Ring HV – Bottom – 40 mm 16 mm Press Board Spacer + 12 mm Permawood Ring + 12 mm Press Board Block LV – Minimum 10 mm Edge Strip

(*Continued*)

TABLE 5.16 (*Continued*)
Winding to Yoke Insulation

kVA	Voltage class of HV and LV	Arrangement of Insulation to Yoke
630	22 kV/433 V	HV – Top – 40 mm 12 mm Press Board Spacer + 12 mm Permawood HV Ring + 15 mm Common Ring with Petals made up by pasting 12 mm Permawood / Press Board on 3 mm press board ring HV – Bottom – 40 mm 16 mm Press Board Spacer + 12 mm Permawood Ring + 12 mm Press Board Block LV – Minimum 10 mm Edge Strip
800, 1000	22 kV/433 V	HV – Top – 40 mm 12 mm Press Board Spacer + 12 mm Permawood HV Ring + 15 mm Common Ring with Petals made up by pasting 12 mm Permawood / Press Board on 3 mm press board ring HV – Bottom – 40 mm 16 mm Press Board Spacer + 12 mm Permawood Ring + 12 mm Press Board Block LV – Minimum 10 mm Edge Strip
1600	22 kV/433 V	HV – Top – 40 mm 12 mm Press Board Spacer + 12 mm Permawood HV Ring + 15 mm Common Ring with Petals made up by pasting 3 mm Press board ring HV – Bottom – 40 mm 16 mm Press Board Spacer + 12 mm Permawood Ring + 12 mm Press Board Block LV – Minimum 10 mm Edge Strip
50, 100, 200, 250	33 kV/433 V	HV - Top and Bottom– 55 mm each 15 mm Press Board Spacer + 25 mm Permawood HV Ring + 15 mm Common Ring made up by pasting 12 mm Permawood / Press Board on 3 mm Press Board Ring LV – Minimum 10 mm Edge Strip
315, 400	33 kV/433 V	HV – Top – 65 mm 15 mm Press Board Spacer + 25 mm Permawood HV Ring + 15 mm Common Ring with Petals made up by pasting 3 mm Permawood / Press Board + 25 mm perma wood ring HV – Bottom – 55 mm 15 mm Press Board Spacer + 25 mm Permawood Ring + 15 mm Common Ring with Petals made up by pasting 12 mm Permawood / Press Board on 3 mm Press Board Ring LV – Minimum 10 mm Edge Strip

(Continued)

TABLE 5.16 (*Continued*)
Winding to Yoke Insulation

kVA	Voltage class of HV and LV	Arrangement of Insulation to Yoke
500	33 kV/433 V	HV – Top – 55 mm 15 mm Press Board Spacer + 15 mm Common Ring with Petals made up by pasting 3 mm press board ring on 25 mm Permawood Ring HV – Bottom – 55 mm 15 mm Press Board Spacer + 25 mm Permawood Ring + 15 mm Common Ring with Petals made up by pasting 12 mm Permawood / Press Board on 3 mm Press Board Ring LV – Minimum 10 mm Edge Strip
630	33 kV/433 V	HV – Top – 55 mm 15 mm Press Board Spacer + 15 mm Common Ring with Petals made up by pasting 12 mm Permawood / Press Board on 3 mm Press Board Ring + 25 mm Permawood Ring HV – Bottom – 55 mm 15 mm Press Board Spacer + 25 mm Permawood Ring + 15 mm Common Ring with Petals made up by pasting 12 mm Permawood / Press Board on 3 mm Press Board Ring LV – Minimum Common Edge Strip
800, 1000	33 kV/433 V	HV – Top – 55 mm 15 mm Press Board Spacer + 25 mm Permawood HV Ring + 15 mm Common Ring with Petals made up by pasting 12 mm Permawood / Press Board on 30 mm press board ring HV – Bottom – 55 mm 15 mm Press Board Spacer + 25 mm Permawood Ring + 15 mm Common Ring with Petals made up by pasting 12 mm Permawood / Press Board on 3 mm Press Board Ring LV – Minimum 10 mm Edge Strip
1600	33 kV/433 V	HV – Top – 55 mm 15 mm Press Board Spacer + 25 mm Permawood HV Ring + 15 mm Common Ring with Petals made up by pasting 12 mm Permawood / Press Board on 3 mm press board HV – Bottom – 55 mm 15 mm Press Board Spacer + 25 mm Permawood Ring + 15 mm Common Ring with Petals made up by pasting 12 mm Permawood / Press Board on 3 mm Press Board Ring LV – Minimum 10 mm Edge Strip

(*Continued*)

TABLE 5.16 (*Continued*)
Winding to Yoke Insulation

kVA	Voltage class of HV and LV	Arrangement of Insulation to Yoke
1600	33 /11 kV	HV and LV Top – 55 mm
		15 mm Press Board Spacer +
		25 mm Individual Permawood Ring +
		15 mm Common Ring with Petals made up by pasting 12 mm Permawood / Press Board on 3 mm press board ring
		HV – Bottom – 55 mm
		15 mm Press Board Spacer +
		25 mm Individual Permawood Ring +
		15 mm Common Ring with Petals made up by pasting 12 mm Permawood / Press Board on 3 mm Press Board Ring
3150	33 /11 kV	HV and LV Top – 65 mm
		15 mm Press Board Spacer +
		45 mm Common Ring made up by pasting 10 mm Press Board / Permawood Block on 35 mm Permawood Ring+
		5 mm free space
		HV and LV - Bottom – 55 mm
		15 mm Press Board Spacer +
		25 mm Individual Permawood Ring +
		15 mm Common Ring made up by pasting 10 mm Press Board/ Permawood Block on 5 mm Press Board ring
5000, 8000	33 /11 kV	HV and LV Top –70 mm
		15 mm Press Board Spacer +
		50 mm Common Ring made up by pasting 10 mm Press Board / Permawood Block on 40 mm Permawood Ring + 5 mm free space
		HV and LV - Bottom – 60 mm
		15 mm Press Board Spacer +
		25 mm Individual Permawood Ring +
		20 mm Common Ring made up by pasting 15 mm Press Board/ Permawood Block on 5 mm Press Board

5.11 CALCULATION OF ELECTRICAL STRESSES FOR DIFFERENT CONFIGURATIONS

5.11.1 Two Bare Uniform Electrodes (Parallel Electrodes) (One Dielectric) (Figure 5.14)

$$E = \frac{V}{a}$$

a = distance between electrodes
V = applied voltage
E = voltage stress

Example: Parallel plate capacitor

TABLE 5.17

Typical Internal Clearances of Oil-Filled Transformers

Sl. No.	Rated Voltage (kV)	11	22	33	66	110 / 132	132	220
1.	Impulse test voltage level (kV peak)	75	125	170	325	550	650	950/1050
2.	AC test voltage (kV rms)	28	50	70	140	230	275	395/400
3.	Insulation of lead wire (increase in diameter)	3	4	4	8	12	12	28/28
4.	Clearance from lead wire to flat plate	30	40	70	120	140	180	200
5.	Clearance from lead wire to square corner to tank	30	40	80	140	170	250	300
6.	Clearance of lead wire to core clamp	30	40	95	175	220	310	380
7.	Clearance of lead wire to winding	20	40	60	120	140	180	200
8.	Clearance of lead wire to Permawood clamping ring	20	40	80	140	170	290	250
9.	Clearance between lead wire and lead wire	20	40	70	120	140	180	200
10.	Clearance from flat plate to winding without outer cylinder	20	40	80	140	170	250	300
11.	Clearance from flat plate to winding with outer cylinder	20	40	75	100	130	190	220
12.	Clearance of winding to square corner of tank	30	40	80	140	175	290	350
13.	Clearance from winding to Permawood ring	20	30	70	130	165	270	330
14.	Clearance of bare part of bushing to flat plate	30	60	80	140	170	250	330
15.	Clearance of lower shield of bushing to flat plate	-	-	-	140	170	250	300
16.	Clearance of lower shield of bushing to core clamp	-	-	-	175	190	290	350

5.11.2   Multidielectric (Parallel Electrodes – Figure 5.15)

$$E_i = \frac{V}{\epsilon_i \left[\dfrac{d_1}{\epsilon_1} + \dfrac{d_2}{\epsilon_2} + \dfrac{d_3}{\epsilon_3} \right]} \qquad i = 1, 2, 3 \qquad\qquad (5.35)$$

V = applied voltage
$\epsilon_1, \epsilon_2, \epsilon_3$ = relative permeability of the dielectric
d_1, d_2, d_3 = thickness of dielectric
E_i = voltage stress in dielectric i

The stress value is constant within each dielectric

5.11.3   Concentric Cylindrical Electrodes (One Dielectric – Figure 5.16)

$$E = \frac{V}{r \, \log_e \dfrac{r_2}{r_1}} \qquad\qquad (5.36)$$

a   V = applied Voltage

FIGURE 5.14   Parallel plate capacitor.

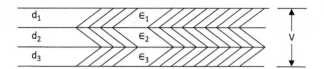

FIGURE 5.15 Parallel electrodes.

V = applied voltage
E = stress at radius r
r_1 = diameter of inner electrodes
r_2 = diameter of outer electrodes

Maximum stress occurs at inner electrode $E_{max} = \dfrac{V}{r_1 \, \log_e \dfrac{r_2}{r_1}}$

5.11.4 Concentric Cylindrical Electrodes with Multiple Dielectrics (Figure 5.17)

$$E_i = \frac{V}{r_i E_i \left[\dfrac{1}{E_1} \log_e \dfrac{r_2}{r_1} + \dfrac{1}{E_2} \log_e \dfrac{r_3}{r_2} + \dfrac{1}{E_3} \log_e \dfrac{r_4}{r_3} \right]} \qquad i = 1,2,3\ldots \qquad (5.37)$$

5.11.5 Cylindrical Conductor to Plane Electrode (Figure 5.18)

(Uninsulated Conductor and Electrode)
   The maximum stress occurs on conductor at "P" (shortest distance from conductor to plane)

R = radius of cylindrical conductor
S = distance from the centre of conductor
V = voltage

$$E_p = \frac{V}{(S-P)} \left[\frac{\sqrt{\left(\dfrac{S}{R}\right)^2 - 1}}{\log_e \left(\dfrac{S}{R} + \left(\dfrac{S}{r}\right)^2 - 1 \right)} \right] \qquad (5.38)$$

E_p = maximum stress

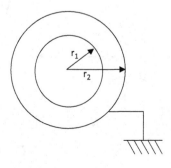

FIGURE 5.16 Concentric cylindrical electrodes.

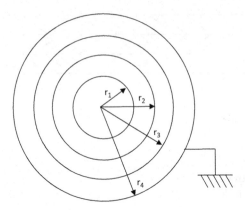

FIGURE 5.17 Concentric cylindrical electrodes with multiple dielectrodes.

5.11.6 INSULATED CYLINDRICAL CONDUCTOR TO PLANE ELECTRODE (FIGURE 5.19)

[Insulated lead to Tank]

V = voltage on the conductor
r_1 = radius of conductor
r_2 = radius of conductor + insulation
(outside radius of insulated conductor)
S = distance from centre line of lead to ground (insulated lead centre to tank)
E_1 = dielectric strength of insulation
E_2 = dielectric strength of insulation between lead and electrode
(dielectric strength of oil for oil-filled transformers)

Maximum stress in oil occurs at A, which is the shortest distance from the outer surface of insulated lead to flat electrode.

Maximum stress in paper occurs at B, which is the shortest distance from the outer surface of conductor without insulation to flat electrode.

$$E_A = \cfrac{V}{r_2 \in_2 \left\{ \cfrac{\left(\dfrac{S}{r_2}-1\right)\log_e\left[\left(\dfrac{S}{r_2}\right)+\sqrt{\left(\dfrac{S}{r_2}\right)^2-1}\right]}{\in_2\sqrt{\left(\dfrac{S}{r_2}\right)^2-1}} + \cfrac{\left(\dfrac{r_2}{r_1}-1\right)\log_e\left[\left(\dfrac{r_2}{r_1}\right)+\sqrt{\left(\dfrac{r_2}{r_1}\right)^2-1}\right]}{\in_2\sqrt{\left(\dfrac{r_2}{r_1}\right)^2-1}} \right\}} \qquad (5.39)$$

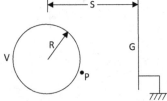

FIGURE 5.18 Uninsulated conductor and electrode.

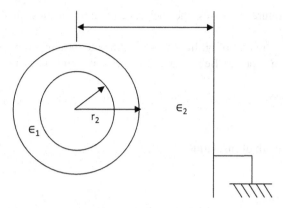

FIGURE 5.19 Insulated cylindrical conductor to plane electrode.

$$E_B = \cfrac{V}{r_1 \in_1 \left\{ \cfrac{\left[\left(\dfrac{s}{r_2}-1\right)\log_e\left[\left(\dfrac{s}{r_2}\right)+\sqrt{\left(\dfrac{s}{r_2}\right)^2 - 1}\right]\right]}{\in_2 \sqrt{\left(\dfrac{s}{r_2}\right)^2 - 1}} + \cfrac{\left[\left(\dfrac{r_2}{r_1}-1\right)\log_e\left[\left(\dfrac{r_2}{r_1}\right)+\sqrt{\left(\dfrac{r_2}{r_1}\right)^2 - 1}\right]\right]}{\in_2 \sqrt{\left(\dfrac{r_2}{r_1}\right)^2 - 1}} \right\}} \tag{5.40}$$

5.11.7 Factors Affecting Insulation Strength

It is a statistically distributed quantity and is affected by
Moisture and Impurities

- Time and frequency (duration of voltage application and frequency)
- Temperature
- Thickness
- Stressed volume
- Voltage wave shape and type of voltage

a. Volt–time characteristics (*V–t* characteristics – Figure 5.20)

The higher the voltage, the lower the time to cause a breakdown. The *V–t* characteristics (typical) are as below:

Some amount of energy is required to cause breakdown, and breakdown voltage and time are interdependent.

The infinite time strength $\cong \left(\dfrac{2}{3}\right)$ one minute strength

One second strength $\cong 1.6$ times one minute strength

b. Effect of frequency

For solid insulation, the strength $f^{-0.137}$

For liquid insulation, the effect is not much.

The effect of frequency on the creepage distance is not high (Figures 5.21 and 5.22).

c. Effect of Impulse Voltage

d. Effect of Temperature

- When temperature increases, the dielectric strength of most of the solid insulation reduces.
- The dielectric strength of oil increases with the increase in temperature in the operating range of transformer. Below −5°C, the dielectric strength increases as the moisture gets frozen.

e. Effect of Thickness

$$E \propto T^n \tag{5.41}$$

E = dielectric strength of insulation
T = thickness
$n \cong 0.5–0.96$

f. Stressed Volume Effect
    Plain oil gap

$$E = 17.9 \, (SOV)^{-0.137} \tag{5.42}$$

E = breakdown strength kV/mm (rms)
SOV = stressed volume (cm³)

g. Creepage Strength

$$E_{creep} = 15 \, d^{-0.37} \tag{5.43}$$

E_{creep} = creepage strength kV/mm (rms)
d = creepage path length (mm)

5.12   INSULATION BETWEEN LAYERS

Kraft paper is normally used for interlayer insulation. If epoxy dotted paper is used, it gives more mechanical strength to the winding. It is recommended to use it when foil windings are used.
    The thickness of insulation between layers is determined on the following basis:

a. Maximum test voltage between layers at the time of induced overvoltage test.
b. Impulse voltage between layers during impulse test. This is significant for the first two layers near the line.
c. Insulation provided on the conductor.

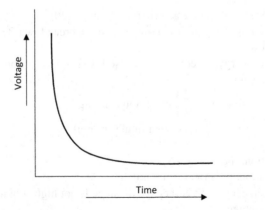

FIGURE 5.20   The V–t characteristics.

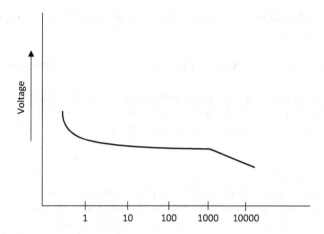

FIGURE 5.21   Impulse volt–time curve of oil gap.

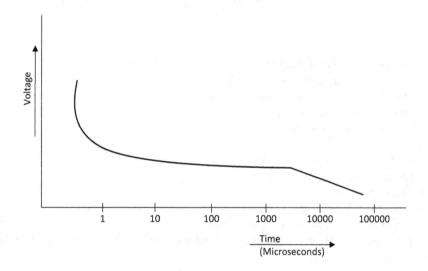

FIGURE 5.22   Volt–time characteristics of oil-impregnated press board.

1. Interlayer insulation required for induced overvoltage test

$$t = \frac{2\, n\left(V/T \right) E_{test}}{E\ K} - t_1 \qquad (5.44)$$

where
t = thickness of interlayer insulation (mm)
n = turns per layer (use maximum turns per layer)
V/T = volts per turn
E_{test} = test voltage (volts)
E = rated voltage of winding (volts)
t_1 = thickness of conductor insulation (mm)

For enamelled conductor, equivalent thickness of paper covering shall be used. For medium covering, use

$t_1 = 0.125$

K = allowable maximum stress on interlayer insulation (V/mm) when induced overvoltage test is done

= 4500 V/mm for conventional transformers without vacuum filling

= 5000 V/mm for sealed/CSP transformers without vacuum filling

= 5500 V/mm for transformers with vacuum filling

Note:

1. Minimum value of "t" shall be 0.15 mm
2. When interlayer insulation thickness as calculated above exceeds 0.4 mm, graded insulation between layers can be tried so that the overall insulation requirement is reduced. Figure 5.23 shows uniform integer insulation and graded integer insulation.
3. Interlayer insulation for impulse strength between first two layers from line end:
    The interlayer insulation thickness for first two line layers is calculated as below:

$$t = \frac{P\,(\text{BIL}) \propto \beta\, n}{20\, T_{min}} - t_1 \tag{5.45}$$

where

t = interlayer insulation thickness (mm)

$P = 1$ for delta connection for three-phase transformers

0.7 for Wye connection for three-phase transformers

1 for single-phase transformers

BIL = impulse test voltage kV$_{peak}$

$\alpha = 1.5$

$\beta = 1.5$ for three-phase transformer

= 1.8 for single-phase transformer

n = turns per layer

T_{min} = turns at lowest tap

t_1 = thickness of conductor insulation (mm)

For enamelled conductor, equivalent thickness of paper covering shall be used. For medium covering, use

$t_1 = 0.125$

Note: The interlayer thickness at the two line end layers shall be the maximum value of "t" calculated as above in (2) for impulse strength or calculated as in (1) for induced overvoltage withstand strength, whichever is higher.

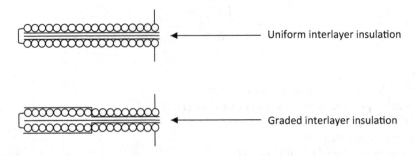

FIGURE 5.23   Layer arrangement.

5.13   CALCULATION OF WINDING DIAMETER AND RADIAL DEPTH

Winding diameters and radial depth of windings are calculated as below:

$$ID = \text{Inside diameter, } OD = \text{Outside diameter}$$

$$RD = \text{Radial depth}$$

$$LV\ ID = \text{Core diameter} + 2(\text{core} - LV\ \text{gap})$$

$$LV\ OD = LV\ ID + 2(m\ n\ t + KD) - \text{for spiral winding with axial duct}$$

$$LV\ ID + 2[m\ n\ t + (m-1)t_1] - \text{for spiral winding without duct (with interlayer insulation only)}$$

where
m = number of layers of winding
n = number of conductors in over (radial) direction per turn
(i.e. if conductor arrangement is 1F × 2 over, $n = 2$)
t = covered thickness of one insulated parallel conductor
K = number of cooling ducts in axial direction
D = size of duct in axial direction
t_1 = thickness of interlayer insulation when no duct is provided

For Helical Winding

$LV\ OD = LV\ ID + 2\ (m\ n\ t + KD)$ – when cooling ducts are provided
$= LV\ ID + 2\ [m\ n\ t + (m-1)\ t_1]$ – when no cooling ducts are used
$RD = \text{radial depth of winding} = \dfrac{OD + ID}{2}$
$MLT = \text{mean length of turn} = (ID + RD)\pi$
$HV\ ID = LV\ OD + 2(HV - LV\ \text{Gap})$
For foil winding
$HV\ OD = HV\ ID + 2[nt + (n-1)t_1]$ – when cooling ducts are not used
$HV\ OD = HV\ ID + 2[nt + KD + (n-1(K+1))t_1]$ – when cooling ducts are provided

where
d = covered diameter of conductor
n = number of layers

For crossover coils:

$$HV\ OD = HV\ ID + 2[dn + (n-1)t_1 + KD]$$

where
d = covered diameter of conductor
n = number of layers
t_1 = thickness of interlayer insulation
D = cooling duct thickness
K = number of cooling ducts

For disc winding:

$$HV\ OD = HV\ ID + 2[pnt + KD]$$

where
n = number of turns per disc
t = covered thickness of one conductor
p = number of parallel conductors
K = number of cooling ducts
D = cooling duct thickness

5.14 WEIGHT OF BARE CONDUCTOR, COVERED CONDUCTOR AND RESISTANCE OF WINDING

1. Weight of bare conductor is calculated using mean length of turn, number of turns, area of cross section of the turn and the density of the material. Weights corresponding to lowest tap, rated tap and maximum tap shall be calculated.
2. When calculating bare weight, covered weight, resistance and I^2R loss, lead length also shall be taken into consideration. For transformers up to 200 kVA with LV voltage of 240 or above, the allowance for lead wire can be taken as 2% of the weight of LV conductor. For higher ratings, length of LV lead wire up to bushing and neutral bar shall be calculated for estimating the weight of LV conductor.
3. Resistance of winding is calculated as below:

$$R = \frac{\rho l}{A}$$

where
$\rho = 2.1 \times 10^{-6}$ Ω.cm for copper at 75°C
$= 3.46 \times 10^{-6}$ Ω.cm for aluminium at 75°C
l = total length of conductor (cm)
$= \dfrac{MLT \times N}{10}$
A = area of cross section of conductor (cm²)

Note: For high-conductivity aluminium, "ρ" will be lower than the above value.

6 Calculation of Load Loss

6.1 CALCULATION OF I^2R LOSS

The I^2R loss is calculated by adding up the DC resistance losses of all windings. The phase current corresponding to the required tap position shall be taken for the calculation. The losses need to be calculated for the reference temperature, and the resistance values corresponding to the applicable reference temperature shall be used. As per IEC 60076, the reference temperature is 75°C for oil-filled transformers, whereas it is 85°C as per ANSI C57.12 standard.

The I^2R loss can be estimated by knowing the weight of conductor and the current density. The calculation using resistance value shall be used for design, and the method as per this section shall be used for quick checking/estimation only.

$$I^2R = K\delta^2W \text{ (Watts)} \tag{6.1}$$

where
  $K = 2.4$ for copper
      13.0 for aluminium
  δ = current density (A/mm^2)
  W = bare conductor weight (kg)
  The above formula is derived by the following equations:

$$I^2R = I^2\frac{\rho l}{A} \tag{6.2}$$

$$lAD = W$$

  ρ = resistivity of conductor material
  W = weight
  l = length
  D = density
  A = area of conductor

$$\therefore l \propto \frac{W}{AD}$$

Substituting in Equation (6.2)

$$I^2R \propto \frac{I^2\rho\ W}{A^2D} \propto \frac{\delta^2\rho\ W}{D} = k\delta^2W$$

When values of ρ and D are substituted for copper and aluminium, the values of K become 2.4 and 13, respectively, at 75°C.

6.2 CALCULATION OF EDDY CURRENT LOSSES AND STRAY LOSSES

Eddy current and stray losses are produced due to leakage fluxes in windings, bushing plates, busbars, tanks, core channels and other metal parts in proximity of the leakage flux. The spatial distribution of the axial and radial leakage flux is required for accurate calculation of these losses, especially when the ampere-turn distribution is asymmetrical and when the leakage flux density is higher. For transformers up to 10000 kVA, 33 kV class, the following formula can be used for the estimation of eddy current and stray losses.

Total eddy current and stray loss = Eddy current loss in HV winding

+ eddy current loss in LV winding + Stray loss in bushing plate + stray loss in tank

+ Stray loss due to circulating current + Stray loss in channels + lead wire loss

+ miscellaneous stray loss

Total load loss = I^2R loss + winding eddy current loss + stray loss

6.2.1 EDDY CURRENT LOSS IN WINDING

$$w_s = 9f^2B^2t^2w \times 10^{-3} \text{ Watts for Copper winding (for rectangular conductor)} \quad (6.3)$$

$$w_s = 9f^2B^2t^2w \times 0.49 \times 10^{-3} \text{ Watts for Copper winding (for round conductor)} \quad (6.4)$$

$$w_s = 19f^2B^2t^2w \times 10^{-3} \text{ Watts for Aluminium winding (for rectangular conductor)} \quad (6.5)$$

$$w_s = 19f^2B^2t^2w \times 0.49 \times 10^{-3} \text{ Watts for Aluminium winding (for round conductor)} \quad (6.6)$$

where
W_s = stray loss in winding
f = frequency (Hz)
B = maximum value of axial leakage flux density (Tesla)

$$B = 1.78 \frac{IT}{H_{av}} \times 10^{-3}$$

I = rated phase current of winding
T = turns at rated tap of the same windings

$$H_{av} = \text{Average axial height (mm)} = \frac{\text{HV axial height} + \text{LV axial height}}{2}$$

t = thickness of conductor (mm)
w = bare weight of winding (kg)

Note
1. Winding stray loss shall be calculated separately for HV and LV windings. If separate tap winding is used, eddy current loss of tap winding shall be calculated using the maximum value of leakage flux linkage to the tap coil.
2. Total weight of winding shall include the weight of tap conductors also which are inside the leakage field. If taps are taken from inside layer/layers of winding, the complete turns will be subjected to leakage flux, even if some turns are not carrying current.

- Eddy current loss of winding (Alternate Formula 1 for copper winding)

$$P_{eddy} = 3.1\left(\frac{f}{50}\right)^2 \left(\frac{t}{10}\right)^{3.84} (n_r m_l)^{2.07} I^2 R \qquad (6.7)$$

where

f = frequency
t = conductor thickness (mm)
n_r = number of parallels in radial direction
m_l = number of layers of winding
$(n_r m_l)$ = total number of conductors in radial direction
$I^2 R = I^2 R$ loss (kW)

Example

$I^2 R$ loss = 19,214 W
f = 50 Hz
t = 3.2 mm
n_r = 5
m_l = 4

$$P_{eddy} = 3.1\left(\frac{50}{50}\right)^2 \left(\frac{3.2}{10}\right)^{3.84} (5\times4)^{2.07} \times \frac{19,214}{100} \qquad (6.8)$$

$$= 3.1 \times 0.01256 \times 493.2 \times 192.14 = 3690 \text{ watts}$$

- Eddy current loss as % of $I^2 R$ loss winding (Alternate Formula 2 for copper winding)

$$W_E = \frac{(IT)^2 k t^2}{(100.H)^2 \times 1.07 \times P^2} \qquad (6.9)$$

W_E = eddy current loss as % of $I^2 R$ loss (copper winding)
I = phase current
T = turns/phase
K = 15 for normal spiral winding, transposed helical winding, disc winding
t = conductor thickness (mm)
P = current density (Amps/sq mm)
H = effective electrical height of winding (mm)

6.2.2 Stray Loss in Bushing Plate

$$W_{Bp} = 0.045\, I^{1.25}\left(\frac{f}{50}\right)^{1.5} (\text{Watts}) \qquad (6.10)$$

W_{Bp} = stray loss on tank plate from bushing current
I = rated current of transformer through bushing (Amps)
When brass or stainless steel is used,

$$W_{Bp} = 0.045\, I^{1.25}\left(\frac{f}{50}\right)^{1.5} \times 0.5 (\text{Watts}) \qquad (6.11)$$

where

f = rated frequency

6.2.3 STRAY LOSSES IN FLITCH PLATE (TIE PLATE)

- If tie rods are provided, these losses are not applicable
- If flitch plates are provided, losses are calculated as per the following formula:

$$\text{Losses of steel flitch Plate} = 2.6 \times 10^4 \, W^{2.4} B^2 \tag{6.12}$$

W = width of flitch plate (m)
B = flux density in flitch plate (Tesla)
For stainless steel flitch plate

$$\text{Losses} = 1.44 \times 10^5 \, W^3 B^2 \, (\text{kW/m}) \tag{6.13}$$

The flux density in flitch plate is calculated as below:

$$\frac{B_1}{B_2} = \frac{\mu_1}{\mu_2} \tag{6.14}$$

B_1 = flux density in core (Tesla)
B_2 = Flux density in flitch plate (Tesla)
μ_1 = relative permeability of core ≈ 5000
μ_2 = relative permeability of flitch plate ≈ 200
If core flux density is 1.7 T, the flux density on 100 mm wide steel plate is calculated as follows:

$$B_2 = 1.7 \times \frac{\mu_2}{\mu_1} = \frac{1.7 \times 200}{5000} = 0.068 \text{ Tesla}$$

$W = 0.1 \, \text{m}$
Length of tie plate $= 1.1 \, \text{m}$

$$\text{Losses in Tie Plate} = 2.6 \times 10^4 \times 0.1^{2.4} \times 0.068^2 \text{ kW/m}$$

$$= 0.4786 \, \text{kW/m}$$

For full length of tie plate loss $= 1.1 \times 0.4786 = 0.526 \, \text{kW}$
This loss is due to the flux in the core. The losses due to the leakage flux will have to be added to the above.

6.2.4 CIRCULATING CURRENT LOSS IN CONTINUOUS DISC WINDING

$$P_{cc} = 2.4 \times 10^6 \times \frac{1}{180} (KR)^4 \left(1 - \frac{5}{n^2} + \frac{4}{n^4} \right) \times I^2 R \tag{6.15}$$

where
R = radial size of the insulated turns
$= n(t + t_i)$ in m
n = number of turns in radial direction
t = thickness of one conductor
t_i = insulation thickness between conductors

$$k = 0.48 \times 10^{-4} \sqrt{\frac{h_c}{h_w}} \left(\frac{t}{t + t_i} \right) \frac{f}{p} \tag{6.16}$$

TABLE 6.1

Increase in Circulating Current Loss %

Number of Conductors in Radial Direction/Disc	Proportionate Increase
1	0
2	0
3	0.4938
4	0.7031
5	0.8064

p = resistivity of conductor material
h_c = total height of copper in axial direction
h_w = total winding height in axial direction

$\left(\dfrac{h_c}{h_w}\right)$ and $\left(\dfrac{t}{t+ti}\right)$ are space factors in axial and radial directions

Example

$f = 50\,\text{Hz}$
$t = 2.4\,\text{mm}$
$t_i = 0.7\,\text{mm}$
$\dfrac{h_c}{h_w} = 0.8$

$$\rho = 2.1 \times 10^{-8}\,\Omega/\text{m}\,(\text{for copper})$$

$$\therefore R = 8(2.4+0.7)\times 10^{-3} = 0.0248\,\text{m}$$

$$K = 0.48\times 10^{-4}\sqrt{0.8\times\left(\frac{2.4}{2.4+0.7}\right)\frac{50}{2.1\times 10^{-8}}} = 1.843$$

$n = 8$

$$P_{cc} = 2.4\times 10^6 \times \frac{1}{180}(1.843\times 0.0248)^4\left(1-\frac{5}{8^2}+\frac{4}{8^4}\right)\times\left(I^2 R\,\text{loss}\right)$$

$$= 0.053\times\left(I^2 R\,\text{loss}\right)$$

Increase of circulating current loss depends on number of conductors in radial direction as given in Table 6.1.

6.2.5 EMPIRICAL FORMULA FOR TANK LOSS CALCULATION

Empirical formula–1

$$\text{Tank Loss} = 6.3\times\left[\left(\frac{IT}{h}\right)^2\times 10^{-6}\times(LVOD + HL\ Gap)\times(LV\ RD + 3\times HL\ Gap + HV\ RD)\right]$$

$$(6.17)$$

Where,

     h = winding axial height (Electrical) (mm)

     I = phase current (A)

     T = Number of turns / phase

All dimensions are in mm

For corrugated fin type (with corrugations all round), multiply the above by 0.70.

Empirical formula–2

$$W = N \times \left(\frac{f}{50}\right)^{1.5} \frac{I^2 T^2}{H} \times 10^{-4} \tag{6.18}$$

W = tank loss (watts)

N = number of winding faces exposed to tank

(N = 6 if winding to tank clearance is almost equal all around)

(N = 5 if tap changer is on short side, etc.)

I = phase current

T = turns/phase

H = window height

HV OD = outer diameter of HV winding

C = clearance from HV to tank

f = frequency

Empirical Formula 3

     Stray loss in tank due to axial component of leakage flux {unshielded steel tank}

$$W_T = 15 \, A\left(\frac{L}{P}\right) K^2 \phi^2 \times 10^{-10} \text{ watts at 50 Hz}$$

$$= 1.3 \times 15 A\left(\frac{L}{P}\right) K^2 \phi^2 \times 10^{-10} \text{ watts at 60 Hz}$$

where

$$A = \frac{\text{Area of leakage flux flow}}{\text{Tank perimeter}} \, \text{m}^2/\text{m}$$

Area of leakage flux flow is calculated as below:

- Area of leakage flux flow is the shaded area in Figure 6.1. Ignore the area occupied by OLTC (if any) and any major additions to tank size other than normal.

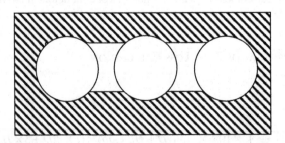

FIGURE 6.1   Area of leakage flux flow.

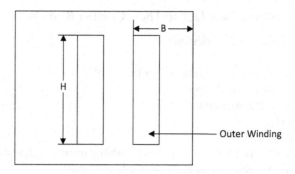

FIGURE 6.2 Parameters for tank loss calculation.

- Tank perimeter is the effective perimeter of the tank inside dimensions (mm). The parameters for tank loss calculations are shown in Figure 6.2.

$L = H + 2B$ (mm)
H = winding axial height (mm)
B = average distance from the tank to the inside edge of outer winding
P = tank perimeter (mm)
$K = 0.6$
ϕ_{max} = leakage flux/limb lines
N = turns in untapped winding
I = rated current in untapped winding

$$B_{max} = \frac{1.78 I\, N}{H} \times 10^{-3}\ \text{Tesla}$$

Leakage flux in lines is calculated as below
The mean circumference of the flux path has calculated from the parameters shown in Figure 6.3

$$\text{Mean circumference of flux path} = \pi \left[\text{LV ID} + 2 \times W_{LV} + W_g \right]$$

LV ID = Inside diameter of LV
W_{LV} = Radial depth of LV
W_g = HV-LV gap

$$\varphi = B_{max} \times 10^2 \times \pi \left[LVID + 2W_{LV} + W_g \right] \times \left[\frac{1}{2} W_{LV} + W_g + \frac{1}{2} W_{HV} \right] \qquad (6.19)$$

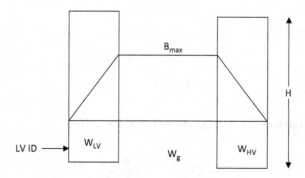

FIGURE 6.3 Windings and HL gap.

6.2.6   LOSS ON TRANSFORMER TANK DUE TO HIGH-CURRENT BUSBARS

Figure 6.4 shows the parameters for calculations

$W = 300\ I^2 d^n$

W = loss on tank due to current carrying busbar (watts (W)/m^2)

I = current through busbar (kilo Amperes)

d = distance between busbar and tank (mm)

$n = -0.4$ for mild steel

$\quad = -0.8$ for non-magnetic steel

The area of the tank affected by the proximity of busbar current is calculated as $1.5 \times L \times w$

$\{L$ = length of busbar exposed to tank and w = width of busbar$\}$

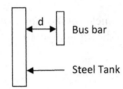

FIGURE 6.4   Busbar clearance.

Example

Current = 2000 A (I = 2)

  Distance between tank and busbar = 100 mm

  Tank material is mild steel (n = −0.4)

$$W = 300 \times 2^2 \times 100^{-0.4}$$

$$= 190\ W/m^2$$

a. If the tank material is stainless steel

$$W = 300 \times 2^2 \times 100^{-0.8}$$

$$= 30\ W/m^2$$

b. If clearance is 50 mm with mild steel tank

$$W = 300 \times 2^2 \times 50^{-0.4}$$

$$= 251\ W/m^2$$

c. If clearance is 200 mm

$$W = 300 \times 2^2 \times 200^{-0.4}$$

$$= 144\ W/m^2$$

6.2.7   EMPIRICAL FORMULA FOR CALCULATING TOTAL STRAY LOSS

$$W_s = K \left[\frac{IT}{1000\,h} \right]^2 D \left[a + \frac{b+c}{3} \right] \qquad (6.20)$$

The parameters of formula 6.20 are shown in Figure 6.5

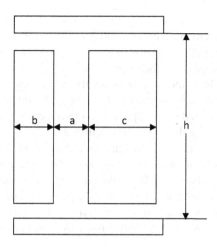

FIGURE 6.5   Core coil assembly.

W_s = stray loss (watts)
K = 25 up to 20 MVA
40 25–40 MVA
50 50–100 MVA
a = H–L gap
b = radial depth of LV winding
c = radial depth of HV winding
D = mean diameter of leakage flux
(mean diameter of HL gap)
h = distance between core clamp flanges (if steel core clamp is used)
= window height (mm) (if steel core clamp is not used)
(All dimensions are in meters)

- If wooden core clamp is used and if top clamp ring is not steel, the stray loss is 70% of the above value.
- If tank shield is used, the stray loss is 50% of the above value.

6.3   CALCULATION OF LOAD LOSS

$$W = 1.015 \left(I^2 R_{\mathrm{LV}} + I^2 R_{\mathrm{HV}} + W_{\mathrm{sLV}} + W_{\mathrm{sHV}} + W_B + W_L + W_{\mathrm{Cir}} + W_{\mathrm{Contact}} + W_{\mathrm{Misc}} \right) \text{watts} \qquad (6.21)$$

where
$I^2 R_{\mathrm{LV}}$ = LV-$I^2 R$ loss (watts)
$I^2 R_{\mathrm{HV}}$ = HV – $I^2 R$ loss (watts)
W_{sLV} = LV eddy current loss (watts)
W_{sHV} = HV eddy current loss (watts)
W_B = bushing plate loss (watts)
W_L = lead wire loss (watts)
W_T = tank loss (watts)
W_{Cir} = circulating loss in parallel conductors
W_{Contact} = contact losses of busbar joints of high-current leads (watts)
(to be considered for current above 500 A)
W_{Misc} = Miscellaneous stray loss

Note

1. If LV leads carrying current in excess of 1000 A run in parallel with tank at close proximity such that the field strength on steel is in excess of 25 A/cm, the tank losses and miscellaneous losses will be higher than the above value.
2. Eddy current and stray loss calculations as per the section are not directly applicable for transformers where harmonics are present. Examples are transformers for rectifier application, drives and inverter-fed solar application; in such cases, the effect of harmonics on eddy current and stray loss will have to be calculated.
3. Winding eddy current calculation and tank loss calculations as above assume that the leakage flux density distribution is primarily axial where the ampere-turns are balanced. If the ampere-turns are imbalanced due to the effect of asymmetry of windings, body tapping especially from disc winding, the effect of radial component of leakage flux is to be calculated. The flux density distribution (both axial and radial) will have to be calculated for accurate calculations. When the asymmetry is high and accurate results are required, FEM will be necessary.
4. For guaranteeing the losses, a safety margin based on previous test result is to be considered in the design. This safety margin will change from manufacturer to manufacturer depending on the production method and process stability. The average ratio of calculated value to the measured results of several transformers will have to be calculated to fix a reliable factor.

Tables 6.2–6.4 show the measured results of load loss of a number of 3.15 MVA, 5 MVA and 8 MVA, 33/11 kV ONAN transformers.

TABLE 6.2
Measured Losses of 3.15 MVA Transformers

Sl. No	HV Resistance/ Phase (Ω)	LV Resistance/ Phase (m/Ω)	HV I^2R Losses (W)	LV I^2R Losses (W)	Amb. Temp °C	Total I^2R Losses at Amb. (W)	Total I^2R Losses at 75°C (W)	Load Loss at Amb. (W)	Stray Loss at Amb. (W)	Stray Loss at 75°C (W)	Load Loss at 75°C (W)
1.	3574	94.5	10855	7749	28.5	18604	21887	20620	2016	1714	23601
2.	3479	94.4	10566	7741	29.0	13307	21497	20690	2016	2029	23601
3.	3568	94.57	10837	7755	24.0	18520	22167	20141	1621	1354	23521
4.	3547	94.17	10773	7722	25.5	18495	22009	19998	1503	1263	23272
5.	3463	93.93	10518	7701	26.5	19219	21593	20315	2096	1768	23366
6.	3486	94.97	10588	7788	27.5	18376	21701	20400	2024	1714	23415
7.	3479	95.23	10566	7809	26.5	18375	21783	20395	2020	1704	23487
8.	3446	93.57	10466	7673	26.5	18139	21717	20402	2263	1908	23626
9.	3566	96.48	10831	7912	30.5	18743	21884	20906	2063	1767	23651
10.	3589	95.95	10900	7865	27.0	15765	22206	20736	1968	1663	23869
11.	3645	96.55	11070	7917	31.5	18987	22086	20785	1798	1546	23632
12.	3573	93.3	10852	7651	25.5	18539	22062	20101	1562	1313	23375
13.	3540	95.33	10752	7817	26.0	18569	22055	20229	1660	1398	23453
14.	3462	92.63	10515	7596	20.5	18111	21974	20202	2081	1723	23697
15.	3537	94.93	10742	7785	25.0	18527	22090	20507	1980	1661	23751
16.	3553	97.93	10791	8031	26.0	18822	22356	20876	2054	1720	24085
17.	3481	93.5	10572	7667	25.0	18239	21746	20332	2093	1755	23501
18.	3579	94.97	10870	7788	32.0	18658	21663	20616	1958	1686	23349
19.	3664	96.87	11123	7944	32.0	19072	22143	20345	1273	1096	23239

TABLE 6.3
Measured Load Loss of 5 MVA Transformer

Sl. No	HV Resistance/ Phase (Ω)	LV Resistance/ Phase (m/Ω)	HV I^2R Losses (W)	LV I^2R Losses (W)	Amb. Temp °C	Total I^2R Losses at Amb. (W)	Total I^2R Losses at 75°C (W)	Load Loss at Amb. (W)	Stray Loss at Amb. (W)	Stray Loss at 75°C (W)	Load Loss at 75°C (W)
1.	2073	54.97	15901	11357	35.0	27258	31296	32555	5297	4614	35910
2.	2103	58.37	16093	12060	38.5	28153	31910	33313	5160	4552	36462
3.	2088	57.93	15978	11969	38.5	27947	31677	33510	5563	4903	36585
4.	2018	59.53	15442	12299	36.5	27844	31792	32779	4935	4322	36114
5.	2009	57.00	15373	11777	34.5	27505	31230	32014	4864	4228	35459
6.	2032	57.87	15549	11956	39.5	27505	31062	32613	5105	4523	35585
7.	1969	56.50	15067	11673	36.0	26740	30588	31810	5070	4432	35020
8.	1971	56.90	15082	11673	33.5	26755	30990	32103	5345	4632	35522
9.	2072	61.67	15855	12742	42.0	28597	32004	33301	4704	4203	36207
10.	2073	56.63	15863	11700	35.5	27563	31586	32929	5365	4682	36270
11.	1988	57.30	15212	11839	37.5	27051	30774	32136	5035	4470	35248
12.	1982	57.93	15167	11969	38.5	27136	30757	32273	5137	4532	35280
13.	1994	57.57	15259	11894	33.0	27153	31408	32203	5050	4366	35774
14.	1950	56.45	14922	11663	33.0	26635	30809	31698	5063	4377	35186
15.	19735	56.71	15102	11717	33.5	26819	30964	31838	5019	4347	35311
16.	19805	56.67	15155	11709	33.0	26864	31074	31858	4994	4317	35391
17.	1975	56.68	15113	11711	34.0	26824	30912	31572	4751	4123	35035
18.	2061	57.68	15771	11917	34.0	27688	31905	32512	4824	4186	36094
19.	1965	56.85	15037	11746	34.0	26883	30980	31554	4671	4053	35033
20.	1935	55.22	14807	11409	27.0	26067	30943	31923	5856	4949	35792
21.	1932	56.12	14784	11595	30.0	26479	30975	31918	5439	4649	35624
22.	2004	56.18	15335	11607	31.0	26942	31399	32319	5377	4613	36013

In Table 6.5, the averages of measured losses of the three ratings and the ratio of calculated values to the measured figures are compared. The calculation methods of stray losses in this section give actual results which are very close.

TABLE 6.4
Measured Load Loss of 8 MVA Transformer

Sl. No	HV Resistance/ Phase (Ω)	LV Resistance/ Phase (m/Ω)	HV I^2R Losses (W)	LV I^2R Losses (W)	Amb. Temp °C	Total I^2R Losses at Amb (W)	Total I^2R Losses at 75°C (W)	Load Loss at Amb. (W)	Stray Loss at Amb. (W)	Stray Loss at 75°C (W)	Load Loss at 75°C (W)
1.	1.1295	32.19	22127	17026	36.5	38796	44297	48761	9965	8727	53024
2.	1.0765	31.575	21079	16701	31.5	37851	44029	47361	9510	8176	52205
3.	1.1015	31.407	21578	16612	27.0	38190	45187	47169	8979	7589	52776
4.	1.0975	31.657	21500	16744	28.5	38244	44993	47858	9614	8172	53165
5.	1.09955	32.095	21540	16976	31.5	38516	44803	48630	10114	8695	53498
6.	1.09315	31.775	21415	16807	31.5	38222	44461	48925	10706	9204	53665
7.	1.09055	31.89	21364	16867	30.0	38231	44723	49306	11075	9467	54190
8.	1.0788	31.297	21133	16554	28.0	37687	44422	49999	12312	10445	54567
9.	1.08165	31.532	21189	16678	30.0	38155	44634	50578	12423	10620	55254
10	1.07775	31.03	21113	16412	31.5	37880	44063	47686	9806	8430	52493
11.	1.0806	31.238	21169	16522	32.0	37762	43844	46411	8649	7449	51293
12.	1.08485	31.35	21252	16582	29.5	37906	44427	46905	8990	7679	52105
13.	1.08235	31.563	21203	16694	29.5	37542	44000	48267	10725	9151	53151

TABLE 6.5
Comparison of Measured and Calculated Values

kVA	Calc. winding Eddy current Losses (W)	Calc. Other Stray Losses (W)	Total Calc. Stray Losses (W)	Total Measured Stray Losses (W) (Average Value)	Calculated/ Measured	Standard Deviation of Measured Value for Stray Losses (W)
3150	1219	381	1600	1621	0.987	306
5000	3933	588	4521	4432	1.02	342
8000	8080	908	8988	8754	1.03	969

7 Calculation of Reactance

7.1 REACTANCE CALCULATION OF TWO-WINDING TRANSFORMER

Percentage reactance of two-winding transformers with symmetrical, balanced ampere-turn distribution is calculated using the following formula:

$$x = 8\pi^2 f \frac{I_{ph}}{V_{ph}} T_{ph}^2 k\Delta \times 10^{-8} \qquad (7.1)$$

x = % Reactance
f = frequency (Hz)
I_{ph} = Phase current (A)
V_{ph} = Phase voltage (V)
T_{ph} = Number of turns per phase

(Current, voltage and turns shall be taken corresponding to the rated tap position, for either LV winding or HV winding).

$$K = \text{Rogowski factor}$$

$$K = \left[1 - \frac{W_1 + W_g + W_2}{\pi \left(\dfrac{h_1 + h_2}{2}\right)}\right] \qquad (7.2)$$

When the winding axial height is low and radial depths of windings are high, the Rogowski factor calculated as above may not give accuracy and the following formula will have to be used.

$$k = 1 - \frac{1 - e^{\frac{-2\pi(w_1 + w_g + w_2)}{(h_1 + h_2)}}}{\dfrac{2\pi(h_1 + h_2)}{2}}\left[1 - \frac{e^{\frac{-4\pi(h_1 + h_2)}{2}}}{2}\left(1 - e^{\frac{2\pi(h_1 + h_2)}{2}}\right)\right] \qquad (7.3)$$

$$\Delta = \frac{W_1 L_1}{3h_1} + \frac{W_g L_2}{\left(\dfrac{h_1 + h_2}{2}\right)} + \frac{W_2 L_3}{3h_2} \qquad (7.4)$$

W_1 = radial depth of LV winding (mm)
W_g = HV–LV gap (mm)
W_2 = radial depth of HV winding (mm)
L_1 = mean length of LV (mm)
L_2 = mean length of HV–LV gap (mm)
L_3 = mean length HV (mm)

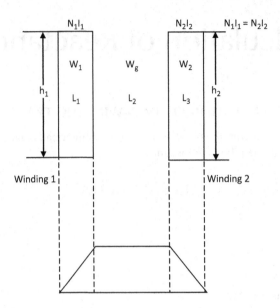

FIGURE 7.1 Winding dimensions.

h_1 = electrical axial height of LV (mm)

h_2 = electrical axial height of HV (mm)

N_1 = number of turns in LV

N_2 = number of turns in HV

I_1 = phase current in LV (A)

I_2 = phase current in HV (A)

The symbols are shown in Figure 7.1.

Note

1. Axial height of helical and spiral winding will be more than the axial electrical height by the height of one turn. Calculation of reactance shall be done using the electrical height by subtracting the width of one turn from the axial height.

The reactance calculated shall be multiplied by the factors as per Table 7.1, to consider the effect of unbalance in ampere-turns due to tapping from the body of the winding.

TABLE 7.1

Multiplication Factors for Unbalance in Ampere-Turns

Sl. No.	Type of Unbalance	Multiplication Factor	Remarks
1.	Winding with taps in the middle – no balancing on other winding	$1 + \dfrac{K}{2}\left(1 + \dfrac{\pi}{2}\dfrac{Kl}{T}\right)$	Other winding can be foil, spiral, layer, disc, etc.
2.	Winding with taps at the middle, 50% balancing on other winding	$1 + \dfrac{0.7K}{2}\left(1 + \dfrac{\pi}{2}\dfrac{Kl}{T}\right)$	Balancing of other winding is not possible if it is foil winding (this factor is not applicable, if other winding is foil)
3.	Winding with taps at one end – no balancing on other winding	$1 + \dfrac{K}{2}\left(1 + \pi\dfrac{Kl}{T}\right)$	Other winding can be foil, spiral, layer, disc
4.	Winding tapped at one end, 50% balancing on other winding	$1 + \dfrac{0.7K}{2}\left(1 + \pi\dfrac{Kl}{T}\right)$	If other winding is foil, balancing of other winding is not possible

K = winding zone with zero current

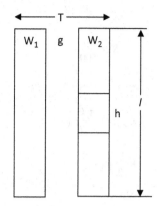

FIGURE 7.2 Winding tapped in the middle.

$$K = \frac{h}{l},$$

where
   h = height of tapped portion with zero current
   l = height of winding (total including h)

$$T = W_1 + g + W_2$$

W_1 = radial depth of winding 1
g = HL gap
W_2 = radial depth of winding 2

The above symbols are shown in Figure 7.2.

Example

Winding tapped in the middle, as shown in Figure 7.3.
   No balancing of other winding
   $l = 500$ mm
   $h = 50$ mm
   $T = 30 + 10 + 50 = 90$ mm
   $X = 5\%$ (reactance of two-winding arrangement with $h = 0$)

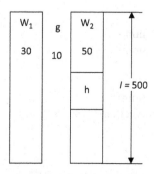

FIGURE 7.3 Dimensions of winding tapped in the middle.

i.e. $h = 0$

$$K = \frac{50}{500} = 0.1$$

Multiplication factor

$$= 1 + \frac{k}{2}\left(1 + \frac{\pi}{2}\frac{kl}{T}\right)$$

$$= 1 + \frac{0.1}{2}\left(1 + \frac{\pi}{2} \times 0.1 \times \frac{500}{90}\right)$$

$$= 1.0936$$

$\therefore$ Reactance with 50 mm tapping on one winding with no balancing on other winding $=$ $5 \times 1.0936 = 5.47\%$

If the winding is tapped at one end and if there is no balancing on the other winding, the reactance with 50 mm tapping will be as below:

$$\text{Multiplication factor} = 1 + \frac{k}{2}\left(1 + \pi\frac{kl}{T}\right)$$

$$= 1 + \frac{0.1}{2}\left(1 + \pi \times \frac{0.1 \times 500}{90}\right) = 1.1372$$

$$\text{Reactance} = 5 \times 1.1372 = 5.68\%$$

7.1.1 ALTERNATE FORMULA FOR REACTANCE CALCULATION

Percentage reactance of two-winding transformers with symmetrical, balanced ampere-turn distribution at rated tap position is calculated using the following formula:

$$X = \frac{1.24 \times \text{kVA} \times D_m\left(\dfrac{W_1 + W_2}{3} + W_g\right)K \, K_1 \, K_2}{3\left(\dfrac{V}{T}\right)^2 h_2 \times 10} \qquad (7.5)$$

$X =$ reactance (%)
kVA $=$ rated power
$D_m =$ mean diameter of leakage flux density distribution

$$= \frac{\text{LV OD} + \text{HV ID}}{2}(\text{mm})$$

$W_1 =$ radial depth of LV winding (mm)
$W_2 =$ radial depth of HV winding (mm)
$W_g =$ HV–LV gap (mm)
$V/T =$ volts/turn
$h_2 =$ electrical axial height of HV (mm)
$K =$ Rogowski's factor (same as in Section 7.1)

$K_1 = 1.04$ (factor based on actual measured results)
$K_2 = f/50$ (frequency factor)
$f =$ frequency (Hz)

7.2 REACTANCE CALCULATION OF ZIGZAG CONNECTED TWO-WINDING TRANSFORMERS

Effective reactance of zigzag connected transformers with core-type concentric winding arrangement shall be calculated as below for the various connection methods. X_{12}, X_{13}, X_{14}, X_{23}, X_{24}, X_{34} are the reactances calculated between the respective winding pair.

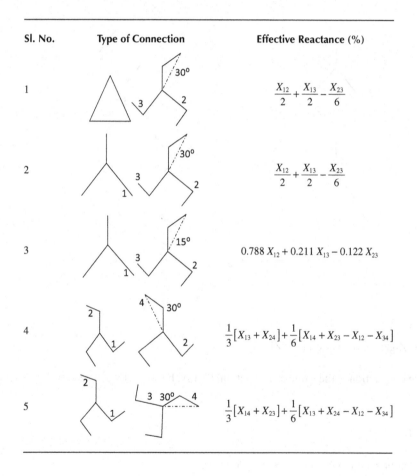

Sl. No.	Type of Connection	Effective Reactance (%)
1		$\dfrac{X_{12}}{2} + \dfrac{X_{13}}{2} - \dfrac{X_{23}}{6}$
2		$\dfrac{X_{12}}{2} + \dfrac{X_{13}}{2} - \dfrac{X_{23}}{6}$
3		$0.788\, X_{12} + 0.211\, X_{13} - 0.122\, X_{23}$
4		$\dfrac{1}{3}\left[X_{13} + X_{24}\right] + \dfrac{1}{6}\left[X_{14} + X_{23} - X_{12} - X_{34}\right]$
5		$\dfrac{1}{3}\left[X_{14} + X_{23}\right] + \dfrac{1}{6}\left[X_{13} + X_{24} - X_{12} - X_{34}\right]$

7.2.1 EFFECTIVE REACTANCE OF ZIGZAG CONNECTED TRANSFORMERS

A. Star–Zigzag Connection

Star–Zigzag Connection	Symmetrical Zigzag (30° Zigzag)

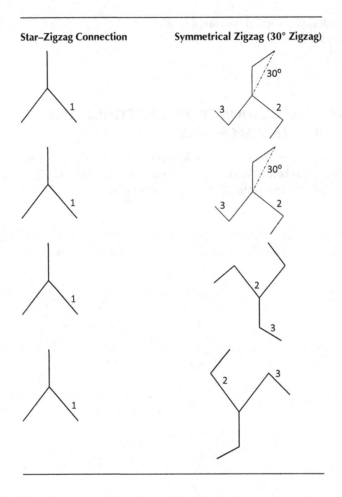

$$\%X = \%\frac{X_{12}}{2} + \%\frac{X_{13}}{2} - \%\frac{1}{6}X_{23} \qquad (7.6)$$

B. Star–Zigzag Connection (15° Zigzag)

The zigzag connection is the same as in Section 7.2.1.A, but with 15° (instead of 30°) in all the four cases.

Example

The first connection in A is changed as below.

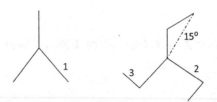

% X for all four cases is

$$\%X = 0.788\ (\%X_{12}) + 0.211(\%X_{13}) - 0.122(\%X_{23})$$

C. Zigzag connection – 30° Zigzag Angle

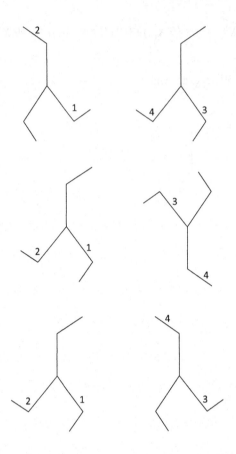

The reactance for all the above eight cases is shown in Section 7.2.1.C.

$$\%X = \frac{1}{3}\left[\%X_{14} + \%X_{23}\right] + \frac{1}{6}\left[X_{13} + X_{24} - X_{12} - X_{34}\right] \tag{7.7}$$

7.3   REACTANCE CALCULATION WITH DIFFERENT AMPERE-TURN DISTRIBUTIONS

7.3.1   LV–HV–LV ARRANGEMENT OR SIMILAR (FIGURE 7.4)

$$D_2' + D_2'' = D_2$$

$$D_2' = D_2 P$$

$$D_2'' = D_2(1-P)$$

h = average winding stack height

$$D = \frac{P^2}{h}\left\{\frac{D_1 L_1}{3} + \delta_1 L\delta_1 + \frac{D_2'}{3}L_2'\right\} + \frac{(1-P)^2}{h}\left\{\frac{D_2 L_2''}{3} + \delta_2 L\delta_2 + \frac{D_3}{3}L_3\right\} \tag{7.8}$$

$L_1, L\delta_1, L_2', L_2'', L\delta_2, L_3$ − mean lengths (mm)

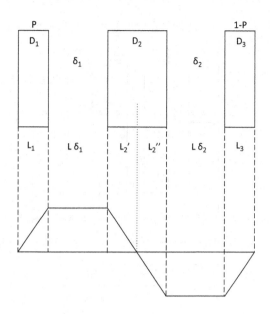

FIGURE 7.4 LV–HV–LV arrangement.

$D_1, \delta_1, D_2', D_2'', \delta_2, D_3$ – radial depths (mm)

7.3.2 WINDING WITH DUCTS INSIDE OR WINDINGS MADE IN TWO SEPARATE LAYERS WITH GAPS (FIGURE 7.5)

h = average height of winding (mm)

$$D = \frac{1}{h} \left\{ \frac{D_1 L_1}{3} + \delta_1 L \delta_1 + D_2 L_2 \left(\frac{1 + P + P^2}{3} \right) + P^2 \delta_2 \ L \delta_2 + P^2 \frac{D_3}{3} L_3 \right\} \qquad (7.9)$$

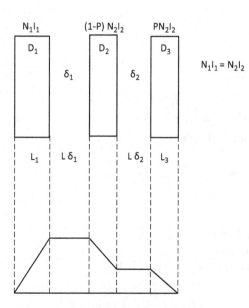

FIGURE 7.5 Winding with separate tap coil.

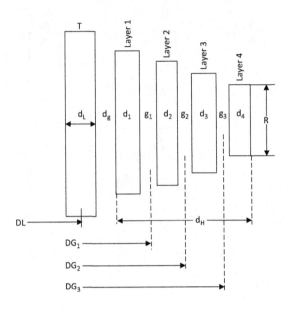

FIGURE 7.6 Layer winding.

7.3.3 Reactance of Layer Winding with Reduced Layer Height towards Outer Layers (Figure 7.6)

Multilayer HV winding with reduced layer heights towards line end.

$$\%X = 1.1938 \times 10^{-4} \times \left(\frac{f}{50}\right) \frac{IT\ \varepsilon D_m}{10^3 \times \dfrac{V}{T} \times L} \tag{7.10}$$

f = frequency
I = phase current
T = turns corresponding to I
$\dfrac{V}{T}$ = volt/turn
L = effective winding height
$$= R\left(\frac{\text{Total turns } = N}{N_1 = \text{Turns in height of } R}\right) + \frac{d_L + d_G + d_H}{3}$$
R = height of layer with lowest height
N = total turns
(if R is the same for all layers, $N_1 = N$)
d_L = radial depth of LV
d_H = radial depth of HV
d_g = HV–LV gap
d_1, d_2, d_3, d_4 = radial depth of layers 1, 2, 3, 4
g_1, g_2, g_3 = radial depth of gap
n_1, n_2, n_3, n_4 = number of turns in layers 1, 2, 3, 4

$$\varepsilon D_m = \left[\frac{D_L d_L}{3} + D_G d_G + Y\right]$$

D_L = mean diameter of LV winding
D_G = mean diameter of HL gap
D_H = mean diameter of HV winding

$$Y = \frac{1}{3} D_H \sum_{d=1}^{d=n} d + DG_1 g_1 \left(\frac{n_2 + n_3 + n_4}{n_1 + n_2 + n_3 + n_4} \right)^2 + DG_2 g_2 \left(\frac{n_3 + n_4}{n_1 + n_2 + n_3 + n_4} \right)^2$$

$$+ DG_3 g_3 \left(\frac{n_4}{n_1 + n_2 + n_3 + n_4} \right)^2 \tag{7.11}$$

$$\text{The quantity in bracket in each case} = \frac{\text{Number of Turns from gap to minimum flux}}{\text{Total turns in the circuit}}$$

$$DG_1, DG_2, DG_3 = \text{Mean dia of gap 1, 2, 3}$$

Example

LV current = 1010 A
LV turns = 165
HV turns = 1000
HV turns at height R (lowest layer height) = 820

V/T	= 135
n_1	= 265
n_2	= 250
n_3	= 245
n_4	= 240
Total	= 1000

Figure 7.7 shows the layer winding dimensions.

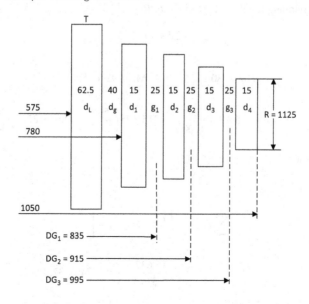

FIGURE 7.7 Layer winding dimensions.

L (effective winding height for reactance purpose)

$$= R \times \left(\frac{1000}{820}\right) + \frac{d_L + d_G + d_H}{3}$$

$$= 1125 \times \frac{1000}{820} + \frac{62.5 + 40 + 135}{3} = 1451$$

D_H = Mean dia of HV winding

$$= 780 + 135 = 915$$

$$Y = \frac{915}{3} \times 15 \times 4 + 835 \times 25 \left(\frac{733}{1000}\right)^2 + 915 \times 25 \left(\frac{485}{1000}\right)^2 + 995 \times 25 \left(\frac{240}{1000}\right)^2$$

$$= 36329.47$$

$$\in D_m = \left[\frac{D_L d_L}{3} + D_G d_G + Y\right]$$

$$= \frac{62.5 \times 637.5}{3} + 40 \times 740 + 36329.47$$

$$= 13281.25 + 29{,}600 + 36329.47$$

$$\%X = 1.1938 \times 10^{-4} \left(\frac{50}{50}\right) \frac{165 \times 1010}{1451 \times 135} \times 79210.72 = 8.04\%$$

A. Calculation assuming that turns are equal in all layers and all layer heights are equal.

$$\therefore n_1 = n_2 = n_3 = n_4 = 250$$

Assume that winding height is 1172 mm (i.e. R)

$$L = 1172 + \frac{62.5 + 40 + 135}{3} = 1251$$

$$D_H = 780 + 135 = 915 \ (\text{Mean dia of HV})$$

$$Y = \frac{915}{3} \times 15 \times 4 + 835 \times 25 \left(\frac{750}{1000}\right)^2 + 915 \times 25 \left(\frac{500}{1000}\right)^2 + 995 \times 25 \left(\frac{250}{1000}\right)^2$$

$$= 18300 + 11742 + 5718.75 + 1554.7$$

$$= 37315.45$$

$$\sum D_m = \left[\frac{D_L d_L}{3} + D_G d_g + Y\right]$$

$$= \frac{62.5 \times 637.5}{3} + 40 \times 740 + 37{,}315.45$$

$$= 80196.7$$

$$\%X = 1.1938 \times 10^{-4} \times \left(\frac{50}{50}\right) \frac{165 \times 1010}{1251} \times \frac{80,196.7}{135}$$

$$= 9.44\%$$

B. Calculation using standard formula for two-winding transformers (Figure 7.8)

$$\%X = 8\pi^2 \times 50 \times \frac{1010}{(135 \times 165)} \times 165^2 (KD) \times 10^{-8}$$

$$= 0.0487339\, KD$$

$$K = 1 - \frac{62.5 + 40 + 135}{\pi \times 1251} = 0.94$$

$$D = \frac{\pi}{1251} \left[\frac{62.5 \times 637.5}{3} + 40 \times 740 + \frac{135}{3} \times 915\right]$$

$$= 211.086$$

$$\%X = 9.7$$

C. Calculation considering the effect of ducts in HV (Figure 7.9)

$$D = \frac{\pi}{1251} = 80,196 \frac{\pi}{1251}$$

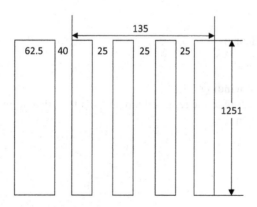

FIGURE 7.8 Layer winding (equal height of layer).

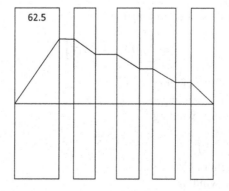

FIGURE 7.9 Layer winding.

$$\%X = 0.0487339 \ K \ X \ 80{,}196 \ \frac{\pi}{1251}$$

$$= 9.81K = 9.22\%$$

$$D = \frac{\pi}{1251}\left[\frac{62.5\times 637.5}{3} + 40\times 740 + \frac{\left(1+0.75+0.75^2\right)}{3} 15\times 795 + 835\times 0.75^2 \times 25 + \left(0.75+0.75\times .5+0.5^2\right) \right.$$

$$\left. \times \frac{875}{3}\times 15 + 915\times 25\times .5^2 + \left(0.5+0.5\times .25+0.25^2\right)\frac{15}{3}\times 955 + 25\times 995\times .25^2 + 0.25^2 \times \frac{15}{3}\times 1035 \right]$$

$$= \pi \times \frac{80710.93}{1251} = 202.68$$

$$\therefore IX = 0.0487339 \ X \ K \ X \ 202.68$$

$$= 9.28\%$$

7.4   REACTANCE CALCULATIONS BASED ON TOTAL INDUCTANCE

In this method, the inductance of HV winding and LV winding and the mutual inductance are calculated. The total inductance is calculated from these three values.

$$L = \left(L_{LV} + L_{HV} + L_{LV-HV}\right)\frac{T_{HV}^2}{3} \tag{7.12}$$

where
   L = total inductance (Henry)
   $X = 2\pi fL$
   f = frequency
   T_{Hv} = turns/phase in HV winding
   [use the turns corresponding to the required tap position for calculation]
   L_{LV} = reactance of LV winding

$$L_{LV} = \frac{A_{LV} \ \mu_0}{\left(\dfrac{\pi^2 H_{Leq}}{4}\right)} \tag{7.13}$$

$$\mu_0 = 4\pi \times 10^{-7}$$

A_{LV} = area of LV winding cross section linking the flux

$$= \frac{\pi}{4}\left[LV_{OD}^2 - LV_{ID}^2\right]$$

   LV_{OD} = outside diameter of LV winding
   LV_{ID} = inside diameter of LV winding
   H_{Leq} = equivalent height of LV winding
$$= \left(H_{LV} + \frac{2\times L_{VRD} + HL}{2}\right)$$
   H_{LV} = axial electrical height of LV
   LV_{RD} = radial depth of LV winding
   HL = HV–LV gap
   L_{HV} = inductance of HV winding

$$L_{HV} = \frac{A_{HV}\mu_0}{\dfrac{\pi H_{Heq}}{4}} \tag{7.14}$$

H_{Heq} = equivalent height of HV winding

$$= \left(H_{HV} + \frac{2\,H_{VRD} + H - L}{2} \right)$$

H_{HV} = axial electrical height of HV
HV_{RD} = radial depth of HV winding

$$L_{HV} = \frac{A_{HV}\mu_0}{\dfrac{\pi^2}{4}\,H_{HV}} \tag{7.15}$$

A_{HV} = area of HV winding cross section linking the flux

$$= \frac{\pi}{4}\left[HV_{OD}^2 - HV_{ID}^2 \right]$$

HV_{OD} = outside diameter of HV winding
HV_{ID} = inside diameter of HV winding
L_{LV-HV} = inductance between LV winding and HV winding

$$L_{LV-HV} = \frac{(A_{LV-HV})\mu_0}{\left(\dfrac{H_{HVeq} + H_{LVeq}}{2}\right)} \tag{7.16}$$

A_{LV-HV} = cross-sectional area of the LV–HV gap

$$A_{LV-HV} = \frac{\pi}{4}\left[HV_{ID}^2 - LV_{OD}^2 \right] \tag{7.17}$$

Example

Figure 7.10 gives the dimensions of winding.
   Frequency (Hz) = 50
   LV phase volts = 250
   HV phase current (R-Tap) = 30.3 A
   LV turns = 16
   HV phase volts = 11000 V

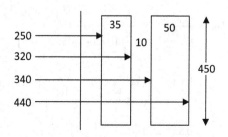

FIGURE 7.10   Dimensions of windings

HV turns = 704 at rated tap
[HV radial depth corresponding to rated tap is 50]
Electrical height of LV = 475
Electrical height of HV = 450

a. Applying the formula as per Equation (7.1)

$$\% \, X = 8\pi^2 f \frac{I_{ph}}{V_{ph}} \, T^2 \, K\Delta \times 10^{-8}$$

$$= 8\pi^2 \times 50 \times \frac{30.3}{11000} \times 704^2 \, K\Delta \times 10^{-8} = 0.0538 \, K\Delta$$

$$K = 1 - \frac{95}{\pi \times \left(\dfrac{450 + 475}{2} \right)} = 0.934$$

$$\Delta = \pi \left[\frac{35}{3} \times \frac{285}{475} + \frac{10 \times 330}{462.5} + \frac{50}{3} \times \frac{390}{450} \right]$$

$$= \pi [7 + 7.135 + 14.44] = 89.7255$$

$$\therefore \% \, X = 4.51$$

b. Applying total inductance formula

$$L = (L_{LV} + L_{HV} + L_{LV\,HV}) \frac{T_{HV}^2}{3}$$

$$T_{HV} = 704$$

$$L_{LV} = \frac{A_{LV} \, \mu_0}{\dfrac{\pi^2 H_{LV\,eq}}{4}} = \frac{\pi/4 \left(0.32^2 - 0.25^2 \right) 4\pi \times 10^{-7}}{\dfrac{\pi^2}{4} \left(H_{LV\,eq} \right)}$$

$$H_{LVeq} = 0.475 + \frac{2 \times 0.035 + 0.01}{2} = 0.475 + 0.04 = 0.515$$

$$\therefore L_{LV} = \frac{\left(0.32^2 - 0.25^2 \right) \times 10^{-7} \times 4}{.515} = 10^{-7} \left[\frac{4}{0.515} \times 0.0399 \right]$$

$$= 0.3099 \times 10^{-7}$$

$$L_{HV} = \frac{A_{HV} \, \mu_0}{\dfrac{\pi^2 \, H_{Heq}}{4}} = \frac{\dfrac{\pi}{4} \left[0.44^2 - 0.34^2 \right] \pi 4 \times 10^{-7}}{\dfrac{\pi^2}{4} [0.45 + 0.05 \times 2 + .01]}$$

$$= \frac{\left[0.44^2 - 0.34^2 \right] 4 \times 10^{-7}}{0.56} = 10^{-7} \left[\frac{0.078 \times 4}{0.56} \right]$$

$$= 0.557 \times 10^{-7}$$

$$L_{HV-LV} = \frac{(A_{LVHV})\mu_0}{\left(\dfrac{H_{HVeq} + H_{LVeq}}{2}\right)} = \frac{4\pi \times 10^{-7}\left[\dfrac{\pi}{4}(0.34^2 - 0.32^2)\right]}{0.5375}$$

$$= \frac{4\pi \times 10^{-7}}{0.5375}\left[\frac{\pi}{4} \times 0.0132\right]$$

$$= 0.24237 \times 10^{-7}$$

$$L = 10^{-7} \times \frac{704^2}{3}[0.3099 + 0.557 + 0.24237] = 0.01833$$

Total Inductance $= 3L = 0.05499$

$$X = 2\pi \times 50 \times 0.05499 = 17.27\,\Omega$$

$$\%X = \frac{17.27 \times 30.3 \times 100}{11,000} = 4.76\%$$

7.5 REACTANCE CALCULATION OF EXTENDED WINDING (FIGURE 7.11)

1 – winding 1 {winding without extension}
2 and 3 – extended winding {winding 3 is the extended part}
X_{12} = reactance between windings 1 and 2
X_{13} = reactance between windings 1 and 3
X_{23} = reactance between windings 2 and 3

$$X_\% = \frac{2}{\sqrt{3}}\left[X_{12}\sin(60-\theta)\cos\theta + X_{13}\cos(60-\theta)\sin\theta\right] - \left(\frac{2}{3}(X_{23})\sin(60-\theta)\sin\theta\right)$$

If $\theta = 0$, this formula will reduce to the formula for two-winding transformer.
When $\theta = 0$,

$$X_\% = \frac{2}{\sqrt{3}}\left[X_{12}\sin 60\cos\theta + X_{13}\cos 60\sin\theta\right] - \left(\frac{2}{3}(X_{23})\sin 60\sin\theta\right)$$

$$= \frac{2}{\sqrt{3}}\left[(X_{12})\frac{\sqrt{3}}{4} + 0\right] - 0 = X_{12}$$

If $\theta = 30°$, the above formula becomes formula for zigzag connection with auxiliary winding.

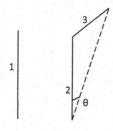

FIGURE 7.11 Extended winding.

When $\theta = 0$,

$$X_\% = \frac{2}{\sqrt{3}}\left[X_{12}\sin(60-30)\cos30 + X_{13}\cos(60-30)\sin30\right] - \left((X_{23})\frac{2}{3}\sin30\sin30\right)$$

$$\%X = \left(\frac{X_{12}}{2} + \frac{X_{13}}{2} - \frac{1}{6}X_{23}\right) \tag{7.18}$$

7.6 ZERO SEQUENCE IMPEDANCE OF ZIGZAG EARTHING TRANSFORMER (FIGURE 7.12)

$$X = 8\pi^2 fN^2 K_1 K_2 \times 10^{-10} \tag{7.19}$$

where
  X = zero sequence impedance (ohms)
  f = frequency
  N = number of turns of one winding {either zig winding or zag winding}

$$K_1 = 1 - \frac{W_1 + W_2 + W_3}{\pi H} \tag{7.20}$$

W_1 = radial depth of winding 1 (mm)
W_2 = gap between winding 1 and winding 2 (mm)
W_3 = radial depth of winding 2 (mm)
H = axial height (electrical) of winding (mm) {if heights are different, the formula will have to be modified}

$$K_2 = \left[\frac{W_1}{3}\frac{L_1}{H} + \frac{W_2 L_2}{H} + \frac{W_3}{3}\frac{L_3}{H}\right] \tag{7.21}$$

L_1 = mean length of zig winding (W_1) (mm)
L_2 = mean length of gap (W_2) (mm)
L_3 = mean length of zag winding (W_3) (mm)

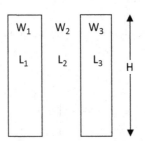

FIGURE 7.12 Zigzag winding.

7.7 REACTANCE CALCULATION OF NEUTRAL EARTHING TRANSFORMER WITH AUXILIARY WINDING (FIGURE 7.13)

$$X = \frac{X_{12}}{2} + \frac{X_{13}}{2} - \frac{X_{23}}{6} \tag{7.22}$$

X_{23} will have +ve tolerance only
   If fault current $= I_f$

$$Z_0 = \frac{3V}{I_f} \text{ ohms}$$

(V = phase voltage (V), I_f = fault current (A)

Leakage Reactance of Zigzag Transformer

$$X_T = \frac{2}{\sqrt{3}} \left(X_{12} \sin(60-\theta)\cos\theta + X_{13} \cos(60-\theta)\sin\theta \right) - X_{23} \times \frac{2}{3} \sin 60 \sin 6 \tag{7.23}$$

When $\theta = 0$,

$$X_T = \frac{2}{\sqrt{3}} \left(X_{12} \sin 60 \cos\theta + X_{13} \cos 60 \sin\theta \right) - X_{23} \times \frac{2}{3} \sin 60 \sin\theta$$

$$= \frac{2}{\sqrt{3}} \left[X_{12} \frac{\sqrt{3}}{2} + 0 \right] - 0$$

$$= X_{12}$$

When $\theta = 120°$,

$$X_T = \frac{2}{\sqrt{3}} \left[X_{12} \sin(60-120)\cos 120 + X_{13} \cos(60-120)\sin 120 \right] - X_{23} \times \frac{2}{3} \sin(60-120)\sin 120$$

$$= \frac{2}{\sqrt{3}} \left[X_{12} \sin(-60)\cos 120 + X_{13} \cos(-60)\sin 120 \right] - X_{23} \frac{2}{3} \sin(-60)\sin 120$$

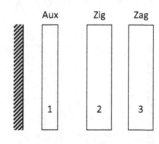

Aux   Zig   Zag

1   2   3

FIGURE 7.13   Neutral grounding transformers with auxiliary winding.

$$= \frac{2}{\sqrt{3}} \left[X_{12} \frac{\sqrt{3}}{2} (-1/2) + X_{13} \times 0.5 \times \frac{\sqrt{3}}{2} \right] - X_{23} \times \frac{2}{3} \left(\frac{-\sqrt{3}}{2} \right) \frac{\sqrt{3}}{2}$$

$$= \frac{2}{\sqrt{3}} \left[X_{12} \frac{\sqrt{3}}{4} + X_{13} \frac{\sqrt{3}}{4} \right] + \frac{X_{23}}{2}$$

When $\theta = 30°$,

$$X_T = \frac{2}{\sqrt{3}} [X_{12} \sin 30 \cos 30 + X_{13} \cos 30 \sin 30] - \frac{2}{3} X_{23} \sin 30 \sin 30$$

$$= \frac{2}{\sqrt{3}} \left[X_{12} \times 0.5 \times \frac{\sqrt{3}}{2} + X_{13} \frac{\sqrt{3}}{2} \times \frac{1}{2} \right] - \left(\frac{2}{3} \times 0.5 \times 0.5 \right) X_{23}$$

$$= \frac{2}{\sqrt{3}} \left[(X_{12}) \frac{\sqrt{3}}{4} \right] + \left(X_{13} \times \frac{\sqrt{3}}{4} \right) \frac{2}{\sqrt{3}} - X_{23} \left(\frac{1}{6} \right)$$

$$= \frac{X_{12}}{2} + \frac{X_{13}}{2} - \frac{X_{23}}{6}$$

7.8 REACTANCE OF AUTOTRANSFORMER WITH TERTIARY WINDING (FIGURE 7.14)

T = tertiary winding
C = common winding
S = series winding

In an autotransformer consisting of series winding S, common winding C and tertiary winding T, the reactance of $(S + C)$ with T is calculated as follows:

$\%IX_{SC}$ – % reactance between series and common winding
$\%IX_{CT}$ – % reactance between common and tertiary winding
$\%IX_{ST}$ – % reactance between tertiary and series winding

All reactances will have to be on a common base.

$$\%IX_{(SC)T} = K_s \, \%IX_{ST} + K_C \, \%IX_{CT} - K_S K_C \%IX_{SC} \tag{7.24}$$

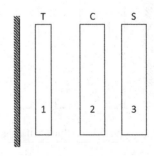

FIGURE 7.14 Autotransformer with tertiary winding.

$$K_s = \frac{\text{Turns in series winding}}{\text{Turns in } (\text{series} + \text{common}) \text{winding}}$$

$$K_C = \frac{\text{Turns in common winding}}{\text{Turns in } (\text{series} + \text{common}) \text{winding}}$$

To get the effective reactance at the external circuit terminals, the reactance so obtained must be multiplied by the ratio $\dfrac{\text{output kVA}}{\text{Base kVA}}$

Alternative method

$$\%IX_{(SC)T} = \%IX_T + \%IX_S K_s^2 - \%IX_C\ K_C^2 \tag{7.25}$$

where
$\%IX_S$, $\%IX_C$ and $\%IX_T$ = percent three-winding transformer reactance on a common base (these are not reactance on autotype)
i.e. first calculate $\%IX_{TC}$, $\%IX_{CS}$ and $\%IX_{TS}$
Then,

$$\%IX_S = \frac{\%IX_{TC} + \%IX_{CS} + \%\ IX_{TS}}{2} - \%IX_{TC} \tag{7.26}$$

$$\%IX_C = \frac{\%IX_{TC} + \%IX_{CS} + \%\ IX_{TS}}{2} - \%IX_{TS} \tag{7.27}$$

$$\%IX_T = \frac{\%IX_{TC} + \%IX_{CS} + \%\ IX_{TS}}{2} - \%IX_{CS} \tag{7.28}$$

Example

$$\%IX_{TC} = 3.33$$

$$\%IX_{CS} = 4.50$$

$$\%IX_{TS} = 5.07$$

$$\%IX_T = \frac{3.33 + 4.5 + 5.70}{2} - 4.5 = 2.265$$

$$\%IX_C = \frac{3.33 + 4.5 + 5.7}{2} - 5.07 = 1.695$$

$$\%IX_S = \frac{3.33 + 4.5 + 5.7}{2} - 3.33 = 3.435$$

$$K_S = \frac{200}{400} = 0.5, K_C = 0.5$$

$$\therefore \%IX(SC)T = \%IX_T + K_S^2 \%IX_S + K_C^2 \%IX_C$$

$$= 2.265 + 0.5^2 \times 3.435 + 0.5^2 \times 1.695$$

$$= 3.5435\%$$

ALTERNATIVE

$$\%IX(SC)T = K_S\%IX_{ST} + K_C\%IX_{CT} - K_S K_C\%IX_{SC}$$

$$= 0.5 \times 5.07 + 0.5 \times 3.33 - 0.5 \times .5 \times 4.5$$

$$= (3.075)$$

7.9 EFFECTIVE REACTANCE OF WINDINGS IN SERIES (AUTOTRANSFORMER)

Coils 2 and 3 are connected in series (Figure 7.15) [in autotransformer series and common windings, in other transformers, 2 and 3 are in series]

$$K_2 = \frac{N_2}{N_2 + N_3} \qquad K_3 = \frac{N_3}{N_2 + N_3}$$

Calculate X_{12}, X_{13}, X_{23} on a common MVA base

$$\%IX_1 = \frac{1}{2}\left(\%IX_{12} + \%IX_{13} - \%IX_{23}\right)$$

$$\%IX_2 = \frac{1}{2}\left(\%IX_{12} + \%IX_{23} - \%IX_{13}\right)$$

$$\%IX_3 = \frac{1}{2}\left(\%IX_{13} + \%IX_{23} - \%IX_{12}\right)$$

$$\%IX_{1-23} = \%IX_1 + K_2^{\,2}\left(\%IX_2\right) + K_3^{\,2}\left(\%IX_3\right)$$

$$= \frac{1}{2}\Big[\left(\%IX_{12} + \%IX_{13} - \%IX_{23}\right)$$

$$+ K_2^{\,2}\left(\%IX_{12} + \%IX_{23} - \%IX_{13}\right) + K_3^{\,2}\left(\%IX_{23} + \%IX_{13} - \%IX_{12}\right)\Big]$$

IF $K_2 = K_3 = \dfrac{1}{2}$

$$\%IX_{1-23} = \frac{\%IX_{12}}{2} + \frac{\%IX_{13}}{2} - \frac{\%IX_{23}}{4}$$

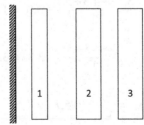

FIGURE 7.15   Effective reactance of winding in series.

The above formula can be used for the calculation of reactance for tapped winding, tertiary winding to series + common winding, etc.

Example

To calculate the reactance of an auto transformer with tertiary as per dimensions shown in Figure 7.16.
Frequency = 50 Hz

$$I_1 = 340 \quad I_2 = 34 \quad I_3 = 34$$

$$V_1 = 250 \quad V_2 = 1250 \quad V_3 = 1250$$

$$N_1 = 40 \quad N_2 = 200 \quad N_3 = 200$$

On common MVA basis

$$\%IX = 8\pi^2 \times 50 \times \frac{340}{250} \times 40^2\, KD \times 10^{-8} = 0.0859050 KD$$

A – Calculation assuming that windings 2 and 3 are one coil, and ignoring the effect of duct on reactance

$$K_{1-23} = 1 - \frac{105}{\pi \times 600} = 0.944$$

$$D_{1-23} = \frac{\pi}{600}\left[\frac{25}{3} \times 235 + 10 \times 270 + \frac{70}{3} \times 350\right]$$

$$= 67.146$$

$$\%IX_{1-23} = 0.944 \times 0.0859050 \times 67.146 = 5.44$$

B – Calculation of reactance using procedure as per 7.9 (Figure 7.17)

$$\%IX_{1-2} = 0.0859050\ KI$$

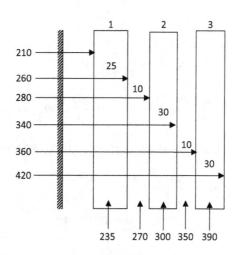

FIGURE 7.16 Dimensions of autotransformer with tertiary.

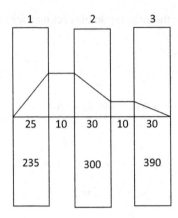

FIGURE 7.17   Ampere-turn diagram.

$$K_{1-2} = 1 - \frac{65}{\pi \times 600} = 0.9655$$

$$D_{1-2} = \frac{\pi}{600}\left[\frac{25}{3} \times 235 + 10 \times 270 + \frac{30}{3} \times 300\right]$$

$$= 40.097$$

$$\%X_{1-2} = 0.0859050 \times 0.9655 \times 40.097 = 3.32569$$

$$\%IX_{2-3}$$

$$K_{23} = 1 - \frac{70}{\pi \times 600} = 0.963$$

$$D_{23} = \frac{\pi}{600}\left[\frac{30}{3} \times 300 + 10 \times 350 + \frac{30}{3} \times 390\right] = 54.45$$

$$\%IX_{2-3} = 0.085905 \times 0.963 \times 54.45 = 4.50$$

$$\%IX_{1-3}$$

$$K_{1-3} = 1 - \frac{105}{\pi \times 600} = 0.944$$

$$D_{1-3} = \frac{\pi}{600}\left[\frac{25}{3} \times 235 + 50 \times 310 + \frac{30}{3} \times 390\right] = 111.83$$

$$\%IX_{1-3} = 0.085905 \times .944 \times 111.83 = 9.0687778$$

$$\%IX_{1-23} = \frac{\%IX_{12}}{2} + \frac{\%IX_{13}}{2} - \frac{\%IX_{23}}{4}$$

$$= \frac{3.32569}{2} + \frac{9.0687778}{2} - \frac{4.5}{4}$$

$$= 5.07$$

All dimensions and parameters are the same as above

$$K_2 = \frac{N_2}{N_2 + N_3} = 0.5$$

$$K_3 = \frac{N_3}{N_2 + N_3} = 0.5$$

$$\%IX_{1-23} = 0.085905\ KD$$

$$K = 1 - \frac{105}{\pi \times 600} = 0.944$$

$$D = \frac{\pi}{600}\left[\frac{25}{3} \times 235 + 10 \times 270 + \frac{30}{3}\left(1 + 0.5 + 0.5^2 K_3 K_2\right) \times 300 + 0.5^2 \times 10 \times 350 + \frac{30}{3} \times 0.5^2 \times 390\right] = 58.619$$

$$IX_{1-23} = 4.75\%$$

B and C give very close results. If the effect of duct in the winding is ignored, the calculated reactance is higher.

7.10 REACTANCE CALCULATION OF SPLIT WINDING

When two secondaries of equal ratings are placed axially, the reactance between primary and secondary is calculated by the vector summation of the axial and radial reactances. The winding arrangement is shown in Figure 7.18.

The reactance between 1 and 4 (2 and 3) can be calculated by the following formula:

$$X_{14} = \sqrt{(X_{14}\text{axial})^2 + (X_{14}\text{radial})^2}$$

The parameters of windings 1 and 4 are shown in Figure 7.19.

$$H_m = \frac{H_1 + H_4}{2}$$

$$= \text{average winding height}$$

$$X_{14}\text{axial} = \frac{\dfrac{\mu_0 \pi N^2}{H_m}\left[\dfrac{B_1 D_{m1}}{3} + \dfrac{B_4 D_{m4}}{3} + B_g D_{mg}\right]}{R_{ogax}}$$

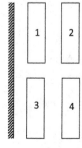

1, 3 – Secondary windings

2, 4 – Primary Windings

FIGURE 7.18 Split winding.

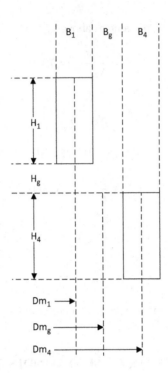

FIGURE 7.19 Split winding parameters.

where

$\mu_0 = 4\pi \times 10^{-7}$

N = number of turns

$H_m = \dfrac{H_1 + H_4}{2}$ or $\dfrac{H_2 + H_3}{2}$ = Average winding height (m)

B_1 = Radial Depth (Thickness) of winding 1 (m)

D_{m1} = Mean diameter of winding 1 (m)

B_4 = Radial depth of winding 4 (m)

D_{mg} = Mean diameter of gap between winding 1 and 4 (m)

D_{m4} = Mean diameter of winding 4 (m)

Figure 7.19 shows the above parameters.

$$R_{\text{oga}} = \frac{1 - e^{\frac{-\pi H_m}{B_1 + B_g + B_4}}}{\dfrac{\pi H_m}{B_1 + B_g + B_4}} \tag{7.29}$$

R_{oga} = Rogowski factor – Axial

$$X_{14} \text{ radial} = \frac{\mu_0 N^2 D_m}{\dfrac{D_m}{K_{\text{rogr}}}} \left[\frac{B_1}{3} + \frac{B_4}{3} + B_g \right] \tag{7.30}$$

where

$D_m = D_{m1} + D_{m4}$ m

$B_m = B_1 + B_4$ m

H_g = Axial gap between H_1 and H_4 m

K_{rogr} = Rogowski factor − radial

$$K_{rogr} = \frac{1 - e^{-\pi \frac{B_m}{H_1 + H_g + H_4}}}{\dfrac{\pi B_m}{H_1 + H_g + H_4}}$$ (7.31)

7.11 CALCULATION OF REACTANCE OF INDIVIDUAL WINDINGS

Equivalent reactance of HV and LV can be calculated on the basis that the sum of the reactances when converted to any reference side is equal to the total reactance calculated as per the procedures explained in Section 7.1.

$$X = X_1 + X_2'$$

where

X = reactance referred to rated voltage of winding 1

X_1 = reactance of winding 1

X_2' = equivalent reactance of winding 2 referred to winding 1

If per unit or percent values are calculated, $X = X_1 + X_2$

X_1 and X_2 are calculated from the ampere-turn diagrams for reactance calculations, and the calculations are done as below:

$$X_1 = X \left[\frac{A_1 + \frac{1}{2} A\delta}{A_1 + A\delta + A_2} \right]$$ (7.32)

where

X = reactance of the transformer

A_1 = area of ampere-turn diagram of winding 1

$A\delta$ = area of ampere-turn diagram of HV–LV gap

A_2 = area of ampere-turn diagram of winding 2

$A_1 + A\delta + A_2$ = total area of ampere-turn diagram

7.12 CALCULATION OF REACTANCE BY FINITE ELEMENT METHOD

Reactance of transformer can be calculated by finite element method (FEM). This method is useful when the windings have irregular ampere-turn distribution, where analytical methods may not give sufficient accuracy.

The reactance is calculated by the stored energy method.

$$W = \frac{1}{2} L I^2$$ (7.33)

where
    W = stored energy
    L = inductance
    I = current

From Equation (7.33),

$$L = \frac{2\,W}{I^2} \tag{7.34}$$

Reactance $X = 2\pi f L$

$$X = 2\pi f\,\frac{2W}{I^2} \tag{7.35}$$

In the above equation, W is the only unknown.

    X = reactance (ohms)
    f = frequency

The windings and gaps are divided into meshes, and the total stored energy is calculated.

$$W = \int_v \frac{1}{2}\,B\,H\,dv \tag{7.36}$$

where
    H = magnetic field strength
    B = magnetic flux density

$$= \mu\,H$$

μ = permeability

$$\therefore H = \frac{B}{\mu}$$

$$\therefore W = \int_v \frac{B^2}{2\mu}\,dv \tag{7.37}$$

$$W = \frac{1}{2\mu}\int_v B^2\,dv \tag{7.38}$$

The stored energy in terms of the magnetic vector potential and current density is given by Equation (7.39):

$$W = \frac{1}{2}\int_v A\,J\,dv \tag{7.39}$$

where
    A = magnetic vector potential
    J = current density

The solution in two-dimensional form is worked out by assuming a triangular shape function, where the corners represent the vector potentials, A_1, A_2 and A_3 (Figure 7.20).
    The shape function is taken as

$$A_1 = S_1 + S_2 x_1 + S_3 y_1$$
$$A_2 = S_1 + S_2 x_2 + S_3 y_2$$
$$A_3 = S_1 + S_2 x_3 + S_3 y_3$$

The solutions of S_1, S_2 and S_3 can be made by matrix method.

$$S_1 = \frac{\begin{bmatrix} A_1 & x_1 & y_1 \\ A_2 & x_2 & y_2 \\ A_3 & x_3 & y_3 \end{bmatrix}}{\begin{bmatrix} 1 & x_1 & y_1 \\ 1 & x_2 & y_2 \\ 1 & x_3 & y_3 \end{bmatrix}} \tag{7.40}$$

$$S_1 = \frac{A_1 \left(x_2 y_3 - x_3 y_2\right) - A_2 \left(x_1 y_3 - x_3 y_1\right) + A_3 \left(x_1 y_2 - x_2 y_1\right)}{\left(x_2 y_3 - x_3 y_2\right) - \left(x_1 y_2 - x_3 y_1\right) + \left(x_1 y_2 - x_2 y_1\right)} \tag{7.41}$$

The denominator is the area of the triangle. So S_1 can be written as

$$S_1 = \frac{\left(x_2 y_3 - x_3 y_2\right) A_1 + \left(x_3 y_1 - x_1 y_3\right) A_e + \left(x_1 y_2 - x_2 y_1\right) A_3}{2\Delta}$$

where

$$\Delta = \frac{\left|\left(x_2 - x_3\right)\left(y_3 - y_1\right) - \left(y_2 - y_3\right)\left(x_3 - x_1\right)\right|}{2}$$

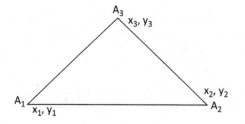

FIGURE 7.20   Shape function.

S_2, S_3 can be similarly written as

$$S_2 = \frac{(y_2 - y_3) A_1 + (y_3 - y_1) A_2 + (y_1 - y_2) A_3}{2\Delta}$$

$$S_3 = \frac{(x_3 - x_2) A_1 + (x_1 - x_3) A_2 + (x_2 - x_1) A_3}{2\Delta}$$

The flux density at any point is B is

$$B = \sqrt{Bx^2 + By^2}$$

B_x = axial component of the flux density
B_y = radial component of the flux density

The axial component B_x is the derivative of the vector potential A.

$$\therefore \frac{\delta_A}{\delta_y} = B_x$$

Similarly,

$$\frac{\delta_A}{\delta_x} = B_y$$

The solution in matrix form is

$$\begin{bmatrix} K_{11} & K_{12} & K_{13} \\ K_{21} & K_{22} & K_{23} \\ K_{31} & K_{32} & K_{33} \end{bmatrix} \begin{bmatrix} A_1 \\ A_2 \\ A_3 \end{bmatrix} = \frac{\mu_0 J \Delta}{3} \begin{bmatrix} 1 \\ 1 \\ 1 \end{bmatrix} \tag{7.42}$$

The values of K_{12}, K_{23}, K_{13}, etc., are calculated as follows:

$$K_{12} = K_{21} = \frac{1}{4\mu r \Delta} \left[(y_3 - y_1)(y_2 - y_3) + (x_3 - x_1)(x_2 - x_3) \right] \tag{7.43}$$

$$K_{23} = K_{32} = \frac{1}{4\mu r \Delta} \left[(y_3 - y_1)(y_1 - y_2) + (x_3 - x_1)(x_1 - x_2) \right] \tag{7.44}$$

$$K_{13} = K_{31} = \frac{1}{4\mu r \Delta} \left[(y_1 - y_2)(y_2 - y_3) + (x_1 - x_2)(x_2 - x_3) \right] \tag{7.45}$$

$$K_{11} = \frac{1}{4\mu r \Delta} \left[(x_2 - x_3)^2 + (y_2 - y_3)^2 \right] = -K_{13} - K_{12} \tag{7.46}$$

$$K_{22} = \frac{1}{4\mu r \Delta} \left[(x_1 - x_3)^2 + (y_1 - y_3)^2 \right] = -K_{12} - K_{23} \tag{7.47}$$

$$K_{33} = \frac{1}{4\mu r\Delta}\left[(x_1 - x_2)^2 + (y_1 - y_2)^2\right] = -K_{23} - K_{13} \tag{7.48}$$

7.13 ZERO SEQUENCE IMPEDANCE OF THREE-PHASE TRANSFORMERS

The zero phase sequence impedance characteristics of three-phase transformers depend on the winding connections and in some cases on the core construction.

For a delta–star connected transformer, the zero phase sequence impedance is approximately 80% of the positive sequence impedance. The zero phase sequence impedance of a transformer can be measured when at least one neutral point of a winding is brought out externally.

Zero phase sequence impedance of delta–star connected transformer is measured by connecting together all three line terminals of star connected winding and applying single-phase voltage at rated frequency between neutral terminal and the shorted line terminals. The applied voltage shall be such that the rated phase currents of any of the windings are not exceeded. The voltage and current are measured, and the percent zero phase sequence impedance is calculated by the following formula:

$$Z = 300\left[\left(\frac{E}{E_r}\right)\left(\frac{I_r}{I}\right)\right] \tag{7.49}$$

where

Z = percent zero sequence impedance
E = measured voltage
E_r = rated phase to neutral voltage of excited winding
I = measured total input current following in the three paralleled phase windings
I_r = rated current per phase of excited winding

The paths available for zero phase sequence currents and the relevant impedances are shown in Table 7.2 for typical transformer connections.

7.14 ZERO SEQUENCE IMPEDANCE CALCULATION

Effective leakage reactance of a system of n winding
I_j, I_k = currents in windings j and k
Z_{jk} = leakage reactance between j and k pair
Total power = $p + jq$

$$P = \frac{-1}{2}\sum_{k=1}^{n}\sum_{j=1}^{n}\overrightarrow{Z_{ik}}\overrightarrow{I_j}\overrightarrow{I_k}$$

Since resistance can be neglected, P can be ignored

$$\text{Total Power} = \frac{-1}{2}\sum_{k=1}^{n}\sum_{j=1}^{n}\overrightarrow{Z_{ik}}\overrightarrow{I_j}\overrightarrow{I_k} \tag{7.50}$$

$$\%X_0 = 2.48\times10^{-5}f\frac{AT}{H_{eq}\left(\dfrac{V}{T}\right)}\left[\frac{1}{3}\,T_w\,D_w + T_g\,D_G\right] \tag{7.51}$$

TABLE 7.2

Zero Phase Sequence Current Paths

Winding Diagram	Zero Sequence Network	Current Diagram

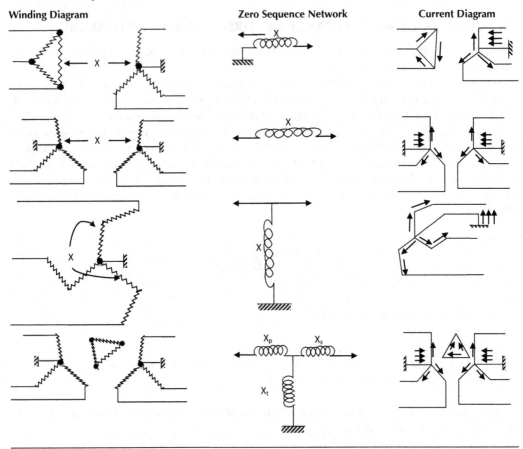

f = frequency

A = rated current

T = number of turns

H_{eq} = effective height of winding

$\dfrac{V}{T}$ = volt/turn

T_w = radial depth of excited winding (cm)

D_w = mean diameter of excited winding (cm)

T_g = gap between tank and excited winding (cm)

D_g = mean diameter of excited winding to tank (cm)

For a two-winding three-limb transformer with star connected primary and delta connected secondary, the windings can be represented as 1, 2 and the tank is represented as 3.

Per unit zero sequence reactance is

$$X_0 = \frac{1}{2}\left[2\,X_{12}I_1(-I_2) + 2X_{23}(-I_2)(-I_3) + 2X_{13}I_1(-I_3)\right] \qquad (7.52)$$

Neglecting magnetizing current

$$I_1 = I_2 + I_3 \qquad \therefore I_3 = I_1 - I_2$$

Substituting for $I_3 = I_1 - I_2$

$$Q = X_{12} \, I_1 \, I_2 + X_{23} \, I_2 \, (I_1 - I_2) + X_{13} \, (I_1 - I_2) \tag{7.53}$$

The currents get distributed in the winding such that the total energy is minimum:

$$\frac{d_q}{dI_2} = 0 = X_{12} - X_{23} + 2 \, X_{23} \, I_2 - X_{13} \tag{7.54}$$

$$\therefore I_2 = \frac{X_{13} + X_{23} - X_{12}}{2X_{23}} \tag{7.55}$$

$$I_3 = \left(\frac{X_{23} + X_{12} - X_{13}}{2X_{23}} \right) \tag{7.56}$$

8 Calculation of Core Frame Size, Core Losses, Efficiency and Regulation

8.1 CORE FRAME SIZE AND CORE WEIGHT CALCULATION

a. For Three-Phase Three-Limb Core

Core window height (LL) = HV axial height + Winding to yoke clearance at bottom +

Winding to yoke clearance at top

Core leg centre (CL) = HV OD + Phase to phase clearance

Total core weight can be calculated by adding the core weight across the grain and the core weight along grain.

$$\text{Total core weight}\,(W_{\text{core}}) = W_{\text{along}} + W_{\text{across}} \tag{8.1}$$

$$W_{\text{along}} = \left[3\text{LL} + 4(\text{CL} - W_{\text{max}})\right]7.65\,A_{\text{eff}} \times 10^{-4}\,(\text{kg}) \tag{8.2}$$

$$W_{\text{across}} = 7.65 \times 6\,W_{\text{max}} \times A_{\text{eff}} \times 10^{-4}\,(\text{kg}) \tag{8.3}$$

where
LL = core window height (mm)
CL = core leg centre distance (mm)
$W_{\text{max.}}$ = maximum step width (mm)
A_{eff} = core effective cross-sectional area (cm²)

b. Single-phase two-limb core (Figure 8.1)

$$L = 2\text{LL} + 2\text{CL} + 2W_{\text{max}}$$

$$\text{Core weight} = 7.65 \times 10^{-4} \times A_{\text{eff}} \times L \tag{8.4}$$

c. Single-phase three-limb core (centre limbs = 100%, side limbs = 50%) (Figure 8.2)

$$L = 2\text{LL} + 2\text{CL} + 2W_{\text{max}} \tag{8.5}$$

$$\text{Core weight} = 7.65 \times 10^{-4} \times A_{\text{eff}} \times L \tag{8.6}$$

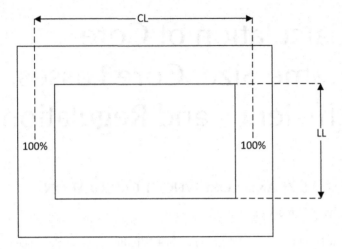

FIGURE 8.1   Single-phase two-limb core.

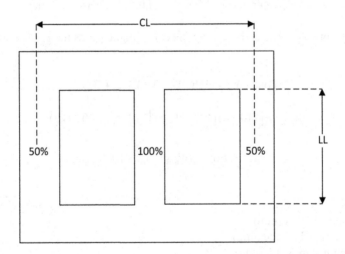

FIGURE 8.2   Single-phase three-limb core.

d. Three-phase five-limb core (centre limbs = 100%, side limbs = 50%) (Figure 8.3)

$$L = 4LL + 2.4CL + 2.4E + 1.2W_{max} \tag{8.7}$$

$$\text{Core weight} = 7.65 \times 10^{-4} \times A_{eff} \times L \tag{8.8}$$

8.2   CORE LOSS CALCULATION

Core losses (no-load losses) at rated frequency and voltage can be calculated by the following two methods.

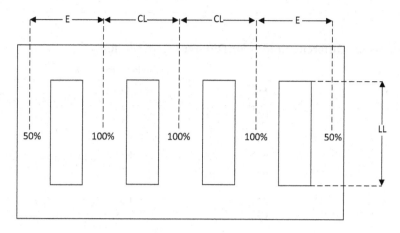

FIGURE 8.3 Three-phase five-limb core.

8.2.1 Loss Calculation Based on Average Building Factor

In CRGO steel, the specific losses along the grain orientation are given in manufacturer's graphs by conducting the Epstein tests. When the core is manufactured, the flux flow at the joints and its surroundings is deviated from the grain-oriented direction. This flow increases the core losses.

In the first method of calculation of core loss, it is assumed that the ratio of core diameter: leg centre: leg length remains more or less constant, and hence, the overall loss is a multiple of the specific core loss for a given core diameter. The building factor for smaller core circles will be higher, and for larger core circles, it will be lower.

Typical building factors used for the calculation of core loss are listed in Table 8.1. These factors require periodic revision based on comparison with tested values. These factors change with the type of material used.

$$W_i = K \, W_{core} S \tag{8.9}$$

where

W_i = total core loss (W)

W_{core} = total weight of core

S = W/kg of material at working flux density as per manufacturer's test certificate or graph, if test certificate is not available (the W/kg of some of the commercially available materials at different flux densities is given in Tables 8.2 and 8.3).

K = building factor (overall)

For three-phase three-limb fully mitred core, the following values of K as per Table 8.1 can be used.

8.2.2 Loss Calculation by Adding of Losses Across the Grain and Along the Grain

The no-load losses are calculated by adding the losses along the grain and the losses across the grain. It is assumed that the losses along the grain are equal to the specific loss as per manufacturer's graph or Epstein test result for the flux density under consideration. The losses across the grain are calculated by assuming a building factor for considering the effect of flux flow at corners.

$$W_1 = S \times W_{along} + K_2 S \left(W_{across} \right) \tag{8.10}$$

where

K_2 = building factor for flux flow deviation from grain-oriented direction at corners.

TABLE 8.1

Building Factors for Different Grades for Different Core Circles

Core Circle (mm)	Building Factor "K"		
	M4	MOH	23ZDKH90
75–95	1.35	1.4	1.45
96–105	1.32	1.35	1.40
106–130	1.32	1.32	1.40
131–160	1.32	1.32	1.40
161–200	1.30	1.30	1.35
201–250	1.28	1.28	1.33
251–300	1.25	1.25	1.30
301–400	1.25	1.25	1.30

For laser grade material, the building factor is about 5% more than the values, for M4 MOH, etc. The building factor for step-lap joints is about 12%–15% lower than the above values.

The value of K_2 shall be established for different types of core materials on the basis of comparing test results with calculated values.

K_2 will increase with improved grain orientation, and as such, K_2 will be higher for MOH when compared to M4, and still higher for laser grades, e.g. 23ZDKH70.

K_2 can be assumed as 1.8 for M4 and 2.0 for MOH and 2.25 for laser grade for fully mitred joints and ¾ mitred joints. K_2 is about 20% reduced when step lap is used.

8.3 SPECIFIC LOSSES OF DIFFERENT GRADES OF CRGO MATERIALS (TABLES 8.2 AND 8.3)

8.4 CORE LOSSES OF SYMMETRICAL CORE TRANSFORMERS

The symmetrical core for three-phase transformers consists of three wound single core frames with the same size to form a 3D triangle. In the stacked core three-limb core construction, the reluctance of the two outer limbs is higher because of longer flux paths and also due to the effect of the joints. When a three-phase voltage is applied, a three-phase balanced flux goes through the unbalanced magnetic path and unbalance of the magnetic circuit takes place. In a wound core, the effect of joints in the magnetic circuit is eliminated. However, the asymmetry of reluctance path is not avoided for a three-phase construction. The symmetrical core eliminates the above shortcomings and gives a balanced magnetic path. A comparison of three-phase stacked core, plane wound core and symmetrical wound core is given in Table 8.4.

8.4.1 ADVANTAGES OF SYMMETRICAL WOUND CORE

The cross section of the stacked core is of stepped shape as it is formed by the assembly of laminations of different widths. The filling factor of the core circle can approach 100% only by using very large number of steps with practical number of steps; the filling factor is 90%–92%.

The cross section of the symmetrical wound core is very nearly circular, and the filling factor can approach 97% to 98%. Hence, for the same cross-sectional area of core, the core circle of symmetrical wound core is reduced by 2.5%–3% and the mean length of winding turns is reduced by 6%–8%; a comparison of building factors for different types of core constructions for three-phase transformer is given in Table 8.5. It is assumed that the grade of core material used is the same for all the three types of constructions.

TABLE 8.2

Specific Losses of Different Grades

Tesla	23 Z DKH 75		23 Z DKH 80		23 Z DKH 85		23 Z H 85	
	50 Hz	60 Hz	50 Hz	60 Hz	50 Hz	60 Hz	50 Hz	60 Hz
1.00	0.24	0.32	0.25	0.33	0.26	0.34	0.27	0.37
1.02	0.25	0.34	0.26	0.34	0.27	0.35	0.28	0.38
1.04	0.255	0.35	0.27	0.35	0.28	0.36	0.29	0.41
1.06	0.26	0.36	0.28	0.36	0.28	0.38	0.30	0.42
1.08	0.27	0.37	0.29	0.38	0.29	0.39	0.31	0.43
1.10	0.28	0.38	0.30	0.40	0.30	0.40	0.33	0.44
1.12	0.29	0.39	0.31	0.41	0.31	0.41	0.34	0.45
1.14	0.30	0.40	0.32	0.42	0.32	0.42	0.35	0.46
1.16	0.32	0.42	0.33	0.43	0.33	0.43	0.36	0.48
1.18	0.33	0.44	0.34	0.45	0.34	0.44	0.37	0.50
1.20	0.34	0.46	0.35	0.46	0.35	0.46	0.38	0.52
1.22	0.35	0.47	0.36	0.47	0.36	0.48	0.40	0.53
1.24	0.36	0.44	0.38	0.49	0.38	0.50	0.41	0.54
1.26	0.37	0.51	0.39	0.51	0.39	0.52	0.43	0.56
1.28	0.38	0.53	0.40	0.53	0.40	0.54	0.44	0.58
1.30	0.40	0.54	0.42	0.55	0.42	0.56	0.46	0.60
1.32	0.41	0.55	0.43	0.57	0.44	0.58	0.47	0.61
1.34	0.41	0.56	0.45	0.58	0.45	0.59	0.49	0.63
1.36	0.42	0.58	0.46	0.60	0.46	0.61	0.50	0.64
1.38	0.43	0.60	0.47	0.62	0.48	0.63	0.51	0.66
1.40	0.44	0.62	0.48	0.64	0.50	0.65	0.53	0.68
1.42	0.46	0.63	0.49	0.65	0.51	0.67	0.54	0.70
1.44	0.48	0.65	0.51	0.67	0.52	0.69	0.55	0.72
1.46	0.50	0.67	0.52	0.69	0.53	0.71	0.56	0.74
1.48	0.52	0.69	0.54	0.71	0.54	0.73	0.59	0.76
1.50	0.54	0.71	0.56	0.73	0.56	0.75	0.60	0.78
1.52	0.55	0.73	0.57	0.75	0.58	0.77	0.60	0.81
1.54	0.56	0.75	0.58	0.77	0.60	0.79	0.62	0.84
1.56	0.58	0.77	0.60	0.79	0.62	0.81	0.64	0.88
1.58	0.60	0.79	0.62	0.82	0.64	0.84	0.67	0.90
1.60	0.62	0.82	0.64	0.85	0.66	0.87	0.70	0.99
1.62	0.64	0.84	0.66	0.88	0.68	0.89	0.72	0.96
1.64	0.66	0.87	0.68	0.91	0.70	0.91	0.74	0.98
1.66	0.68	0.90	0.70	0.94	0.72	0.94	0.76	1.00
1.68	0.70	0.93	0.72	0.97	0.75	0.97	0.79	1.03
1.70	0.72	0.96	0.74	1.00	0.78	1.00	0.82	1.06
1.72	0.75	1.00	0.77	1.04	0.81	1.05	0.85	1.12
1.74	0.79	1.05	0.80	1.08	0.84	1.11	0.89	1.19
1.76	0.82	1.10	0.83	1.12	0.88	1.17	0.93	1.26
1.78	0.86	1.15	0.86	1.16	0.92	1.23	0.97	1.33
1.80	0.90	1.20	0.90	1.20	0.96	1.30	1.01	1.40

8.4.2 MANUFACTURING PROCESS OF SYMMETRICAL WOUND CORE

The core is first slit in a diagonal slitting machine and then wound to bobbins. Then, it is rewound for feeding to the core winding machine on formers of the designed dimensions of the core. The wound core from the winding machine then undergoes a stress relieving annealing process at 820°C.

TABLE 8.3
Specific Losses of Different Grades

B_m	M6 (0.35 mm)		M5 (0.3 mm)		M4 (0.27 mm)		MOH (0.27 mm)		ZH90 (0.23 mm)		23ZDKH90 (0.23 mm)		RGHP (0.23 mm)	
	50 Hz	60 Hz	50 Hz	60 Hz	50 Hz	60 Hz	50 Hz	60 Hz	50 Hz	60 Hz	50 Hz	60 Hz	50 Hz	60 Hz
1.20	0.62	0.81	0.56	0.73	0.51	0.67	0.47	0.61	0.41	0.54	0.36	0.48	0.38	0.51
1.21	0.63	0.83	0.57	0.74	0.52	0.68	0.48	0.62	0.42	0.55	0.37	0.49	0.39	0.52
1.22	0.64	0.84	0.58	0.75	0.53	0.69	0.48	0.63	0.42	0.55	0.38	0.50	0.40	0.52
1.23	0.65	0.85	0.59	0.77	0.54	0.70	0.49	0.64	0.43	0.56	0.39	0.50	0.40	0.53
1.24	0.66	0.87	0.60	0.78	0.55	0.71	0.50	0.65	0.43	0.57	0.39	0.51	0.41	0.54
1.25	0.67	0.89	0.61	0.79	0.56	0.72	0.50	0.66	0.44	0.58	0.40	0.52	0.41	0.55
1.26	0.68	0.90	0.62	0.80	0.56	0.73	0.51	0.68	0.44	0.59	0.40	0.53	0.42	0.56
1.27	0.69	0.92	0.62	0.81	0.57	0.74	0.52	0.69	0.45	0.60	0.41	0.54	0.43	0.57
1.28	0.70	0.94	0.63	0.83	0.58	0.75	0.53	0.70	0.46	0.60	0.42	0.55	0.44	0.58
1.29	0.71	0.96	0.64	0.84	0.59	0.77	0.54	0.71	0.47	0.61	0.43	0.56	0.44	0.59
1.30	0.72	0.97	0.65	0.85	0.60	0.78	0.54	0.72	0.47	0.62	0.44	0.57	0.45	0.60
1.31	0.73	1.00	0.66	0.87	0.61	0.79	0.55	0.73	0.48	0.63	0.44	0.58	0.46	0.61
1.32	0.74	1.01	0.67	0.88	0.62	0.79	0.56	0.74	0.49	0.64	0.45	0.59	0.46	0.62
1.33	0.75	1.02	0.68	0.89	0.63	0.82	0.57	0.75	0.50	0.65	0.46	0.60	0.47	0.63
1.34	0.76	1.04	0.69	0.91	0.64	0.83	0.58	0.76	0.50	0.66	0.46	0.61	0.48	0.64
1.35	0.78	1.05	0.70	0.92	0.65	0.84	0.58	0.77	0.51	0.67	0.47	0.62	0.49	0.64
1.36	0.79	1.07	0.71	0.94	0.66	0.84	0.59	0.78	0.52	0.68	0.48	0.63	0.50	0.65
1.37	0.80	1.09	0.72	0.96	0.67	0.87	0.60	0.79	0.52	0.68	0.49	0.64	0.50	0.66
1.38	0.81	1.11	0.74	0.98	0.68	0.88	0.61	0.80	0.54	0.70	0.49	0.65	0.51	0.67
1.39	0.83	1.13	0.75	0.99	0.69	0.90	0.61	0.81	0.54	0.70	0.50	0.66	0.52	0.68
1.40	0.84	1.15	0.76	1.01	0.70	0.91	0.62	0.83	0.55	0.72	0.51	0.67	0.52	0.69
1.41	0.87	1.17	0.77	1.02	0.71	0.93	0.63	0.89	0.56	0.72	0.52	0.68	0.53	0.70
1.42	0.88	1.19	0.78	1.04	0.72	0.96	0.64	0.85	0.56	0.74	0.53	0.69	0.54	0.71
1.43	0.90	1.21	0.79	1.06	0.73	0.98	0.65	0.86	0.57	0.75	0.54	0.70	0.55	0.72
1.44	0.91	1.23	0.80	1.07	0.75	1.00	0.66	0.88	0.58	0.76	0.54	0.71	0.56	0.74
1.45	0.93	1.25	0.81	1.08	0.77	1.01	0.67	0.89	0.58	0.76	0.55	0.72	0.57	0.75

(Continued)

TABLE 8.3 (Continued)
Specific Losses of Different Grades

B_m	M6 (0.35 mm)		M5 (0.3 mm)		M4 (0.27 mm)		MOH (0.27 mm)		ZH90 (0.23 mm)		23ZDKH90 (0.23 mm)		RGHP (0.23 mm)	
	50 Hz	60 Hz	50 Hz	60 Hz	50 Hz	60 Hz	50 Hz	60 Hz	50 Hz	60 Hz	50 Hz	60 Hz	50 Hz	60 Hz
1.46	0.94	1.27	0.82	1.10	0.78	1.03	0.68	0.90	0.60	0.78	0.56	0.73	0.58	0.76
1.47	0.97	1.30	0.83	1.12	0.79	1.04	0.68	0.91	0.60	0.79	0.57	0.74	0.58	0.77
1.78	0.98	1.31	0.85	1.13	0.80	1.05	0.69	0.93	0.61	0.80	0.58	0.75	0.59	0.78
1.49	1.00	1.33	0.87	1.15	0.82	1.07	0.70	0.94	0.62	0.81	0.58	0.76	0.59	0.79
1.50	1.02	1.35	0.89	1.18	0.83	1.10	0.71	0.96	0.62	0.82	0.59	0.78	0.60	0.80
1.51	1.03	1.38	0.90	1.19	0.85	1.12	0.72	0.98	0.63	0.83	0.60	0.79	0.61	0.81
1.52	1.05	1.40	0.92	1.21	0.86	1.13	0.73	1.00	0.64	0.84	0.61	0.80	0.62	0.82
1.53	1.07	1.42	0.93	1.22	0.89	1.15	0.74	1.01	0.65	0.85	0.62	0.81	0.63	0.83
1.54	1.09	1.45	0.95	1.25	0.90	1.18	0.75	1.02	0.66	0.86	0.63	0.82	0.64	0.84
1.55	1.10	1.49	0.97	1.27	0.92	1.20	0.76	1.04	0.67	0.88	0.64	0.83	0.65	0.85
1.56	1.13	1.50	0.99	1.29	0.93	1.22	0.77	1.06	0.68	0.89	0.65	0.84	0.66	0.86
1.57	1.15	1.51	1.01	1.30	0.96	1.25	0.78	1.07	0.69	0.91	0.65	0.85	0.67	0.88
1.58	1.17	1.55	1.02	1.34	0.98	1.27	0.79	1.08	0.70	0.93	0.66	0.87	0.68	0.89
1.59	1.20	1.59	1.04	1.35	1.00	1.30	0.81	1.10	0.71	0.94	0.67	0.88	0.69	0.90
1.60	1.21	1.60	1.05	1.39	1.01	1.32	0.82	1.11	0.72	0.95	0.68	0.90	0.70	0.91
1.61	1.23	1.63	1.09	1.40	1.04	1.35	0.83	1.13	0.73	0.97	0.69	0.91	0.71	0.93
1.62	1.25	1.67	1.10	1.42	106	1.38	0.84	1.15	0.74	1.00	0.70	0.92	0.73	0.94
1.63	1.29	1.70	1.12	1.46	1.08	1.41	0.86	1.17	0.76	1.01	0.71	0.94	0.74	0.96
1.64	1.30	1.72	1.15	1.49	1.10	1.42	0.88	1.20	0.77	1.03	0.73	0.96	0.75	0.97
1.65	1.31	1.77	1.17	1.51	1.13	1.45	0.89	1.21	0.78	1.05	0.74	0.98	0.76	0.98
1.66	1.35	1.80	1.20	1.54	1.15	1.49	0.91	1.22	0.79	1.07	0.75	1.00	0.78	1.00
1.67	1.37	1.81	1.21	1.59	1.18	1.53	0.93	1.25	0.81	1.08	0.76	1.01	0.79	1.02
1.68	1.40	1.85	1.25	1.60	1.20	1.55	0.95	1.28	0.82	1.10	0.78	1.04	0.81	1.04
1.69	1.42	1.89	1.26	1.62	1.22	1.60	0.98	1.29	0.84	1.11	0.79	1.05	0.82	1.06
1.70	1.45	1.90	1.29	1.68	1.25	1.64	0.99	1.31	0.85	1.13	0.81	1.07	0.84	1.08
1.71	1.49	1.95	1.31	1.70	1.28	1.69	1.01	1.33	0.87	1.17	0.82	1.09	0.85	1.10

(Continued)

TABLE 8.3 (Continued)
Specific Losses of Different Grades

B_m	M6 (0.35 mm)		M5 (0.3 mm)		M4 (0.27 mm)		MOH (0.27 mm)		ZH90 (0.23 mm)		23ZDKH90 (0.23 mm)		RGHP (0.23 mm)	
	50 Hz	60 Hz	50 Hz	60 Hz	50 Hz	60 Hz	50 Hz	60 Hz	50 Hz	60 Hz	50 Hz	60 Hz	50 Hz	60 Hz
1.72	1.51	1.99	1.35	1.73	1.31	1.70	1.03	1.38	0.90	1.20	0.84	1.10	0.87	1.13
1.73	1.54	2.00	1.36	1.78	1.33	1.75	1.06	1.39	0.91	1.22	0.86	1.12	0.88	1.15
1.74	1.58	2.05	1.40	1.81	1.37	1.79	1.08	1.41	0.95	1.25	0.88	1.15	0.90	1.17
1.75	1.60	2.10	1.41	1.84	1.40	1.82	1.10	1.45	0.96	1.28	0.89	1.18	0.92	1.20
1.76	1.62	2.11	1.45	1.90	1.42	1.86	1.12	1.50	0.99	1.30	0.91	1.20	0.94	1.23
1.77	1.68	2.15	1.48	1.92	1.47	1.90	1.14	1.52	1.01	1.32	0.93	1.22	0.96	1.26
1.78	1.70	2.20	1.50	1.97	1.50	1.95	1.18	1.56	1.02	1.35	0.96	1.25	0.98	1.28
1.79	1.74	2.25	1.55	2.00	1.55	2.00	1.21	1.59	1.05	1.39	0.99	1.29	1.00	1.32
1.80	1.78	2.30	1.59	2.05	1.59	2.05	1.23	1.64	1.07	1.45	1.01	1.31	1.02	1.35
1.81	1.80	2.32	1.62	2.10	1.62	2.10	1.27	1.69	1.10	1.48	1.04	1.35	1.04	1.38
1.82	1.84	2.38	1.68	2.13	1.65	2.15	1.30	1.72	1.12	1.55	1.07	1.39	1.07	1.42
1.83	1.89	2.40	1.70	2.19	1.70	2.20	1.33	1.79	1.16	1.55	1.09	1.41	1.08	1.44
1.84	1.91	2.46	1.75	2.22	1.75	2.25	1.38	1.85	1.20	1.62	1.13	1.46	1.12	1.49
1.85	1.95	2.53	1.78	2.29	1.80	2.30	1.41	1.90	1.22	1.65	1.16	1.50	1.14	1.52
1.86	1.99	2.60	1.81	2.32	1.85	2.40	1.47	1.95	1.25	1.70	1.20	1.54	1.16	1.58
1.87	2.04	2.65	1.88	2.39	1.90	2.45	1.51	2.00	1.30	1.75	1.22	1.59	1.21	1.61
1.88	2.09	2.72	1.91	2.42	1.95	2.50	1.58	2.10	1.35	1.82	1.26	1.62	1.25	1.66
1.89	2.13	2.80	1.97	2.49	2.00	2.60	1.63	2.20	1.39	1.85	1.30	1.69	1.28	1.70
1.90	2.18	2.82	2.00	2.52	2.05	2.65	1.70	2.25	1.45	1.90	1.32	1.72	1.35	1.74

TABLE 8.4
Comparison of Stacked Core, Wound Core and Symmetrical Core

Type of Core Construction	Material Utilization	No-Load Current	No-Load Loss	Noise Level	Flux Symmetry	Harmonics
Stacked core	Normal	High	High	High	Unbalanced	Normal
Plane wound core	Better	Normal	Lower	Lower	Unbalanced	Normal
Symmetrical wound core	Highest	Low	Lowest	Lowest	Balanced	Reduced

TABLE 8.5
Comparison Stacked Core Step Lap, Plane Wound Core and Symmetrical Wound Core

Core Building Factor	Stacked Core Step-Lap Construction	Plane Wound Core	Symmetrical Wound Core
50 kVA	1.3–1.32	1.35	1.05
100 kVA	1.27–1.29	1.35	1.03
200 kVA	1.23–1.25	1.35	1.02
315 kVA	1.18–1.20	1.32	1.01
500 kVA	1.13–1.15	1.32	1.01
1000 kVA	1.11–1.12	1.30	1.01

The annealed cores are assembled to form the symmetrical core and then bound with webbing tape.

The LV and HV windings are made on the core using special winding machines similar to the torroidal winding machine. The core is mounted on the winding machine, and the coils are wound on formers inserted into the core limbs.

8.4.3 SYMMETRICAL WOUND CORE DESIGNS

The design process is similar to the three-phase transformer design, except for the calculation of core loss, no-load current, core weight, etc. The clamping design of the core coil assembly is different due to the symmetrical spacial disposition of the three limbs and the coils. The shape of the tank can be made to fit the core coil assembly to save oil quantity, or alternatively, conventional arrangement of tank can be used. Calculated values of core and core weights and cost of typical stacked core design and symmetrical wound core design are compared in Table 8.6. The transformers compared are 11/0.415 kV three-phase 50 Hz ONAN units with the same guaranteed performance figures.

TABLE 8.6
Comparison of Calculated Values of Core and Core Weights and Cost of Typical Stacked Core Design and Symmetrical Wound Core Design

kVA	Stacked Core		Symmetrical Core		Cost of Copper & Core		Saving (%)
	Core (kg)	Copper (kg)	Core (kg)	Copper (kg)	Stacked	Symmetrical	
100	150	135	134	108	1380	1138	17.5
315	457	225	372	192	2714	2280	16

8.5   CALCULATION OF NO-LOAD CURRENT (EXCITATION CURRENT)

The current that flows in any winding used to excite the transformer when all other windings are open circuited is called no-load current or excitation current. This current maintains the magnetic flux excitation in the core of the transformer.

The excitation current is usually expressed in per unit or in percentage of the rated current of the winding in which it is measured.

The excitation current is calculated as below:

$$I_0\% = \frac{\left[W_{core} \times \left(\frac{VA}{kg} \right) \times K \right] \times 100}{\sqrt{3} \ V_L \ I_L} \qquad (8.11)$$

$I_0\%$ = percent no-load current
W_{core} = weight of core (kg)
VA/kg = excitation VA of core material at working flux density (from manufacturer's graph)
K = building factor for excitation VA (for mitred core)
= 3.0 for core circles from 70 mm diameter to 130 mm diameter
2.5 for core circles above 130 mm but below 250 mm
2.0 for core circles above 250 mm
For step-lap core, K can be taken as 60% of the above value

Note

1. When excitation current is to be guaranteed, the tolerance allowable is +30%. If customer has specified an upper limit for excitation current, the same limit shall be considered while furnishing guarantee.
2. Excitation current also forms a factor for capitalization in some customer specifications.
3. The excitation current at flux densities above 1.75 Tesla increases very high. Table 8.7 gives the formulas for calculating the specific core loss in watts/kg of different grades at 50 Hz for a flux density range of 1.2 T to 1.9 T. For 60 Hz the results can be multiplied by 1.32 for estimation purposes.

TABLE 8.7

Formulas for Specific Core Loss for CRGO Materials between 1.2 Tesla and 1.9 Tesla, W = Watts/kg, B = Flux Density in Tesla

Core Grade	Core Loss W/kg at 50 Hz
M5	$W = 5.1634 - 15.94B + 19.41B^2 - 10.355B^3 + 2.15B^4$
M4	$W = 4.766 - 15.736B + 20.151B^2 - 11.21B^3 + 2.405B^4$
MOH	$W = 45.43 - 127.96B + 135.27B^2 - 63.2B^3 + 11.1B^4$
23ZH90	$W = 31.47 - 89.4B + 95.5B^2 - 45B^3 + 7.9B^4$
23ZDKH90	$W = 14.4 - 43.56B + 49.26B^2 - 24.35B^3 + 4.516B^4$
23ZH85	$W = -8.17 + 17.24B - 11.6B^2 - 2.5B^3 + 0.959B^4$
23ZDKH85	$W = 3.31 - 13.15B + 18.2B^2 - 10.3B^3 + 2.1B^4$
23ZDKH80	$W = 23.92 - 67.23B + 70.82B^2 - 32.8B^3 + 5.72B^4$
ZDKH75	$W = 30.77 - 86.14B + 90.49B^2 - 41.96B^3 + 7.31B^4$
20ZDKH75	$W = 12.468 - 38.75B + 44.84B^2 - 22.59B^3 + 4.25B^4$
20ZDKH70	$W = 13.721 - 42.48B + 48.9B^2 - 24.55B^3 + 4.59B^4$

8.6 SUGGESTED CHANGES OF DESIGN PARAMETERS TO GET DESIRED LOSSES (TABLE 8.8)

When the losses are specified, and the calculated no-load losses and/or load losses are not within the desired limits, the following logical strategies can be used. The calculated values of no-load loss and load loss will have to be less than the guaranteed maximum value, to take into account the tolerances in manufacturing and also to consider the effect of allowable tolerances of the performance parameters of the core steel (CRGO) and conductor material. The suggested strategies are on the assumption that these variations will be less than 2% of the guaranteed values. Therefore, calculated values of losses up to 0.98 times guaranteed values are considered as acceptable.

TABLE 8.8
Suggested Changes of Design Parameters

1	Calculated No-Load Loss	Calculated Load Loss (Times Guaranteed Value)	Suggested Changes of Design Parameters
1.1	Within 0.98 of guaranteed value	0.97–0.94	• No change in core diameter and flux density • Change HV and LV conductor cross section to 0.97–0.94 times of previous design
1.2	Within 0.98 of guaranteed value	0.94–0.89	• No change in core area and flux density • Change HV and LV conductor cross section to 0.94–0.89 of previous design
1.3	Within 0.98 guaranteed value	1.03–1.06	• No change in core area and flux density • Change HV and LV conductor cross section to 1.03–1.06 of previous design
1.4	Within 0.98 of guaranteed value	1.06–1.15	• Change core circle to 1.03 times of previous design • Change HV and LV conductor cross section to 1.06–1.12 of previous design
2	Calculated No-Load Loss	0.98 of guaranteed	• Use same conductor
2.1	Within 0.98–0.94 of the guaranteed value	value	• Reduce core circle to 0.98 times of previous design
2.2	Within 0.98–0.94 of the guaranteed value	0.98–0.94 of guaranteed value	• Change HV and LV conductor area to 0.97 of previous design • Reduce core circle to 0.98 times of previous design
2.3	Within 0.98–0.94 of the guaranteed value	0.94–0.89 of guaranteed value	• Increase core circle to 1.03 times of previous design • Change LV and HV conductor area to 0.97 of previous value
2.4	Within 0.98–0.94 of the guaranteed value	1.02–1.06 of guaranteed value	• Reduce core circle to 0.97 times of previous design • Use same conductor
2.5	Within 0.98–0.94 of the guaranteed value	1.06–1.11 of guaranteed value	• Reduce core circle to 0.95 times of previous design • Changes LV and HV conductor area to 1.05–1.08 times of previous design
3.	Calculated no-load loss	Calculated load loss	
3.1	Within 0.94–0.89 the guaranteed value	0.98 of guaranteed value	• Reduce core circle to 0.97–0.95 times of previous design • Change LV and HV conductor area to 0.98–0.96 times of previous design
3.2	Within 0.94–0.89 the guaranteed value	0.98–0.94 of guaranteed value	• Reduce core circle to 0.97–0.95 of previous design • Change LV and HV conductor area to 0.96–0.93 of previous design

(Continued)

TABLE 8.8 (*Continued*)
Suggested Changes of Design Parameters

1	Calculated No-Load Loss	Calculated Load Loss (Times Guaranteed Value)	Suggested Changes of Design Parameters
3.3	Within 0.94–0.89 the guaranteed value	0.94–0.89 of guaranteed value	• Reduce core circle to 0.97–0.95 of previous design • Change LV and HV conductor area to 0.93–0.88 of previous value
3.4	Within 0.94–0.89 the guaranteed value	0.98–1.03 of guaranteed value	• No change in LV and HV conductor area • Reduce core circle to 0.97–0.95 of previous design
3.5	Within 0.94–0.89 the guaranteed value	1.03–1.11 of guaranteed value	• Increase HV and LV conductor area to 1.03–1.08 times of previous design • Reduce core circle to 0.97–0.95 of previous design
4	Calculated no-load loss	Calculated load loss	
4.1	>0.98–1.03 of guaranteed value	0.98 times guaranteed value	• Increase core circle to 1.01–1.02 times of previous design • Increase LV and HV conductor area to 1.02 of previous design
4.2	>0.98–1.03 of guaranteed value	<0.98–0.94 of guaranteed value	• Increase core circle to 1.01–1.02 times of previous design • Change LV and HV conductor area to. 98.96 of previous design
4.3	>0.98–1.03 of guaranteed value	<0.94–0.89 of guaranteed value	• Increase core circle to 1.02–1.04 of previous design • Change LV and HV conductor area to 0.96–0.93 of previous design
4.4	>0.98–1.03 of guaranteed value	>0.98–1.03 times guaranteed value	• Increase core circle to 1.02–1.04 of previous design • Change HV and LV conductor area to 1.02 of previous design
4.5	>0.98–1.03 of guaranteed value	>1.03–1.10 times guaranteed value	• Increase core circle to 1.02–1.04 of previous design • Change LV and HV conductor area to 1.02–1.05 of previous design
5	Calculated no-load loss	Calculated load loss	
5.1	>1.03–1.1 times guaranteed value	0.98 of guaranteed value	• Increase core circle to 1.03–1.05 of previous design • Change HV and LV conductor area to 1.03–1.04 of previous design
5.2	>0.98–1.03 of guaranteed value	<0.98–0.94 of guaranteed value	• Increase core circle to 1.02–1.03 of previous design • Change HV and LV conductor area to 0.98–0.96 of previous design
5.3	>0.98–1.03 of guaranteed value	0.94–0.89 of guaranteed value	• Increase core circle to 1.02–1.03 of previous design • Change HV and LV conductor area to 0.96–0.93 of previous design
5.4	>0.98–1.03 of guaranteed value	>0.98–1.03 times guaranteed value	• Use core circle of 1.03–1.05 of previous design • Change HV and LV conductor area to 0.98–0.96 of previous design
5.5	>0.98–1.03 of guaranteed value	>1.03–1.10 times guaranteed value	• Increase core circle to 1.03–1.05 of previous design • Increase LV and HV conductor to 1.03–1.07 of previous design

8.7 CALCULATION OF EFFICIENCY AND REGULATION

8.7.1 CALCULATION OF EFFICIENCY

8.7.1.1 Calculation of Efficiency as per ANSI Standard

The efficiency of a transformer is the ratio of its useful power output to its total power input.

$$\eta\% = \frac{100 P_o}{P_i} = \frac{(P_i - P_L)100}{P_i} = \left(1 - \frac{P_L}{P_i}\right)100$$

$$\eta = 100\left(1 - \frac{P_L}{P_o + P_L}\right) \tag{8.12}$$

where

$\eta\%$ = efficiency (%)
P_o = output
P_i = input
P_L = losses

For power factor of "S" and load of "x" per unit, the formula for efficiency is

$$\% = 100\left(1 - \frac{W_i + x^2 \, W_c}{x \, S \, P_o + W_1 + x^2 \, W_c}\right) \tag{8.13}$$

where

$\eta\%$ = efficiency (%)
W_i = no-load loss (W)
x = per unit load
W_c = load losses at full load and at reference temperature
S = power factor per unit

The calculation of efficiency differs from IEC 60076 and ANSI C57.12 Standards because of the difference in defining rated kVA by these standards.

8.7.1.2 IEC 60076 Standard and ANSI C57.12 Standard

The rated power is defined by IEC 60076 as "When the transformer has rated voltage applied to a primary winding, and rated current flows through the terminals of a secondary winding the transformer receives the relevant rated power for that pair of windings".

This implies that it is a value of apparent power input to the transformer, including its own absorption of active and reactive power. The apparent power that the transformer delivers to the circuit connected to the terminals of the secondary winding under rated loading differs from the rated power. The voltage across the secondary terminals differs from the rated voltage by the voltage drop (or rise) in the transformer.

This is different from the definition as per ANSI C57.12.00 standard, where the rated kVA is defined as

> The rated kVA of a transformer shall be the output that can be delivered for the time specified at rated secondary voltage and rated frequency without exceeding the specified temperature rise limitations under prescribed conditions of test, and within the limitations of established standards.

Calculation of efficiency as per IEC 60076 standard

The voltage drop (or rise) of the secondary will have to be considered, while the efficiency is calculated, as per IEC standard.

$$= 100 - \left(\frac{P_o + n^2 P_k}{n \; S_n \; \cos\phi \left(1 - \dfrac{U\phi n}{100} \right) + P_o + n^2 \; P_k} \right) 100 \qquad (8.14)$$

where

$\eta\%$= efficiency (%)

P_o = no-load loss (kW)

$n = \dfrac{\text{Load}}{\text{Rated Power}}$ kVA

P_k = load loss (kW)

Cosϕ = power factor

$U_{\phi n}$ = voltage drop (rise) %

$$U_{\phi n} = n\left(r\cos\phi + x\sin\phi\right) + \frac{\left[n\left(r\sin\phi - x\cos\phi\right)\right]^2}{200} \qquad (8.15)$$

8.7.1.3 Calculation of Efficiency as per IEC 60076 and ANSI C57.12

Transformer parameters

kVA – 1500

Voltage ratio – 11/0.433

% Impedance – 6.0

No-load loss – 2.3 kW

Load loss – 9.0 kW

Table 8.9 shows the efficiency calculated as per IEC standard.

TABLE 8.9
Efficiency Calculation as per IEC Standard

	Unity Power Factor		0.8 Power Factor Lagging	
	100% Load	50% Load	100% Load	50% Load
No-load loss (kW)	2.3	2.3	2.3	2.3
Load loss (kW)	9.0	2.25	9.0	2.25
Voltage drop (%)	0.778	0.3002	4.159	2.055
Impedance (%)	6.0	6.0	6.0	6.0
Resistance (%)	0.6	0.6	0.6	0.6
Reactance (%)	5.97	5.97	5.97	5.97
Efficiency (%)	99.246	99.395	99.027	99.232

At unity power, factor (full load)

$$r\cos\phi + x\sin\phi = 0.6 \times 1 + 5.97 \times 0 = 0.6$$

$$r\sin\phi + x\cos\phi = 0.6 \times 0 - 5.97 \times 1 = -5.97$$

$$\text{Voltage drop} = 0.6 + \frac{(-5.97)^2}{200}$$

$$= 0.778\%$$

At 0.8 pt lag (full load)

$$r\cos\phi + x\sin\phi = 0.6 \times 0.8 + 5.97 \times 0.6 = 4.062$$

$$r\sin\phi - x\cos\phi = 0.6 \times 0.6 - 5.97 \times 0.8 = -4.416$$

$$\text{Voltage drop} = 4.062 + \frac{(-4.416)^2}{200}$$

$$= 4.159\%$$

Unity power factor (50% load)

$$\text{Voltage drop} = 0.5 \times 0.6 + \left(\frac{-5.97 \times 0.5}{200}\right)^2 = 0.3445\%$$

Voltage drop at 0.8 power factor lag (50% load)

$$= 0.5 \times 4.062 + \left(\frac{-4.416 \times 0.5}{200}\right)^2 = 2.055\%$$

As per IEC

Efficiency at 100% load unity power factor

$$\eta = 100 - \left(\frac{2.3 + 9.0}{1500 \times 1\left(1 - \frac{0.778}{100}\right) + 2.3 + 9}\right)100$$

$$\eta = 100 - \left(\frac{11.3}{1499.63}\right)100 = 99.246\%$$

As per ANSI

$$\eta = 100 - \left(1 - \frac{2.3 + 9}{1500 + 2.3 + 9}\right) = 99.252\%$$

TABLE 8.10

Comparison of Efficiency as per IEC and ANSI

	IEC	ANSI
	Efficiency (%)	Efficiency (%)
Full load UPF	99.246	99.252
50% load UPF	99.395	99.397
Full load 0.8 pt lag	99.027	99.067
50% load 0.8 pt lag	99.232	99.247

Table 8.10 shows the comparison of efficiency calculated as per IEC and ANSI standards.

8.7.2 CALCULATION OF REGULATION

The exact formula for calculation of regulation is as follows:

a. When the load is lagging:

$$\text{Reg} = \sqrt{\left(R + F_p\right)^2 + \left(x + q\right)^2} - 1 \tag{8.16}$$

b. When the load is leading:

$$\text{Reg} = \sqrt{\left(R + F_p\right)^2 + \left(x - q\right)^2} - 1 \tag{8.17}$$

where

F_p = power factor of load
$q = \sqrt{1 - F_p^2}$
R = resistance factor of transformer

$$R = \frac{\text{Load Loss in kW}}{\text{Rated kVA}} \tag{8.18}$$

x = reactance factor of transformer

$$x = \sqrt{Z^2 - R^2} \tag{8.19}$$

Z = impedance factor

$$Z = \frac{\text{Impedance kVA}}{\text{Rated kVA}} \tag{8.20}$$

The quantities F_p, q, x and R are on a per unit basis, so the result must be multiplied by 100 to obtain the regulation in percentage.

8.8   CALCULATION OF EQUIVALENT CIRCUIT PARAMETERS OF TRANSFORMER

The parameters of the equivalent circuit can be calculated from the design details of the transformer.
   The procedure for calculation is shown by an example based on the design results of a 1500 kVA three-phase transformer.

Design details of the transformer
    kVA – 1500
    Frequency – 50 Hz
    Phases – 3
    Cooling – ONAN
    Primary volts – 11000
    Secondary volts – 433
    Primary connection – Delta
    Secondary connection – Star
    Percent impedance – 6 %
    No-load loss – 1800 W
    Load loss at 75°C – 12000 W
    No-load current – 0.9% of full load current
Equivalent circuit (Figure 8.4)

The parameters referred to primary are calculated as follows:

I_p = primary rated current

$$= \frac{(\text{kVA})}{\sqrt{3}\ \text{kV}} = \frac{1500}{\sqrt{3} \times 11} = 78.73 \text{ Amps}$$

R = equivalent resistance referred to HV

$$= \frac{\text{Load Loss}}{3 \times I_p^2}$$

$$= \frac{12,000}{3 \times 78.73^2} = 0.645 \text{ ohms / phase}$$

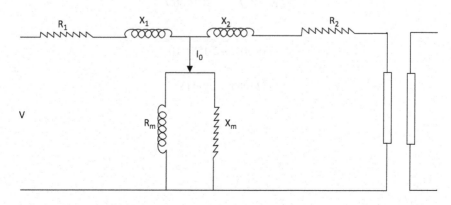

FIGURE 8.4   Transformer equivalent circuit.

Z_- = impedance in ohms referred to primary

$$= \frac{V_{ph}\,(\%Z)}{I_p \times 100} = \frac{11000 \times 6}{78.73 \times 100}$$

= 8.38 Ω (referred to HV side)

X = reactance in ohms referred to primary

$$= \sqrt{Z^2 - R^2} = \sqrt{8.38^2 - 0.645^2}$$

= 8.355 Ω

Z_m = magnetizing impedance referred to HV

I_o = no-load current referred to HV

$$= \frac{0.9}{100} \times 78.73 = 0.708 \text{ Amps}$$

Z_m = magnetizing impedance referred to HV

$$= \frac{11000}{0.708} = 15536 \text{ ohms/phase}$$

R_m = magnetizing resistance referred to HV

$$= \frac{3 \times V^2}{W} = \frac{3 \times 11000^2}{1800}$$

= 201666 Ω referred to HV

X_m = magnetizing reactance referred to HV

$$= \frac{1}{\sqrt{\left(\dfrac{1}{Z_m}\right)^2 - \left(\dfrac{1}{R_m}\right)^2}}$$

$$= \frac{1}{\sqrt{\left(\dfrac{1}{15536}\right)^2 - \left(\dfrac{1}{201666}\right)^2}}$$

$$= \frac{1}{\sqrt{4.143 \times 10^{-9} - 2.458 \times 10^{-11}}}$$

= 15536 Ω referred to HV

9 Lightning and Switching Surges on Transformers

9.1 INTRODUCTION

Lightning is a natural phenomenon, and several millions of it at varying magnitudes strike the earth. When a lightning strikes a power line or a conductor, an impulse voltage travels through the line and enters the terminals of the transformer connected to it.

Being a natural phenomenon, it is difficult to predict the exact magnitude or wave shape of a lightning impulse. However, by analysing the data collected from various parts of the world, three typical wave shapes are identified.

- Full wave
- Chopped wave
- Front of wave

It is to be noted that the actual wave shape of a lightning strike may not resemble any one of the above shapes. However, the above wave shapes are defined by standards in order to ensure that the tests are carried out uniformly and response can be compared on equal basis. Another type of surge can occur in the power system from any one of the following causes.

- Switching on of the transformer
- When a system fault occurs, the circuit breaker opens and the line trips. This generates surges, due to the capacitive, inductive oscillations
- When a line is switched on, after a fault is cleared

The switching surges can create overvoltages and flashover. Typical amplitude of a switching surge is 1.5 pu of the system voltage.

9.2 EFFECT OF SURGES ON TRANSFORMER WINDING

The response of a power frequency voltage is linear on the transformer windings. The voltage is proportionally distributed everywhere in the winding from one end to the other end, and the electro-magnetically transferred voltages on the other windings are also linearly distributed.

The behaviour of surge on a transformer winding is not similar to the response from a power frequency. The reason is that the surges are high-frequency waves and, therefore, the inductive path offers high impedance to its flow. The capacitive reactance of the winding at high frequency becomes low, and the surge travels through the capacitive path. Thus, the voltage distribution is determined by the capacitive equivalent circuit of the transformer and the wave shape of the surge. It is necessary to consider that once the initial stage is over, the damping due to the resistive loss and the inductive effect comes into picture.

9.3   CAPACITIVE EQUIVALENT CIRCUIT OF TRANSFORMER

In order to analyse the surge response of the transformer winding, its equivalent capacitance network is to be derived. This is done by calculating the various capacitances of the winding under consideration as follows:

- Capacitance between turns
- Capacitance between layers
- Capacitance between discs
- Capacitance of winding to tank (earth)
- Capacitance between two windings

When a turn of a disc or layer is considered, it has a capacitance between adjacent turns and capacitance to other discs/layers and ground. Ignoring the stray capacitance from one turn to non-adjacent turns, the capacitive circuit can be approximated as a combination of series capacitance and a shunt (earth) capacitance. The approximate circuit is shown in Figure 9.1, which is called "ladder circuit".

K_1 = capacitance between adjacent turns
C_1 = capacitance of one turn to earth
Since the earth capacitances are in parallel,
$C = \in C_1$, where C is the total capacitance to earth
$K = \in \dfrac{1}{K_1}$, where K is the total capacitance along the winding length

9.4   CALCULATION OF CAPACITANCES

The capacitive equivalent circuit can be derived from the capacitances between turns, layers, discs, etc., and the methods of calculation of capacitances are given in this section.

a. Capacitance between turns of layer winding or disc winding

$$K_1 = 0.974 D_m \left(\frac{h+e}{e} \right) \tag{9.1}$$

where
K_1 = capacitance between turns (pF)
D_m = mean diameter of the conductor of the layer (cm)
h = width of the bare conductor (cm)
e = insulation thickness between conductors (cm)
PF = pico-farads

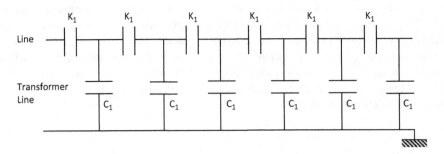

FIGURE 9.1   Ladder circuit.

b. Capacitance between two concentric windings

$$C_1 = \frac{0.378 \ (D_1 + D_2) \ (L + \Delta)}{\Delta - 0.43 \ t}$$ (9.2)

where

C_1 = capacitance between two concentric windings or two concentric layers (pF)
D_1 = outer diameter of inner winding or layer (cm)
D_2 = inner diameter of outer winding or layer (cm)
Δ = radial gap between windings or layer (cm)
t = total insulation thickness between windings/layer (including insulation of conductor) (cm)
L = winding height (cm)

c. Capacitance between discs

$$K_d = 0.851 \times D_m \ \frac{(B+t)}{(t+0.874e)}$$ (9.3)

where

K_d = capacitance between discs (pF)
D_m = mean diameter of discs (cm)
e = insulation thickness between turns (cm)
B = radial thickness of disc (cm)
(width of disc)
t = distance between two discs (cm)
(including conductor insulation)

d. Series capacitance of disc winding (without interleaving)

$$K_s = K_1 \frac{(n-1)}{n^2} + \frac{4}{3} K_d$$ (9.4)

where

K_s = series capacitance of one disc (pF)
K_1 = capacitance between turns as per formula (9.1) (pF)
n = number of turns in one disc
K_d = capacitance between discs as per formula (9.3) (pF)

e. Series capacitance of interleaved disc windings

$$K_{si} = K_1 \left\{ n \times p - \frac{n}{2} + \frac{(n-1)^2 \times \left(\frac{n}{2} - 1\right)}{n^2} \right\}$$ (9.5)

where

K_{si} = series capacitance of interleaved disc winding
n = number of turns/discs
p = number of parallel conductors in one turns

f. Series capacitance of layer-type tap winding with n turns and p parallels

$$K_p = K_1 \frac{(4P-6)n-1}{P^2} \tag{9.6}$$

where
K_p = series capacitance of layer winding with P parallel conductors (pF)
K_1 = capacitance between adjacent conductors as per (9.1) (pF)
P = number of parallel conductors
n = number of turns

g. Series capacitance of shielded disc winding

$$K_{ds} = mK_t \left[8 - \frac{12n-5}{n^2} \right] + \frac{4}{3} K_d \tag{9.7}$$

where
K_{ds} = series capacitance of shielded disc winding (pF)
K_t = capacitance between winding conductor and shield conductor (pF)
n = number of turns per disc
K_d = capacitance between discs
m = number of turns of shield conductor

h. The total series capacitance of a disc winding

$$C_s = \frac{C_t}{N} + \frac{4}{3} C_d \tag{9.8}$$

where
C_s = total series capacitance (pF)
C_t = capacitance between turns (pF)
C_d = capacitance between discs (pF)
N = number of turns/disc

9.5   CALCULATION OF INITIAL VOLTAGE DISTRIBUTION OF A CAPACITIVE LADDER CIRCUIT

The initial voltage distribution is dependent on the series capacitance and earth capacitance. The distribution is different for a solidly earthed neutral or floating neutral.

The simple solution is given in this section.

Figure 9.2 shows the ladder circuit where C_1 is the earth capacitance of an element and K_1 is the series capacitance between two adjacent elements of the winding.

The impulse distribution is decided by the series and earth capacitances of the winding.

Let unity be the length of winding = 1 Pu
Total capacitance to earth = C
Total capacitance along winding length K

$$\therefore C = \in C_1 \quad \text{i.e. } C_1 + C_1 + \ldots$$

All C_1 capacitances are shunt connected
All K_1 capacitors are series connected

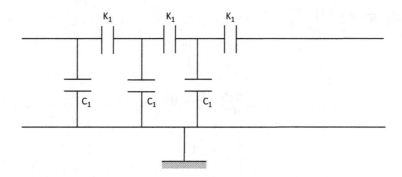

FIGURE 9.2   Capacitive ladder.

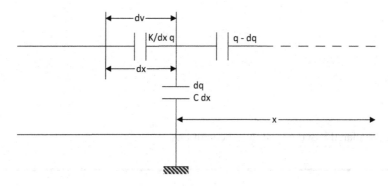

FIGURE 9.3   Winding element.

$$\therefore \frac{1}{K} = \Sigma \frac{1}{K_1} \qquad ie \; \frac{1}{K} = \frac{1}{K_1} + \frac{1}{K_1} + ------$$

Consider a winding element dx, which is x distance from the neutral end of the winding shown in Figure 9.3. Let the voltage at this point be V (voltage relative to earth) and the charge q.

The earth capacitance of element $dx = cdx$

[C is the earth capacitance for the full length which is 1 Pu]

Series capacitance $= \dfrac{K}{dx}$

If the charge of the adjacent winding element is $d\text{-}dq$, the charge of capacitance cdx is equal to dq.

$$q = CV \tag{9.9}$$

For the capacitance cdx, the charge is capacitance x voltage

$dq = VC \, dx$

$$V = \frac{1}{C} \frac{dq}{dx} \tag{9.10}$$

For series capacitance $\dfrac{K}{dx}$

Similarly, charge is capacitance x voltage

$$\frac{dv}{dx} = \frac{q}{k} \tag{9.11}$$

Differentiating equation 9.11

$$\frac{d^2V}{dx^2} = \frac{dq}{k}$$

$$\frac{d^2V}{dx^2} - \frac{dq}{k} = 0$$

$$\frac{d^2V}{dx^2} = \frac{c}{k}V = 0 \tag{9.12}$$

Solving

$$V = A_1 e^{\alpha x} + A_2 e^{-\alpha x} \tag{9.13}$$

where $\alpha = \sqrt{\dfrac{c}{k}}$

A_1 and A_2 are calculated based on the initial conditions.

Case 1

Solidly grounded neutral. In this case, the voltage at neutral end is zero and voltage at the line end is the applied voltage V_{in}

$$\text{when } x = 1, V = V_{in} \tag{9.14}$$

$$\text{When } x = 0, V = 0 \tag{9.15}$$

Substituting Equation (9.13) in Equation (9.15)

$$A_1 = -A_2$$

By using Equation (9.14)

$$V_m = A_1 \left(e^{\alpha} - e^{-\alpha} \right) \tag{9.16}$$

$$\therefore A_1 = \frac{V_m}{e^{\alpha} - e^{-\alpha}} \tag{9.17}$$

$$V = V_m \left(\frac{e^{\alpha} - e^{-\alpha}}{e^{\alpha} - e^{-\alpha}} \right) \tag{9.18}$$

$$V = V_m \frac{\sinh \alpha x}{\sinh \alpha} \tag{9.19}$$

The voltage gradient at the line end of the winding is $\frac{dv}{dx}$ when $x = 1$

$$\therefore \frac{dv}{dx} = V_m \, \alpha \coth \alpha$$

For values of $\alpha > 3$, $\coth \alpha \sim 1$

$$\frac{dv}{dx} = V_m \, \alpha$$

i.e. the voltage gradient at line end is α times more than average gradient

---

Case 2

Isolated neutral

In this case, the boundary conditions are

at $x = 1$ $V = V_m$

at $x = 0$ $q = 0$

$$\therefore \frac{dv}{dx} \, x = 0$$

Differentiating Equation (9.11) and putting $x = 0$ and $\frac{dv}{dx} = 0$

$$A_1 = A_2$$

$$\therefore A_1 = -A_2 = \frac{V_m}{e^\alpha + e^{-\alpha}} \tag{9.20}$$

$$\therefore V = V_m \left[\frac{e^\alpha + e^{-\alpha x}}{e^\alpha + e^{-\alpha}} \right]$$

$$V = V_m \frac{\cosh \alpha x}{\cosh \alpha} \tag{9.21}$$

When $\alpha = 0$, $V = V_m$

The voltage gradient at the line end is

$$\frac{dv}{dx} \, x = 1 \, V_m \frac{\alpha \sinh}{\cosh \alpha}$$

$$V_m \, \alpha \mathrm{Tan} \, h\alpha \cong V_m \, \alpha$$

{for values of $\alpha > 3$}

∴. The voltage gradient at line end is approximately equal for both grounded neutral and isolated neutral.

From the above, an approximate expression for calculating the voltages on the discs or turns of a layer winding can be calculated.

The maximum voltage drop is on the first disc or first turn.

The initial voltage distribution of a layer winding with grounded neutral is given by the following formula:

$$V(o,\, x) = \left[\frac{\sinh(1-x)}{\sinh\alpha} \right] \tag{9.22}$$

where

$V(o,\, x)$ is the initial voltage distribution

BIL = basic impulse level

$$\alpha = \sqrt{\frac{C}{K}} \tag{9.23}$$

x = per unit distance from neutral end

9.6   IMPULSE AND SWITCHING SURGE WAVES AS PER IEC 60076-4

The lightning impulse and switching surge waves are defined in IEC 60076-4 for uniformity of application in testing. Table 9.1 gives the characteristics of the different waves.

TABLE 9.1

Impulse and Switching Surge Waves

No.	Description of Wave	Wave Shape as per IEC Standard	Remarks
1	Full-wave lightning impulse (LI)	Wave front 1.2 µs±30% Wave tail 50 µs±20%	Tolerance of test voltage ± 3%
2	Chopped-wave lightning impulse (LIC)	Wave shape as in (1). But, chopping of the wave tail between 3 and 6 ms	Overswing of voltage in the opposite direction shall be maximum 30%
3	Lightning impulse on the neutral (LIN)	Duration of the front can be maximum of 13 µs	-
4	Switching impulse (SI)	Time to peak of at least 100 µs Time above 90% of the amplitude for at least 200 µs Time to zero of minimum 1000 µs	This waveform is different from 250/2500 µs as per IEC 60060-1

9.7   SIMULATION OF WAVEFORM FOR ANALYTICAL CALCULATIONS

The lightning impulse wave shapes can be defined by mathematical expressions to enable the analytical solution of the response of the impulse wave on the equivalent circuit of the winding. The full-wave impulse is approximated as the difference between two exponential waves.

$$V_t = V_o\left[e^{-at} - e^{-bt}\right] \tag{9.24}$$

where
  V_t = voltage at any time "t"
  V_o = 1.04 PU
  $a = 0.0146$
  $b = 2.467$
  t is the time in microseconds
  The formula for the impulse chopped wave is as follows:

$$V_t = V_o\left[e^{-at} - e^{-bt}\right] - 0.986\left[e^{-a(t-tc)} - e^{(y-tc)}\right] \tag{9.25}$$

where
  V_t = voltage at any time "t"

$V_o = 1.04$

$a = 0.0146$
  $b = 2.467$
  tc = chopping time [as per standard 2–6 ms]
  $y = 16$
  V_t is equal to $V_o\left\lfloor e^{-\alpha t} - e^{-\beta t} \right\rfloor$ for the time interval $o \le t \le tc$

9.8   DESIGN TECHNIQUES TO REDUCE NON-LINEAR IMPULSE VOLTAGE DISTRIBUTION

Calculations of Section 9.7 have shown that the impulse voltage is distributed non-linearly across the winding and the line end has a stress concentration of α times the average stress where $\alpha = \sqrt{\dfrac{C}{K}}$.

In order to make the most efficient use of the insulation of the transformer, the winding needs to be designed to reduce the initial voltage concentration at the line end. As the stress concentration is controlled by the ratio of the shunt and series capacitance, reducing the shunt capacitance (ground capacitance) or increasing the series capacitance would improve the voltage distribution. Figure 9.4 shows the impulse voltage distribution of a winding with $\alpha = 0$ and $\alpha > 0$.

A number of design techniques were introduced in the past several decades to reduce the ground capacitance or increase the series capacitance. Some of these design methods are as follows:

- Static shields
- Intershield windings
- Different types of interleaved windings, etc.

Static shields connected to the line end turn reduce the ground capacitance. However, it is not capable of increasing the series capacitance.

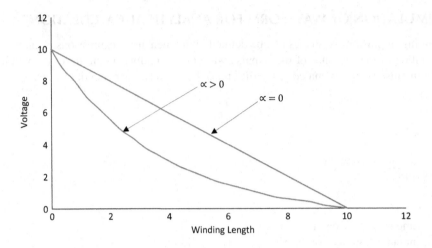

FIGURE 9.4 Impulse voltage distribution with $\alpha > 1$.

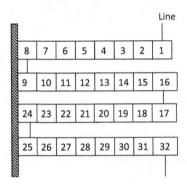

FIGURE 9.5 Conventional disc winding.

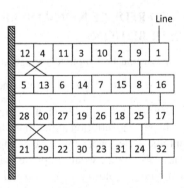

FIGURE 9.6 Interleaved disc winding.

Interleaved winding increases the series capacitance. This technique was first introduced in the 1950s.

Figure 9.5 shows the conventional disc winding, and Figure 9.6 shows an interleaved disc winding.

Different types of interleaved windings are being used including sequential interleaved, continuous interleaved, double interleaved and non-sequential interleaved.

The series capacitance is high for interleaved winding; it is less than interleaved disc for inter-shield-type disc winding.

9.9   POWER FREQUENCY BREAKDOWN AND IMPULSE BREAKDOWN

The breakdown mechanism from power frequency voltage and impulse voltage is not the same. The power frequency breakdown mechanism requires a time lag to operate. It also has the following characteristics in oil paper insulation.

- Electrophoretic particle migration
- Large bubble formation
- Electrohydrodynamic motions producing flow-induced cavitation

The impulse breakdown has fast processes like streamer formation. The electrode configuration, creepage surface of insulation, geometry of electrodes, thickness of solid dielectric, etc., are factors influencing the impulse breakdown. A common design procedure is to use a universal ratio of impulse breakdown and power frequency breakdown voltage of oil – cellulose insulation. However, practical test results on a large number of samples and electrode configurations have shown that this criterion is not accurate.

9.10   SELECTION OF SURGE ARRESTER FOR TRANSFORMER

Surge arresters are installed on the equipment like transformers or a system to protect it from impulses and surges. In the past, the term "lightning arrester" was used to designate surge arrester. The National Electric Code (NEC) defines a "surge arrester" as "a protective device for limiting surge voltage by discharging or bypassing surge current, and it also prevents the flow of follow current while remaining capable of repeating these functions".

The simple form of a surge arrester is the arcing horn or spark gaps. The older surge arresters were made by stacking silicon carbide discs in series with air gaps and housed inside ceramic insulators. There is no current flow through these arresters under normal operating conditions. However, when an overvoltage strikes the arrester, the air gaps between the discs discharge and the silicon carbide elements discharge the current through the neutral. The flow of current increases non-linearly with the increase in voltage.

Metal oxide surge arresters are most commonly used today, and these have normally zinc oxide elements in series without any air gap to work as a non-linear element. These are most commonly connected between the line and neutral and carry very low current in the normal operating conditions. This current is predominantly capacitive.

When a surge occurs, the arrester immediately limits the overvoltage by conducting the current to the ground.

Depending on the application, arresters are given different classes.

- Station class arresters

These are installed to protect important and expensive equipments where high-energy surges are expected. These are generally used in high-voltage systems and substations.

- Intermediate class arresters

These are generally used for medium-voltage applications.

- Distribution class arresters

These are most commonly used for protecting distribution transformer on the primary side where the surge energy levels expected are not high.

TABLE 9.2

Factors (k_e) for Calculating the Temporary Overvoltage

System Earthing	Factor K_e
Solidly grounded four wires	1.25
Solidly grounded three wires	1.40
Impedance grounded	1.73
Ungrounded/isolated neutral	1.73/2.3

9.10.1 SURGE ARRESTERS PARAMETERS

- Maximum continuous operating voltage (MCOV)

    MCOV is the maximum permissible sinusoidal power frequency voltage which can be applied between the arrester terminals. This is selected based on the highest system voltage. Arresters are normally connected between the phase and neutral, and therefore, MCOV is calculated for the same by the following formula.

$$\text{MCOV} = \frac{U_m}{\sqrt{3}} \times 1.05 \qquad (9.26)$$

    where
    U_m = highest system voltage
    1.05 = factor for harmonic distortion

- Rated Voltage U_r

    U_r is defined as the maximum permissible 10-second power frequency rms – voltage that can be applied to the arrester based on the operating duty and temporary overvoltage test.

    U_r is selected after considering the temporary overvoltage of the system, which depends on the neutral grounding of the system. The typical factors (k_e) for calculating the temporary overvoltage are given in Table 9.2.

9.10.2 CALCULATION OF ARRESTER RATING OF SOLIDLY GROUNDED THREE-WIRE SYSTEM

$$U_r = \frac{U_m \times K_e}{\sqrt{3}} \qquad (9.27)$$

$K_e = 1.4$
   MCOV = $0.8\ U_r$
   For 11 kV system, $U_m = 12\,\text{kV}$

$$U_r = \frac{12 \times 1.4}{\sqrt{3}} = 9.6\,\text{kV}$$

Nearest rating of 10 kV can be selected
   MCOV = $0.8 \times 10 = 8\,\text{kV}$

- Temporary overvoltage (TOV) rating

    The arrester can withstand specified overvoltage for specified duration, and it is defined in relation to Ur with prior energy or without prior energy. The arrester-guaranteed TOV ratings shall be greater than the expected TOV at the location of installation. If it is not so,

TABLE 9.3

TOV with Prior Energy

Time Seconds	TOV
1	1.15 Ur
10	1.10 Ur
100	1.05 Ur

the next higher Ur and MCOU ratings shall be selected. Table 9.3 gives the TOV values for time duration up to 100 second.

Nominal discharge current (In) of an arrester is the peak value of the lightning impulse current with a wave shape of 8/20 ms, which is the basis of calculating the lightning impulse protection level of an arrester. The standard values of In are 2500, 5000, 100000 and 20000. A IEC 60099-4 ed 3.0 classifies the arresters based on the application, the service duty, the thermal energy rating and thermal transfer rating. Selection of thermal energy rating of the arrester (W_{th}).

The metal oxide arresters shall absorb the energy due to transient overvoltage of the system which is mainly from the switching on of the system. The energy rating of an arrester will have to be above the system discharge energy. The prospective system discharge energy can be calculated by transient analysis of the system. The following formula can be used if transient analysis data is not available.

$$W = U_{rv} \frac{(U_{su} - U_{rv})}{Z_s} \frac{(2L)}{C}$$ (9.28)

where
W = system discharge energy kJ/switching
L = line length (m)
C = speed of light (m/s)
U_{rv} = arrester residual voltage (kV)
U_{sv} = maximum switching voltage (kV)
Z_s = surge impedance of the line (ohms)
W_{th} shall be greater than W
Protection Levels and protective margins of Arrester

- Lightning impulse protection level is the maximum residual voltage of the arrester for the nominal discharge current
- Switching impulse protection level is the maximum residual voltage of the arrester for switching impulse discharge current
- Steep current impulse protection level is the maximum residual voltage of the arrester for a steep current impulse of the magnitude equal to the nominal discharge current magnitude.

The nominal discharge currents and energy ratings of arresters are given in Table 9.4.
The symbols in the table have the following meaning:
I_n – normal current
H – heavy duty
M – medium duty

L – light duty
D – distribution class
S – station class
kA – kilo amperes

Typical distribution class arrester ratings are given in Table 9.5.

TABLE 9.4
Discharge Current and Energy Rating

	Class of Arrester					
	Distribution Class			Station Class		
Particulars						
Duty	DH	DM	DL	SH	SM	SL
In kA	10	5	2.5	20	10	10
Charge transfer rating (coulombs)	≥0.4	≥0.2	≥0.1	≥2.4	≥1.6	≥1.0
Thermal rating (coulombs)	≥1.1	≥0.7	≥0.45	≥10	≥7	≥4
Approximate 8/20 μs current for thermal rating (Amps)	14	22	34	-	-	-

TABLE 9.5
Typical Distribution Class Arrester Rating

Nominal System Voltage – kV L-L voltage	Solidly and Grounded Neutral (kV)	High Impedance Grounded or Ungrounded Neutral (kV)
6.6	6.0	7.5
11	9 or 10	12 or 15
13.8	10 or 12	15 or 18
24	18 or 21	24 or 27
33	27 or 30	36 or 39

10 Inrush Current in Transformers

10.1 INTRODUCTION

When a transformer works normally, the flux produced in the core is orthogonal to the applied voltage. When the voltage is zero, the flux waveform is at its negative peak. When a transformer is switched on from the primary by keeping the secondary open circulated, it acts as a simple inductance and the flux will start from zero as shown in Figure 10.1

If the switching on is at the instant of voltage zero, the flux is initiated from the same origin of the voltage. The flux is the integral of the voltage, and the value of flux at time π can be written as

$$\phi = \phi_m \int_0^\pi \sin wt \; dwt$$

$$= 2\phi_m$$

The excessive demand of flux would saturate the core and result in a sharp increase of the magnetizing current. This increased current is termed as inrush current, and its magnitude can become several times of the full load current.

When the transformer is switched off, magnetizing current will follow a hysteresis loop and some residual flux may be retained in the core. If the residual flux, ϕr, is in the direction of the flux build-up, the peak flux would become $2\phi m + \phi r$, resulting in much higher inrush current. The waveform of the inrush current is a half-cycle sinusoidal wave superimposed by a DC component.

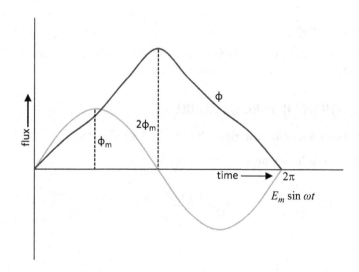

FIGURE 10.1 Voltage and flux.

10.2 PROBLEMS OF TRANSFORMER INRUSH CURRENT

10.2.1 MECHANICAL STRESSES

The amplitude of the first peak of the inrush current can reach high value, and the consequent mechanical forces on the winding can cause excessive stresses on the windings. If there is frequent switching on of the transformer, the recurring surges can cause mechanical weakness of the winding.

10.2.2 OVERVOLTAGE DUE TO HARMONIC RESONANCE

The inrush current has very high harmonic content where the even harmonics are predominant. When any of the harmonics is close to the natural frequency of the system, there is a possibility of harmful resonant overvoltage.

A power system can experience harmonic resonance in some of the following situations:

- A black start of a system to a remote transformer
- Energization of HV converter transformers
- Energization of off-shore wind farms

10.2.3 NUISANCE TRIPPING OF TRANSFORMER

When the transformer is energized, the inrush current flows through one winding only and there is no balancing equivalent current on the energized winding. Even if second harmonic restraint is available on the differential relay, extreme core saturation can cause tripping of a healthy transformer.

10.2.4 TEMPORARY VOLTAGE DIP

The inrush current may cause temporary voltage dips in connected system, which can take few cycles to return back to normalcy.

10.2.5 SYMPATHETIC INRUSH

When a transformer is switched on to the system, another transformer connected to the system can experience unexpected saturation and "sympathetic" inrush.

10.3 CALCULATION OF INRUSH CURRENT

10.3.1 APPROXIMATE VALUE OF THE FIRST PEAK OF INRUSH CURRENT

The first peak of the inrush current can be estimated by

$$I_{in} = \frac{1.15 \sqrt{2}\ V_{ph}}{X_{air}} \tag{10.1}$$

where

I_{in} = first peak of the inrush current (A)
V_{ph} = phase voltage
X_{air} = air core reactance (ohms)

$$= \frac{8\pi^2 f \ AN^2 \times 10^{-7}}{H}$$

f = frequency (Hz)

$A = \frac{\pi}{4} \ D_m^2$

D_m = mean diameter of winding to which power is switched on (m)
(For step-down transformer D_m = mean diameter of HV winding
and for step-up transformer D_m = mean diameter of LV winding)
N = number of turns of winding to which power is switched on
H = axial height of winding to which power is switched on (m)

- Alternate formula for calculation of air core reactance

$$X_{air} = 2\pi f L \qquad (10.2)$$

where

$$L = \frac{6.4 \times 4\pi \times 10^{-7} \times N^2 \ D^2}{(3.5 \ D + 8 \ H)} \left(\frac{D - 2.25 \ T}{D} \right)$$

N = number of turns
D = outside diameter of winding m
H = height of winding m
T = radial depth of winding m

Estimation of transformer air core inductance from short-circuit inductance

CIGRE Study Committee 33 gives guidelines for estimating air core inductance from transformer short-circuit inductance by the following formula:

- Step-down transformer (outer winding)

$$L_{aircore} = 2 \sim 2.5 \ L_{sc}$$

- Step-up transformer (inner winding)

$$L_{aircore} = 1 \simeq 1.5 \ L_{sc}$$

- Autotransformer (high-voltage side)

$$L_{aircore} = 4 \sim 5 \ L_{sc}$$

where
$L_{aircore}$ = air core inductance
L_{sc} = short-circuit inductance of transformer
From L_{sc}, X_{air} in ohms is calculated by

$$X_{air} = 2\pi f L_{air\ core} \qquad (10.3)$$

10.3.2 FIRST PEAK OF INRUSH CURRENT CONSIDERING SWITCHING ANGLE AND CIRCUIT RESISTANCE

An improved estimate of the first peak of inrush current is calculated by considering the effect of the switching angle and the resistance of the circuit to which it is switched on.

$$I_{\text{First peak}} = \frac{\sqrt{2}\, V_{\text{ph}}\, K_2}{\left[R^2 + X_{\text{air}}^2\right]^{\frac{1}{2}}} (1 - \text{Cos}\,\theta) \tag{10.4}$$

where

V_{ph} = phase voltage
R = resistance of winding and system (ohms)
X_{air} = air core reactance (ohms)
B_r = residual flux density (Tesla)
(For CRGO, this can be taken as 0.8 B_{max})
B_s = saturation flux density (Tesla)
(2.03 for CRGO)
B_m = peak working flux density (Tesla)
θ = initial phase angle of the voltage (radians)

$$= K_1\, \text{Cos}^{-1}\left[\frac{B_s - B_r - B_1}{B_{\text{max}}}\right]$$

K_1 = correction factor for saturation angle = 0.9
K_2 = correction factor for peak value = 1.15

10.3.3 ESTIMATION OF INITIAL FEW PEAKS OF INRUSH CURRENT

The first peak and the subsequent peaks of the inrush current can be calculated by the following procedure.

Step 1
    Calculate the air core reactance of the winding to which power is switched on in ohms
Step 2
    Calculate saturation angle θ in radians

$$\theta_{\text{sat}} = K_1\, \text{Cos}^{-1}\left[\frac{B_s - B_{\text{max}} - B_r}{B_{\text{max}}}\right] \tag{10.5}$$

    K_1 = correction factor for saturation angle
        = 0.90
Step 3
    Estimate the first peak of inrush current

$$I_{\text{peak}} = \frac{\sqrt{2}\, V_{\text{ph}}\, K_2}{\left[R^2 + X_{\text{air}}^2\right]^{\frac{1}{2}}} [1 - \text{Cos}\,\theta_{\text{sat}}] \tag{10.6}$$

    K_2 = correction factor for peak value = 1.15

Step 4
    Calculate the new residual flux

$$B_r \text{(new)} = B_r \text{(old)} - B_{max} \frac{R \, K_3}{X_{air}} \times 2 \left(\text{Sin } \theta_{sat} - \theta_{sat} \text{ Cos } \theta_{sat} \right) \tag{10.7}$$

K_3 = correction factor for inrush decay
With the updated B_r, repeat steps 2, 3 and 4 to get the subsequent peaks.

The above procedure calculates the inrush current peaks of a single-phase transformer. For three-phase transformer, the peaks calculated as above can be multiplied by 2/3.

10.3.4 CALCULATION EXAMPLE

To calculate the inrush current peaks of the first 5 cycles of a 25 MVA 132/33 kV star/star connected 50 Hz transformer when energized from 132 kV side.

- Design data of the transformer

Turns/phase of 132 kV winding = 1020
    Mean diameter of 132 kV winding = 0.95 m
    Area of cross section of core = 0.20 m²
    Peak working flux density = 1.7 Tesla
    Saturation flux density of core = 2.03 Tesla
    Residual flux density of core = 0.8×1.7 = 1.36 Tesla
    Axial height of winding = 1.1 m
    System + winding resistance = 2 Ω
    $K_1 = 0.9$, $K_2 = 1.15$, $K_3 = 2.26$

Step 1
    Calculate the air core reactance in ohms

$$X_{air} = \frac{8\pi^2 \, f \, AN^2 \times 10^{-7}}{H}$$

$$A = \frac{\pi}{4} \times D^2 = \frac{\pi}{4} \times 0.95^2 = 0.7088 \ m^2$$
$$H = 1.1 \, m$$

$$X_{air} = \frac{8\pi^2 \times 50 \times 0.7088 \times 1020^2 \times 10^{-7}}{1.1}$$

$$= 264.6 \ \Omega$$

Step 2
    Calculate the core saturation angle of the first peak

$$\theta = K_1 \text{ Cos}^{-1} \left[\frac{B_s - B_m - B_r}{B_s} \right]$$

$$\theta = K_1 \text{ Cos}^{-1} \left[\frac{2.03 - 1.7 - 1.7 \times .8}{1.7} \right]$$

$$= 2 \text{ radians}$$

Step 3

First peak of inrush current

The first peak $= \dfrac{K_2 \, V_{ph} \, \sqrt{2}}{X_{air}} [1 - Cos\theta]$

$$= 1.15 \times 1.15 \times \dfrac{13200}{\sqrt{3}} \times \dfrac{\sqrt{2}}{264.6} [1 - Cos\theta]$$

$$= 663.28 \; Amps$$

Step 4

Residual flux density at end of first cycle

$$B_r \, (new) = B_r \, (old) - B_m \, \dfrac{K_3 R}{X_{air}} \Big[2 \, (Sin\theta - \theta sat \, Cos\theta) \Big]$$

$$= 1.36 - \dfrac{1.7 \times 2.26 \times 2}{264.6} \Big[2 (Sin \, 2 - 2 \, Cos \, 2) \Big]$$

$$= 1.26 \; Tesla$$

Steps 2, 3 and 4 are now repeated to get the peak values of next cycles.

These results as per the above calculation are single-phase values. The three-phase values are $\dfrac{2}{3} \times$ single-phase values calculated.

The first peak for three phases is $\dfrac{2}{3} \times 663.28 = 442.2 \; A$

Rated line current of transformers $= \dfrac{25,000}{\sqrt{3} \times 132} = 109.35 \; A$

$\therefore$ The first peak is $\dfrac{442}{109.35} = 4$ Times rated full load current

10.4 FREQUENCY RANGE OF INRUSH CURRENT AND OTHER TRANSIENTS

The transients on the transformers can be from switching and other sources, which can broadly be divided into the four frequency ranges.

- Low-frequency transients – 0.1 Hz to 3 kHz
- Slow front transients – from power frequency to 20 kHz
- Fast transients – 10 kHz to 3 MHz
- Very fast transients – 100 kHz to 50 MHz

The source of various transients and the frequency range are given in Table 10.1.

10.5 INFLUENCE OF DESIGN ON INRUSH CURRENT

The main parameters of the design affecting the inrush current are as follows:

a. Working flux density of core

Peak inrush current increases and the ratio of the second harmonic to the peak inrush decreases, with increase in working flux density.

b. Core material

TABLE 10.1

Origin of Transients and Frequency Range

Origin of Transients	Frequency Range
Transformer switching	0.1 Hz–1 KHz
Transients due to VCB switching	Up to 1 MHz
Load rejection	0.1 Hz–3 KHz
Fault clearing	Power frequency – 3 kHz
Fault initiation	Power frequency – 20 kHz
Line switching on	Power frequency – 20 kHz
Line reclosing	Power frequency – 20 kHz
Multiple restrikes of circuit breaker	10 kHz–1 MHz
Lightning surges	10 kHz–3 MHz

The Hi-B core material and domain refined (DR) material have higher values of saturation flux density and lower remnant flux density when compared to the normal CRGO steels similar to M4 grade. The Hi-B and DR materials have low inrush currents when working at the same flux density as M4.

c. Core joint geometry

Core joints give higher reluctance and reduce the residual flux when compared to the remnant flux of the core material. The step-lap joints have less reluctance when compared to mitred joints and therefore produce higher inrush current and lower second harmonics.

d. Whether power is switched on to inner winding or outer winding

The inrush current is inversely proportional to the air core reactance which is low for the inner winding. Thus, the inrush current will be higher, if the power is switched on to the inner winding.

10.6 METHODS FOR REDUCTION OF INRUSH CURRENT

Some of the methods employed to reduce the magnitude of switching inrush currents are as follows:

- Pre-insertion resistors

  Pre-insertion resistors are installed in parallel to the main circuit breaker. The resistance value and the pre-insertion time need to be optimally selected.
- Controlled switching

  Controlled switching is used to energize transformers to reduce the inrush current. The strategy involves energizing the first phase at its voltage peak and delays the energization of the two phases by $\frac{1}{4}$ th of a cycle.

10.7 EFFECT OF SYSTEM AND SWITCHING PARAMETERS ON INRUSH CURRENT

10.7.1 SOURCE RESISTANCE

Increase of the source resistance will decrease the amplitude of inrush current, and it also causes faster decay of the amplitude of inrush current. When the system fault level at the transformer primary side is high, the source impedance and therefore the source resistance are low. Such a transformer will have higher magnitudes of inrush current lasting much longer than a transformer installed where the fault level is low.

10.7.2 SWITCHING ANGLE

The higher amplitude of inrush current is at 0°. Increasing the switching angle will decrease the amplitude of inrush current. Also, increasing of the switching angle decreases the percentage second harmonic content of the inrush current. Though the highest amplitude of the inrush current is in the first cycle, the highest percentage second harmonic does not necessarily appear in the first cycle.

10.7.3 EFFECT OF REMNANT FLUX ON THE FIRST CYCLE PEAK CURRENT

When the remnant flux changes, the first peak changes considerably. Switching at the zero residual flux and 90° switching angle necessarily need not reduce the magnitude of the inrush current. The optimum angle will have to be selected, based on the remnant flux direction and magnitude.

11 Calculation of Core and Coil Assembly Dimensions, Tank Size and Tank Weight

11.1 CALCULATION OF THE DIMENSIONS OF CORE AND COIL ASSEMBLY (CCA)

$$\text{Length of CCA} = \begin{cases} 3 \times \text{HV OD} + 2 \left(\text{phase to phase clearence}\right) \\ \text{or} \\ 2 \times LC + \text{HV OD} \end{cases} \quad (11.1)$$

where

HV OD = Outer dimension of outer coil $\left(\text{If LV is outer coil, then use LV OD}\right)$

LC = Core Leg Center distance

Width of CCA = HV OD [If the outer winding is LV or HV tap, use it instead of HV coil]

Height of CCA = window height + 2 × maximum step width of core +

distance from core bottom to tank

a. When off-circuit tap changer is mounted on top of the core, the height of tap changer shall be added to CCA height.
b. When on-load tap changer is mounted on the short side, the length or width of the CCA shall include the overall diameter of tap changer including projections for lead connection and the clearance to the winding (Figure 11.1).

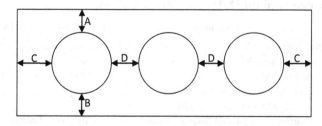

FIGURE 11.1 Tank clearances from CCA.

11.2   CALCULATION OF DIMENSIONS OF TANK

$$\text{Tank length} = 3\,OD + 2\,D + 2\,C \qquad (11.2)$$

OD = HV outside diameter
D = HV–HV interphase clearance
C = winding to tank (short side clearance)
Tank width = OD + A + B
A = winding to tank (long side) clearance on HV side
B = winding to tank (short side) clearance on LV side
A, B, C and D shall be taken from Table 11.1.

For Conventional Transformers

$$\text{Tank height} = LL + 2W + H_{TC} + K_1 + K_2 \qquad (11.3)$$

LL	=	core window height
W	=	maximum step width
H_{TC}	=	height of tap changer from core top yoke surface when wooden core beam (core clamp) is used, and where cable-operated tap changer is used, this can be zero (when tap changer is not provided, $H_{TC} = 0$)
K_1	=	Oil height above tap changer level required at the time of oil filling
K_2	=	Tank bottom to core clearance

Note

When HV layer winding is used, the tap leads can be taken from the top and the tank clearance on HV side will be reduced. Also, when wooden beam is used for top core clamp, the LV leads can be routed directly to top side and LV side clearance can be reduced.

The typical clearances required for voltages up to 33 kV are as per Table 11.1.

Clearances applicable for different design options and ratings are given in Section 5.10.

a. Sealed-type transformer with gas cushion (radiator type) with HV layer winding and wooden beam at top clearances in mm (cable-operated tap changer mounted on side of wooden beam)

TABLE 11.1
Typical Clearances Required for Different Voltage Classes

	Clearances	11 kV Class	22 kV Class	33 kV Class
A	HV OD to tank long side (HV side)	40	75	100
B	HV OD to tank long side (LV side)	40	50	70
C	HV OD to tank short side	30	50	70
D	Core yoke top surface to top oil level (minimum oil level at lowest ambient temperature and no load)	25	25	25
E	Core yoke top surface to top oil level [level at normal filing temperature 30°C]	To be calculated as per Chapter 13		
F	Gas cushion (N₂ cushion) at normal temperature. The gas cushion corresponding to D and E to be calculated	To be calculated as per Chapter 13		
G	Tank bottom (inside) to core bottom (minimum)	60	60	75

$$\text{Tank height} = LL + 2W + D + E + G \tag{11.4}$$

$$\text{Tank height} = 2OD + 2PP + 2C \tag{11.5}$$

$$\text{Tank Width} = OD + A + B \tag{11.6}$$

PP = phase-phase winding clearance

- For sealed transformer

$$\text{Tank height} = LL + 2W + H_{TC} + K_1 + K_2 + K_3 \tag{11.7}$$

K_3 = free air space required over oil level at the time of filling
a. This air space shall be calculated such that the pressure at extreme conditions inside the tank is limited to 0.35 kg/cm^2 below atmosphere or as per the specified value. The calculation of pressure is given in Chapter 13.
b. Where customer has specified the pressure ranges, the same shall be used for calculation of free space.
c. Where customer has specified the minimum volume of free space, the same amount of free space shall be provided.

- For CSP transformers

$$\text{Tank height} = LL + 2W + H_{TC} + K_1 + K_2 + K_3 \tag{11.8}$$

H_{TC} = height of CSP breaker
Tank length and width are calculated as per formula 11.5 and 11.6.
Tank height is calculated by the following formula.
Tank height = $LL + 2W + D + G$
If tap changer is mounted on top of core yoke/core clamp, the tap changer height including the terminals will have to be added to the above.
Also, an extra of K mm is to be provided to take into account for the effect of terminals of the tap changer.

	K (mm)
11 kV	15
22 kV	25
33 kV	30

11.3   CALCULATION OF THE SIZE OF WOODEN BEAM (CORE CLAMP)

Wooden Beam width – Top

A. For Flat yoke (Top Yoke)

$$\text{Width of Beam} = \left[(\text{Core Diameter} \times 0.78) + 5 \right] \tag{11.9}$$

To be rounded off to the lower multiple of 25 mm.

Example

Core diameter = 189 mm

$$189 \times 0.78 + 5 = 152.4$$

Width of wooden beam = 150 mm

B. For normal yoke – Top

$$\text{Width of Wooden Beam} = \text{Max Step Width of core} - 3 \qquad (11.10)$$

For the above core circle of 189 mm, if maximum step width in 180 mm,
Width of wooden beam = 180 – 30 = 150 mm

LENGTH OF WOODEN BEAM – TOP

Minimum of the following is taken as the length of wooden beam.
 CCA length + 70 or tank inside length – 20
 where CCA length = 2 × leg centre + HV OD
 Tank length = CCA length + 2 × clearance from winding to short side of tank.

- If leg centre = 353
- HV winding OD = 343
- Tank short side clearance = 30

Substituting the values,
 CCA length = (353 × 2 + 343) = 1049
 Tank length = CCA length + 2 × 30
 = 1049 + 60
 = 1109
 #(Rounded off to 1110 mm).

$$\text{CCA Length} + 70 = 1049 + 70 = 1119$$

$$\text{Tank Length} - 20 = 1110 - 20 = 1090$$

: Length of top wooden beam = 1090
(Lower value of the above)

LV bushing height from tank bottom
 = Core top level from tank bottom + 60 + H/2
 = (tank bottom to bottom of yoke) + window height + 2 × maximum step width + 60 + H/2.
 H = 63 for monoblock rating less than 3150 Amp
 H = 70 for monoblock rating > 3150 Amp
 Tank bottom to bottom of yoke value to be taken as 20, if the actual is less than 20.

Example

Maximum step width 180
 Monoblock capacity < 3150 Amps
 Window height = 1384
 Gas cushion = 180
 Core top to oil level = 214 [This changes with the dimensions of tap switch]

Tank height = 20 + 1384 + 2 × 180 + 214 +180
= 2158 [round off to 2160]

LV Bushing height [monoblock]

= Core top yoke level + 60 + H/2
= 20 + 1384 +2 × 180 + 60 + 63/2
= 1856

11.4 CALCULATION OF WEIGHT OF TANK [RADIATOR-TYPE TANK]

- Tank plate thickness

  - If customer has specified the thickness, the same can be used after verifying that the stress during pressure test is within limit.
  - If the test pressure of the tank is 0.7 kg/cm² for radiator-type design, the typical minimum thickness of tank plates for distribution transformers up to 2500 kVA is given in Table 11.2.

- Calculating weight of tank

The tank weight is the sum of the weights of top covers, side walls, bottom plate, tank curb, horizontal stiffener, vertical stiffener, base channel, welded pipes, lifting lugs, etc.

11.4.1 WEIGHT OF TOP COVER

Round off dimensions to the higher multiple of five length of tank cover	=	inside length of tank + 2 × Tank curb
Width of tank cover	=	inside width of tank + 2 × Tank curb
Top cover weight	=	length × width × thickness × density

11.4.2 WEIGHT OF SIDEWALLS

$$\text{Weight of sidewalls} = 2\,(\text{Tank width} + \text{Tank Length})\,\text{Tank Height} \times \text{Density}$$

TABLE 11.2
Minimum Thickness of Tank

Description of Tank Plate	Thickness (mm)
Tank wall	4 up to 315 kVA
Tank wall	5 above 315 kVA up to 2500 kVA
Tank cover	6
Tank bottom	6
Tank curb	10 – up to 500 kVA
	12 – Above 500 kVA up to 2500 kVA

11.4.3   BOTTOM PLATE

Length = Tank inside length + 2 × Tank wall Thickness

Width = Tank Inside width + 2 × Tank wall Thickness

Weight = Length × width × thickness × Density

11.4.4   TANK CURB

Use 50×10 mm for rating up to 500 kVA

50×12 mm for rating between 500 and 2500 kVA

Total length of curb = 2 (Tank length + Tank width) + 200

Weight of Curb = Total length of curb × Thickness of curb × Density

11.4.5   HORIZONTAL STIFFENERS

$$\text{Number of horizontal stiffeners} = \left[\frac{\text{Tank height}}{K} - 1 \right] (\text{nearest higher integer})$$

$K = 500$ for transformer pressure test up to 0.35 kg/cm².
= 360 for transformer pressure test up to 1 kg/cm².
Stiffener size = 65 × 10 flat for rating up to 2000 kVA
Length of stiffener = (tank width + tank length) 2 × number of stiffeners.

11.4.6   VERTICAL STIFFENERS

$$\text{Number of vertical stiffeners} = \left[\frac{\text{Tank height}}{K} - 1 \right] (\text{Round off to the nearest higher even number})$$

Total length of vertical stiffeners = tank height × number of vertical stiffeners.
The stiffeners to be adjusted for accommodating cable boxes.

11.4.7   BASE CHANNEL

Use two base channels.
Length of one base channel = tank length + 250

11.5   DESIGN OF CONSERVATOR

Conservator is provided to accommodate the expansion and contraction of oil in transformer from the load and ambient temperature variation. The maximum range of temperature variation to be considered is from the lowest ambient temperature to the maximum oil temperature expected at continuous full load when the ambient is maximum. For a transformer designed for the normal service condition as per IEC 60076, the expansion of oil from −25°C to 110°C needs to be considered.

Conservator volume for conventional transformer can be fixed as 10% of the oil volume. The conservator can be circular, rectangular or oval. For distribution transformer and medium power transformer, circular conservator is used.

The design of the minimum oil level shall correspond to the oil level at lowest ambient temperature specified at no-load condition, and the maximum oil level shall be calculated for the oil volume corresponding to the maximum top oil temperature. The difference in the above oil volume is about 7% of the total quantity of oil.

11.5.1 SIZE OF THE CONSERVATOR

The oil inside the transformer expands and contracts when the temperature changes with load and/ or ambient temperature. In sealed-type transformer with gas cushion, this expansion is absorbed by the gas where the pressure increases or decreases with load and ambient temperature variation. In completely filled transformer, the elasticity of corrugated fin absorbs the expansion or contraction of the oil.

In a conservator-type transformer, the expansion and contraction of the oil are absorbed by the conservator. The size of the conservator is decided by the oil quantity in the transformer, the minimum and maximum ambient temperature and the maximum allowed oil temperature rise.

The maximum change in oil volume is calculated as

$$\Delta V = V\alpha \left[T_2 - T_1\right] \tag{11.11}$$

where ΔV is the maximum change in oil volume

V = oil volume in transformer
α = coefficient of volumetric expansion of oil
  = 0.000718 for mineral oil
T_1 = lowest ambient temperature (°C)
T_2 = average oil temperature at maximum ambient temperature
  = $T_{avo} + T_{ambmax}$
(T_{avo} = average oil temperature rise)
T_{ambmax} = maximum ambient temperature

For normal service conditions and temperature rise as per IEC 60076 standard, conservator volume of 10% of the oil quantity in the tank is sufficient.

The minimum oil level marking on the oil sight window shall be the oil level at the minimum ambient temperature specified. The maximum oil level marking shall be the level corresponding to level at full load at maximum ambient temperature. Conservator is normally designed as a circular tank mounted on the tank. Tanks of rectangular shape also can be used. The mounting height of the conservator shall be calculated on the basis that the minimum oil level in conservator shall always be at a higher level than the oil level in the bushings or any other oil-filled components.

11.6 AIR CELL FOR CONSERVATOR

The oil inside the conventional breathing-type conservator is in contact with the atmospheric air. In order to prevent the atmospheric air from contacting the oil, an airbag (air cell) is fitted inside the conservator. The top side of the air cell is vented outside through a breather. When the oil in the conservator expands or contracts, the air cell deflates or inflates. When the air cell inflates, the outside air comes into the air cell through the conservator. This prevents the transformer oil from getting into contact with the atmospheric air. The conservator has no free space anytime.

The air cell is manufactured using polyamide fabric or similar material with thick coating of elastomers. The inside of the air cell is suitable for exposure to atmospheric air, and the outside has materials suitable for continuous exposure to hot transformer oil/fluid.

The air cell shall withstand heat ageing and shall have low-temperature flexibility, ozone resistance and high bursting strength.

The volume of the air cell is calculated to accommodate the oil volume change within the operating temperature range. The nearest higher standard size available is normally selected. The empty volume of the conservator is 10%–15% more than the air cell volume.

The size of the air cell is specified by the length and width in "flat" condition (when empty).

The sizes can be calculated by the following formula:

D = inside diameter of conservator
L = length of conservator
X = flat length of air cell
Y = flat width of air cell
$Y = \dfrac{\pi D}{2} + K$, where $K \approx 50\,\text{mm}$
$X = L - D + Y$

The above is applicable for conservator of circular shape.

Air cells suitable for conservator designs of rectangular, elliptical cross sections are available.

11.7 DEHYDRATING BREATHERS FOR TRANSFORMERS

Dehydrating breathers are required for transformers with conservator. When the oil in the main tank expands or contracts from the variation of the oil temperature (resulting from load and ambient temperature changes), the oil level in the conservator increases or decreases. The air inside the conservator goes out or comes in through a breather filled with a moisture absorbent to prevent the moisture in the atmospheric air to come into the direct contact with the oil in the conservator.

The breathers are generally transparent hollow cylindrical tubes, which contain chemically pure silica gel with a coloured indicator. The air is breathed in or out of the transformer through the silica gel.

A completely dry silica gel is "orange" colour, and it changes to "green" colour when saturated with moisture.

The major property of silica gel is its high absorption power of moisture. The absorption rate is almost constant till it absorbs moisture up to 15% by weight. Saturation level of moisture can be considered as the absorption of 30% by weight of moisture.

11.7.1 REGENERATION OF SATURATED SILICA GEL

When the colour of silica gel becomes green, it can be regenerated by heating it at 120°C to 150°C to remove the moisture. The heating is to be continued till the orange colour is fully retained.

11.7.2 DESIGN PARAMETERS OF BREATHER

The top and bottom parts of the breather are made from non-corrosive metal, and the vertical hollow cylinder is made from transparent material.

In order to avoid continuous direct contact of the atmospheric air with the silica gel, an oil cup is provided to act as a hydraulic valve. The air inlet happens only if there is pressure reduction inside the transformer. Similarly, the air goes out to the atmosphere if there is a pressure increase inside. The typical pressure values are as follows:

Air inlet: 0.003 kg/cm^2
Air outlet: 0.005 kg/cm^2
Applicable standard of silica gel
DIN 42562 – 1 and 2 – silica gel breathers for transformers
DIN 42567 – dehydrating breathers for transformers
EN 50216-5 – dehydrating breathers for distribution transformers

11.7.3 CALCULATION OF QUANTITY OF SILICA GEL REQUIRED

The quantity of silica gel required depends on the following factors.

- Quantity of oil inside the transformer = V litres
- Average air temperature at the place of installation = $t°C$
- Average humidity of the environment where the transformer is installed = % RH
- Average thermal cycle of the transformer

This is the difference between the minimum temperature and maximum temperature reached by the transformer in a time period = (Δt) °C

- - Average duration of thermal cycle = T hours
    (This is the time interval between two thermal cycles)
- Maintenance interval in days = M
- Quantity of moisture (water) in one m^3 of the air entering the transformer = (Ag/m^3)

Quantity of silica gel required is calculated by the following formula:

$$Q = 1.683(\text{RH})\left(\frac{\Delta t M}{T}\right)\frac{V}{273+t}10^{-6}e^{\left(\frac{17.6t}{243.7+t}\right)} \tag{11.12}$$

where
Q = quantity of silica gel required (kg)

From this equation, it is clear that the size of silica gel breather and the quantity of silica gel required can be reduced by reducing the maintenance interval.

Example

Oil quantity = 10000 L = V
    Average ambient temperature = 30°C = t
    Average humidity = 90% = RH
    Average thermal cycle = 20°C = (Δt)
    Average duration of thermal cycle = 8 hours = T
    Maintenance interval = 90 days = M

$$Q = 1.683 \times 10^{-6} \times 90 \times \left(\frac{10000}{303}\right)\frac{20 \times 90}{8}e^{\left(\frac{17.6 \times 30}{243.7+30}\right)} = 7.7\,\text{kg}$$

The nearest higher size of silica gel breather is selected.

11.7.4 SELF-DEHYDRATING BREATHER

The conventional breather requires regular inspection and maintenance. In extreme humidity conditions and load cycling, the silica gel may get saturated much faster than expected.

Self-dehydrating breathers are developed to avoid frequent maintenance and reactivation of the silica gel.

This type of breather has two chambers filled with silica gel. In normal operating conditions, the moist air passes through the first chamber and the silica gel absorbs the moisture. A weighing cell installed inside the chamber continuously measures the weight of the silica gel. When the weight of the silica gel exceeds a preset value, a solenoid valve blocks the passage of air through chamber 1 and the air is diverted through chamber 2. A heating system inside chamber 1 is activated simultaneously. The moisture expelled from the silica gel is removed by a fan. The temperature of the silica gel is monitored by a probe, and the maximum temperature is set at 130°C–140°C.

The heating in chamber 1 continues till the weight of the silica gel is reduced to the initial value as measured by the load cell.

When the regeneration of the silica gel is completed, the solenoid valve is de-energized which stops the air flow through chamber 2. At this stage, the silica gel in chamber 2 is heated and regeneration is initiated. The process goes on cyclically, and the system can work without external intervention for the replacement/regeneration of the silica gel. Continuous auxiliary power supply and high reliability solenoid values, heater coil, weighing cell and temperature probes are required for the maintenance free operations of the system.

11.7.5 CALCULATION OF DESICCANT IN BREATHER ALTERNATE METHOD

The size of the breather (i.e. the quantity of desiccant required) is calculated by the following formula.

Quantity of oil inside the transformer

$$K = 1.27 \times \frac{V \Delta t A M}{T}$$

where

  K = weight of desiccant (kg)
  V = volume of oil inside tank (litres)
  Δt = average thermal cycle (°C)
  = difference between maximum and minimum temperature reached by oil within a time interval T (hours)
  T = average duration of thermal cycle (hours)
  M = maintenance interval (days)
  A = quantity of water in each cubic meter of air which enters the breather from the atmosphere

A is a function of air humidity and average temperature of the air. Table 11.3 gives values of A.

TABLE 11.3
Values of A (g/mm³)

Temperature °C	Relative Humidity									
	10	20	30	40	50	60	70	80	90	100
0	0.49	0.98	1.47	1.96	2.45	2.94	3.43	3.92	4.4	4.9
5	0.68	1.36	2.04	2.72	3.4	4.08	4.76	5.44	6.1	6.8
10	0.94	1.87	2.82	3.76	4.7	5.64	6.58	7.52	8.5	9.4
15	1.28	2.56	3.84	5.12	6.4	7.68	8.96	10.2	11.5	12.8
20	1.72	3.44	5.15	6.88	8.6	10.3	12.0	13.8	15.5	17.2
25	2.29	4.58	6.87	9.16	11.45	13.7	16.0	18.3	20.6	22.9
30	3.02	6.04	9.05	12.1	15.1	18.1	21.1	24.2	27.2	30.2
35	3.94	7.88	11.8	15.8	19.7	23.6	27.6	31.5	35.4	39.4
40	5.08	10.2	15.3	20.4	25.4	30.5	35.6	40.7	45.8	50.9
50	8.27	16.5	24.8	33.1	41.4	49.6	57.8	66.2	74.4	82.7

Example

Average air temperature	30°C
Average humidity	90%
Average thermal cycle (Δt)	20°C
Average duration of thermal cycle (T)	8 hours
Maintenance interval (M)	90 days
Oil quantity (V)	10000 L

$$K = \frac{1.27 \times 10^{-7} \times V \times \Delta t}{T} AM$$

A = 27.2 g

$$K = 1.27 \times 10^{-7} \times 10000 \times \frac{20 \times 27.2}{8} \times 90$$

= 7.77 kg
(Select 10 kg breather)

12 Calculation of Winding Gradient, Heat Dissipation Area and Oil Quantity

12.1 RADIATION AND CONVECTION FROM SURFACE

The losses generated in winding and core are transferred to the atmosphere from the tank surfaces and radiator/cooling tube surfaces. For ONAN cooled transformers, the heat transfer from tank and radiator/cooling tube surface is by convection and radiation. Convection depends on the temperature rise of the surface, and radiation depends on the surface temperature rise and emissivity constant of the surface.

For an average oil temperature rise of $\theta_{av}°C$ and an ambient temperature of θ_{amb}, the heat transfer by convection and radiation per m^2 area is calculated by the following formula.

$$W = 2\theta_{av}^{1.25} + 5.7 \times 10^{-8} \Sigma \left[(273 + \theta_{amb} + \theta_{av})^4 - (273 + \theta_{amb})^4 \right] \tag{12.1}$$

$$\Sigma = \text{Emissivity factor}$$

(for grey paint $\Sigma = 0.95$)

Assuming an ambient temperature of 40°C and average oil temperature rise of 40°C, the loss dissipation by convection and radiation for grey paint per m^2 area is calculated as below.

$$W = 2 \times 40^{1.25} + 5.7 \times 10^{-8} \left[(273 + 40 + 40)^4 - (273 + 40)^4 \right] \times 0.95$$

$$= 201.2 + 321 = 522.2$$

$$W/m^2 °C = \frac{522.2}{40} = 13.05$$

For other surfaces and colours, the W/m^2/°C can be calculated using above formula.

Radiation from round cooling tube is less than the radiation from flat surface of tank. Similarly, radiation and convection from fin-type radiator are lower than flat surface because of the heat absorption from adjacent fins.

The W/m^2/°C increases with increased temperature rise and decreases with lower temperature rises.

The following dissipation factors can be used for the calculation of heat dissipation, if the mean oil temperature rise is between 32°C and 40°C.

Loss dissipation from tank can be calculated as per IEEE C57.91

$$W_{\text{Tank}} = 5.884 \, S \, \phi^{1.21}$$

W_{Tank} = loss dissipated by tank watts
δ = surface area of tank (m^2)
[tank bottom plate area is not included]
φ = average oil temperature rise

For 40°C average oil temperature rise, the dissipation from tank from 1 m² area is

$$W_{Tank} = 5.884 \times 1 \times 40^{1.21}$$

$$= 510.7 \text{ W}$$

$$\text{W/m}^2 \, °\text{C} = 12.76$$

When the average oil temperature rise is 36°C (corresponding to a top oil temperature rise of 45°C for ONAN transformer below 2500 kVA), the W/m²°C works out to 12.48.

12.2   HEAT DISSIPATION DATA FOR RADIATOR

Heat dissipation data of radiator is given in Table 12.1.

TABLE 12.1

Heat Dissipation Table for "ON" Cooling 520 mm Width Radiator (watts per section)

Centre Distance (L)	Cooling Surface Per Section	Oil Temperature						Weight Per	Oil Content Per
mm	m²	35°C	40°C	45°C	50°C	55°C	60°C	Section in kgs	Section in L
500	0.58	176	224	278	332	395	456	5.55	1.93
600	0.7	203	251	304	355	415	465	6.55	2.22
700	0.84	231	280	333	387	450	510	7.55	2.52
800	0.96	259	308	361	415	478	520	8.55	2.82
900	1.08	288	328	378	426	477	532	9.55	3.12
1000	1.2	316	364	411	462	522	548	10.55	3.42
1100	1.32	342	391	444	498	561	626	11.55	3.7
1200	1.45	364	422	477	537	606	673	12.55	3.99
1300	1.56	391	452	511	572	646	719	13.55	4.28
1400	1.68	414	479	540	608	684	760	14.55	4.55
1500	1.8	434	507	570	621	723	801	15.55	4.83
1600	1.93	459	532	600	673	759	840	16.55	5.11
1700	2.05	479	557	628	706	792	880	17.55	5.4
1800	2.16	502	582	656	734	829	915	18.55	5.68
1900	2.3	523	610	684	766	862	953	19.55	5.98
2000	2.41	545	631	709	795	896	987	20.55	6.26
2100	2.53	563	654	736	825	921	1024	21.55	6.54
2200	2.64	582	978	762	851	959	1057	22.55	6.82
2300	2.77	603	702	786	881	991	1091	23.55	701
2400	2.9	628	724	814	907	1017	1127	24.55	7.39
2500	3.01	645	744	837	936	1046	1156	25.55	7.67
2600	3.13	666	764	816	963	1069	1190	26.55	7.95
2700	3.25	682	786	886	990	1105	1223	27.55	8.25
2800	3.38	704	807	911	1021	1135	1256	28.55	8.53
2900	3.49	721	829	933	1044	1163	1286	29.55	8.81
3000	3.62	744	815	988	1072	1193	1355	30.55	9.09
3100	3.75	762	868	1015	1100	1224	1385	31.55	9.37
3200	3.87	784	888	1044	1128	1252	1419	32.55	9.65
3300	4.0	804	907	1072	1145	1282	1449	33.55	9.95
3400	4.12	826	928	1100	1173	1311	1483	34.55	10.23
3500	4.25	842	948	1128	1203	1341	1511	35.55	10.51

TABLE 12.2
Radiator Correction Factors

		Transformer Radiators – Standard Correction Factors								
1.	Vertical distance between core centre line and radiator centre line (Ref. Figure 12.1)	mm	0	100	200	300	400	500	600	800	1000
	Correction factor (kV)		0.8	0.85	0.89	0.925	0.95	0.975	1.0	1.05	1.1
2.	Horizontal distance between radiators (kH)	520 mm	575	595	620	645	685	735	785	-	-
	Correction factor		0.76	0.82	0.86	0.91	0.95	0.985	1.0	-	-
3.	No. of sections per radiator		2	4–5	6–8	9–11	12–14	15–17	18–20	21–24	-
	Correction factor (kN)		1.11	1.06	1.02	1.0	0.99	0.98	0.97	0.96	-

The heat dissipation from radiator can be calculated by the empirical formula as per Section 12.6. The heat transmission depends on the actual disposition of the radiators. The correction factors as per Table 12.2 will apply for different dispositions.

Effective heat dissipation/fin in watts/m^2 = heat dissipation $\times K_v \times K_H \times K_N$

K_v : Vertical distance correction factor

K_H : Horizontal distance correction factor

K_N : Correction factor due to number of radiator fins

12.3 CALCULATION OF WINDING GRADIENT

The winding gradient is a function of the surface area exposed to the flow of oil, thickness of insulation on conductors and interlayer, type of winding, the disposition of winding (i.e. whether outer or inner) and the radial depth of the winding. The winding gradient is calculated as a function of W/m^2.

Calculate watts dissipated by winding (W); $W = I^2R$ loss + eddy current loss + circulating current loss, if any.

Lead wire loss and stray losses in bushings and other metal parts will not affect the winding gradient, and hence, these losses need not be considered for the calculation of winding gradient.

When temperature rise test is required at lowest tap, the losses corresponding to the lowest tap shall be taken.

Calculate the area exposed to the flow of oil. If the winding is next to core and if runners are not provided between core and winding, the surface next to core should not be considered for cooling area calculation.

$$\text{Cooling area, } A = \sum_{i}^{m} (\pi D - NW)H$$

H = winding height (m)

D = diameter of cooling surface considered (m)

N = number of runners

W = width of runner (m)

m = number of cooling surfaces

If two-layer winding is used with cooling duct between layers and if runners are inserted between core and winding, four cooling surfaces are to be considered. Similarly, if no cooling duct is provided between layers in the above case, two cooling surfaces are to be considered.

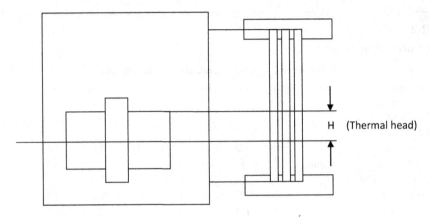

FIGURE 12.1   Thermal head.

If "w" is the calculated loss for three phases of the spiral winding or layer winding under consideration, the heat dissipated per area of cross section in meters is given as (q),

$$\frac{w}{m^2} = \frac{w}{3 \sum_{i}^{m} H(\pi D - NW)} = q \tag{12.2}$$

Then the heat dissipation per decimeter of area (q');

$$\frac{w}{dm^2} = \frac{q}{100} = q' \tag{12.3}$$

Winding gradient is calculated by different methods as given by the version empirical formulas in this section.
- Empirical Formula 1

$$g = 10.5 \frac{A}{T} \delta^2 \frac{W}{D} \times K \times 10^{-3} \tag{12.4}$$

g = winding gradient °C
   A = area of cross section of conductor (mm²)
   T = insulated thickness of conductor (mm)
   δ = current density (A/mm²)
   W = radial thickness of disc (mm)
   D = duct between discs (mm)
   K = axial duct factor
   {= 1 if there is no axial duct in the disc
   = 0.7 if there is axial duct}
- Empirical Formula 2

$$\theta_w = \left[\frac{t_q}{0.15} + \frac{q^{0.75}}{15.1} \right] K \tag{12.5}$$

$$\theta_w = \text{winding gradient} \left({}^\circ C \right)$$

t = thickness of paper insulation on one conductor (m)

TABLE 12.3

Correction Factors

Radial Depth of Winding between two Adjacent Vertical Cooling Surfaces (mm)	Up to 50	51 ~ 62.5	63 ~ 87.5
K_2	1.00	1.1	1.2

K = 1.0 for outer winding

= 1.15 for inner winding

• Empirical Formula 3

$$\theta_w = (0.0016\ H + 1.78)(q')^{0.8}\ K_1 K_2 \tag{12.6}$$

θ_w = winding gradient (°C)

H = winding stack height (mm)

K = 1.0 for outer winding

= 1.15 for inner winding

K_1 = 0.7, if axial duct is provided inside the winding for cooling

= 1 if no axial duct is provided

K_2 = factor for radial depth

(if the radial depth of winding between two adjacent vertical cooling surfaces exceeds 50mm, the values of K_2 in Table 12.3 shall be used).

12.4   CALCULATION OF WINDING GRADIENT: ALTERNATE METHOD

Application: Oil-filled distribution transformer with natural oil flow

Disc winding, layer winding and foil winding

$$G = G_1 + G_2$$

G = winding gradient

G_1 = gradient due to heat flux density

G_2 = temperature drop in the insulation

Calculation of G_1

$$G_1 = \frac{q^{0.75}}{15.1} \text{ for Disc winding} \tag{12.7}$$

$$G_1 = \frac{q^{0.75}}{25} \text{ for layer winding \& Foil winding} \tag{12.8}$$

$$G_1 = \frac{q^{0.8}}{31.4} \text{ for layer winding with one exposed face } \{\text{one side may have a thick cylinder, core etc}\}$$

$$\tag{12.9}$$

q = W/m² (average heat flux)

If the spacers occupy considerably lower or higher value, use actual value for the calculation:

$$q = \frac{\left(1+\dfrac{e}{100}\right)\dfrac{J^2}{\sigma}nab}{\left[2n(a+2\delta)+2(b+2\delta)\right]k} \tag{12.10}$$

J = current density (A/m^2)

e = eddy current loss as percentage of winding loss

σ = conductance of copper = 48.083 X 10^6 at 75 °C

δ = insulation thickness of conductor at one side m.

a = Thickness of conductor m

b = width of conductor m

$n(a+2\delta)$ = radial thickness of disc/winding m

n = number of conductors in radial direction between ducts

l = winding height m

k = oil exposed area factor $\left(\text{This is typically } 0.65\right)$

Example

$$a = 3 \times 10^{-3}\,\text{m} \qquad b = 12 \times 10^{-3}\,\text{m}$$

n = 12 (turns/disc) $\delta = 1 \times 10^{-3}\,\text{m}$
e = 4.5% $J = 3 \times 10^6$ A/m^2

$$q = 920 \text{ W/m}^2$$

Calculation of G_2

$$G_2 = q\frac{\delta}{p}$$

δ = Thickness of one side of turn insulation $\left\{\text{For disc winding}\right\}$ m

= half (total thickness of interlayer insulation + conductor insulation) between two adjacent ducts for foil winding and layer winding)

p = Thermal conductivity of paper

= 0.15 w/m^{-1}c^{-1}

(this is the value for kraft paper at 70°C)

12.5 CALCULATION OF MEAN OIL TEMPERATURE RISE

Mean oil temperature rise (θ_m) = 0.8×guaranteed top oil temperature rise or $(\theta - \theta w)$ (whichever is lower)

θ = average winding temperature rise

θ_w = gradient

a. When round cooling tubes are used and paint colour is light grey or dark grey:

$$W_T = 11\theta_m \ A_T + 0.11 \ L \times 10 \times \theta_m \tag{12.11}$$

where

W_T = losses to be dissipated (watts)

= no-load loss + load loss

(when temperature rises are guaranteed corresponding to maximum losses, load losses corresponding to tap position, for maximum losses, shall be used for calculation)

A_T = cooling surface area of tank (m^2)

θ_m = mean oil temperature rise (°C)

L = length of round cooling tube (m)

b. When pressed steel radiators are used and paint colour is light grey or dark grey:

$$W_T = 11\theta_m \ A_T + 0.16 \ C \times 7.0 \ K_1 \ K_2 K_3 \ \theta_m \tag{12.12}$$

W_T, θ_m, A_T are same as in (a) above

C = effective cooling length of pressed steel radiator

K_1 = correction factor for thermal head

K_2 = correction factor for horizontal distance between radiators

K_3 = correction factor for number of sections per radiator

c. When corrugated fins are used and the paint colour is light grey or dark grey:

$$W_T = 11\theta_m A_T + 7.95 D^{-0.2} \theta_m^{1.2} A_f \tag{12.13}$$

W_T, θ_m and A_T are same as in (a) above.

A_f = total surface area of fin (m²)

$$A_f = 2 \times D \times h \times 10^{-6}$$

D = fin depth (mm)

H = fin height (mm)

12.6 CALCULATION OF WEIGHT OF RADIATOR PANELS

$$\text{Heat dissipation from radiator panel} = 22\theta^{1.1} \ L^{-0.2} \text{w/m}^2 \tag{12.14}$$

θ = top oil temperature rise (°C)

L = radiator panel center to center length (mm)
W = losses to be dissipated by radiator (watts)
T = thickness of radiator fin (mm)

Weight of total radiator panels

$$\text{Weight of total radiator panels} = \frac{0.3568 \ L^{0.2} \ tw}{\theta^{1.1}} \text{kg} \qquad (12.15)$$

$$\text{Weight of Radiator} = \text{Weight of panel} + \text{weight of Headers} \qquad (12.16)$$

$$\text{Number of panels} = \frac{43.7 \ W}{L^{0.8} \ \theta^{1.1}} \ (\text{Integer}) \qquad (12.17)$$

$$\text{Number of Headers} = \frac{\text{Number of panels}}{\text{Number of radiators}} \ (\text{Round to integer}) \qquad (12.18)$$

Example

Top oil temperature rise = 40°C
    Total losses to be dissipated by radiator = 11300 watts
    Panel height (radiator center to center) (L) = 1000 mm
    Thickness of radiator fin = 1.2 mm

$$\text{Total weight of panels} = \frac{0.3568 \ L^{0.2} \ t \ w}{\theta^{1.1}}$$

$$= \frac{0.3568 \times 1000^{0.2} \times 1.2 \times 11300}{40^{1.1}} = 333 \text{ kg}$$

$$\text{Number of panels} = \frac{43.7 \times 11300}{1000^{0.8} \times 40^{1.1}} = 34$$

12.7   CALCULATION OF WEIGHT OF CORRUGATED FINS

W = total loss to be dissipated by fin (watts)
t = fin thickness (mm)
D = fin depth (mm)
θ = maximum top oil temperature rise (°C)
A = surface area of fin (m^2)

$$\text{Specific heat dissipation from fin} = 7.95 \ D^{-0.2} \ \theta^{1.2} \ \text{W/m}^2$$

$$\text{Surface area of fin required } A = \frac{0.126 \ W \ D^{.2}}{\theta^{1.2}} \ \text{m}^2$$

$$\text{Fin weight required} = \frac{0.987 \ D^{0.2} \ t \ w}{\theta^{1.2}} \text{kg}$$

From the fin weight required and the fin area, the number of fins can be calculated after fixing the fin height from the tank dimensions.

Example

Total loss (watts)	11300	11300	7500	7500
t (mm)	1.2	1.2	1.2	1.2
D (mm)	400	220	300	200
θ (°C)	40	40	40	40
Fin weight (kg)	530	470	332	306

12.8 EFFECT OF AMBIENT TEMPERATURE ON THE TOP OIL TEMPERATURE RISE

The viscosity of oil changes with the temperature and therefore the frictional effect of the oil flow reduce with increase in temperature and vice versa.

This effect is more predominant for synthetic liquids like silicone fluid, midel and FR3 fluid.

The approximate change in oil temperature rise with ambient temperature can be expressed by the following formula.

$$\theta_2 = \theta_1 \left[\frac{e^{\frac{2797.3}{T_1+273}}}{e^{\frac{2797.3}{T_2+273}}} \right]^{-0.125} \tag{12.19}$$

where

θ_2 = top oil temperature rise at an ambient temperature at T_2 (°C)

θ_1 = top oil temperature rise at an ambient temperature of T_1 (°C)

The National or International standards on transformer do not consider this effect. However, if the temperature rise test is performed at two different ambient temperatures, the difference can be observed, especially when high viscous liquids are used and the difference in ambient temperatures of tests has a difference of 20°C or more.

Example

Top oil temperature rise at 20°C ambient temperature = 41°C (θ_1)

$T_2 = 50°C$

$$\theta_2 = \theta_1 \left(\frac{e^{\left(\frac{2797.3}{T1+273}\right)}}{e^{\left(\frac{2797.3}{T2+273}\right)}} \right)^{-0.125}$$

$$= 41 \left(\frac{e^{\left(\frac{2797.3}{293}\right)}}{e^{\left(\frac{2797.3}{323}\right)}} \right)^{-0.125}$$

$$= 41 \left(\frac{e^{9.547}}{e^{8.660}} \right)^{-0.125}$$

$= 36.7°C$ at 50°C ambient

Effect of Ambient Temperature on Temperature Rise Test Results

Due to change of viscosity of the transformer oil, the temperature rise of oil changes with change of ambient temperature. The following formula can be used for estimating the temperature rise of oil at different ambient temperature, if the test result at any ambient temperature is known.

$$\theta_2 = \theta_1 \left(\frac{e^{\left(\frac{2797.3}{T_1+273}\right)}}{e^{\left(\frac{2797.3}{T_2+273}\right)}} \right)^{-0.125}$$

where
θ_2 = top oil temperature rise at ambient temperature (T_2)
θ_1 = Top oil temperature rise at ambient temperature (T_1)

Examples

Ambient temperature $T_1 = 35°C$ (T_1)
Top oil temperature rise measured = 45°C (θ_1)
Calculated values of top oil temperature rise at different ambient temperature are as below:

Ambient Temperature	Top Oil Temperature Rise	Top Oil Temperature (Amb + Top Oil Rise)	Remarks
0	52.05	52.05	
10	49.745	59.745	
35	45.00	80	
50	42.70	92.7	

Approx: 0.2°C/degree change in ambient temperature.

12.9   HEAT DISSIPATION BY FORCED AIR COOLING

$$K_w = V_L \; C_{PL} \; P_L \; \Delta\theta$$

K_w = heat dissipated by air flow
V_L = airflow rate (m³/s)
(air flow is the effective quantity of air flowing through hot fins)
C_{PL} = thermal capacity of air = 1.015 kW/kg K

P_L = density of air = 1.18 kg/m³
$\Delta\theta$ = temperature rise of air

Power rating of fans

$$P = \frac{K \; V_L}{3.6 \times 10^6 \; \eta}$$

K = pressure difference due to air flow
(frictional head loss of radiators.

$\approx 50\,\text{N/m}^2$ for 1 m length of radiator)

V_L = airflow rate (m³/h)

η = fan efficiency $(0.7 \sim 0.8)$

12.10 REFERENCE AMBIENT AS PER IEC

The rating of a transformer is specified on the reference ambient temperatures as per Table 12.4.

The temperature rise limits for oil-immersed transformers with Class A insulation are given below in Table 12.5.

TABLE 12.4
Reference Ambient Temperatures

Reference Ambient	IEC 60076
Maximum ambient air temperature (°C)	40
Maximum daily average air temperature (°C)	30
Maximum yearly weighted air temperature (°C)	20
Maximum cooling water temperature (°C)	25
Average cooling water temperature in an year (°C)	20

TABLE 12.5
Temperature Rise Limits for ONAN Class A Insulation

	°C IEC 60076
A. Air cooled	
1. Average winding rise by resistance method	
• ONAN	65
• ONAF	65
• OFAF (non-directed)	65
• ODAF (directed)	70
2. Top oil temperature rise	
• Sealed or with conservator	60
• Not sealed or without conservator	55
B. Water cooled	
1. Average winding rise by resistance method	
• OFWF (non-directed)	65
• ODWF (directed)	60
2. Top oil temperature rise	
• Sealed or with conservator	60
• Not sealed or without conservator	55

If the transformer is designed for service where the temperature of cooling medium exceeds one of the maximum values shown above, by not more than 10°C, the allowable temperature rises of windings and oil shall be reduced as below.

- If the rated power is 10 MVA or greater, the reduction shall correspond to the excess temperature
- For transformers of rating less than 10 MVA, the allowable temperature rises shall be reduced as follows:
  a. By 5°C, if the excess ambient temperature is less than or equal to 5°C and
  b. By 10°C, if the excess temperature is greater than 5°C and less than or equal to 10°C.

12.11   CALCULATION OF WEIGHTED AVERAGE AMBIENT TEMPERATURE

$$\theta_{wat} = \frac{\left(\log_e \int_o^t e^{0.1155\ \theta at}\,dt\right) - \log_e t}{0.1155} \tag{12.20}$$

$$\theta_{wat} = 8.65\ \log_e \frac{1}{n}\sum_{e=1}^{n} e^{0.1155\ \theta at}$$

θ_{at} = Yearly average ambient temperature
θw_{at} = Weighted average ambient temperature
n = total number of months
t = time
Average ambient – months

Example

Average ambient – months

30°C – 2
20°C – 6
10°C – 2
0°C – 2

$$\text{Yearly average ambient temperature} = \frac{30 \times 2 + 20 \times 6 + 10 \times 2 + 0 \times 2}{12} = 16.67°C$$

Yearly weighted ambient temperature

$$= 8.65\log_e \left[\frac{1}{12}\left(2e^{0.115\times30} + 6e^{0.1155\times20} + 2e^{0.1155\times10} + 2e^{0.1155\times0}\right)\right]$$

$$= 20.8°C$$

12.12   CALCULATION OF OIL QUANTITY

Oil quantity = oil in main tank + oil in cooling tube/radiator + oil in conservator + oil in cable box chamber (when applicable) + oil in OLTC (if OLTC is provided)
   For sealed and CSP transformers, oil in conservator is not applicable.
   Oil in main tank is calculated by subtracting the displacement volumes of core, conductor and channel.

$$\text{Oil in main tank} = L \times W \times H \times 10 - W_c/7.65 - W_w/D - W_i/7.5$$

where L = length of tank (inside dimension) (mm)

W = width of tank (inside dimension) (mm)
H = height of oil level (inside dimension) (mm)
For sealed and CSP transformer.
= tank height (inside dimension) (mm)
For conventional transformer.

W_c = weight of core (kg)
W_w = weight of conductor (kg)
D = density of conductor (gm/cc)
= 2.7 for aluminium, 8.9 for copper
W_i = weight of core channel and other steel structures used in core coil assembly (kg)

Displacement volume of paper, press board and other insulating materials are ignored in the above calculation on the basis that the oil absorption in insulation material and core is approximately equal to the displacement volume of the insulation materials. Due to this assumption, the calculated oil volume will be slightly higher than the actual quantity required for filling.

This is applicable for distribution transformers only.

- Oil quantity in radiator/cooling tube
    Oil quantity in 1 m length of round cooling Tube = 1 l
    Oil quantity in 1 m length of pressed steel radiator = 0.9 l
- Oil quantity in conservator, where applicable
    Oil quantity in conservator is 3% of the oil quantity in main tank and radiator/cooling tube.

13 Calculation of Pressure Rise, Stresses and Strength of Tank

13.1 CALCULATION OF PRESSURE RISE IN SEALED TANKS

The oil in transformer tank expands and contracts due to variation of load and ambient temperature. In sealed-type transformers, this expansion and contraction will lead to increase or decrease of internal pressure. The volume of air cushion required above the oil is decided on the basis of limiting the pressure rise inside the tank.

The maximum and minimum pressures are calculated by the following procedure:

V_{oil} = oil volume in tank + radiator at the time of filling (m³)
θ_{oil} = top oil temperature rise at full load or at specified maximum load (°C)
θ_{fill} = filling temperature (°C)
$T_1 = 273 + \theta_{fill}$
P_i = pressure at the time of filling (kg/cm²)
(This is generally 1 kg/cm²)
V_1 = volume of air/gas space at the time of filling
[For rectangular tank, this is tank length×tank width x height of space above oil level at the time of filling]
T_2 = top oil temperature at the time of maximum ambient temperature at full load or specified overload
V_2 = volume of air/gas space at the time of maximum ambient temperature and full load/specified overload

$$V_2 = V_1 - 0.00718 \times 0.8(\theta_{oil} + \theta_{amb} - \theta_{fill}) \tag{13.1}$$

$$P_2 = \frac{P_1 V_1 (273 + \theta_{oil} + \theta_{amb\,max} - \theta_{fill})}{(V_1 - 0.000718 \times 0.8\,(\theta_{oil} + \theta_{amb} - \theta_{fill}))(273 + \theta_{fill})} \tag{13.2}$$

P_2 shall be less than the setting of the pressure relief device, and the setting of the pressure relief device shall be less than the test pressure of transformer.

For calculation of minimum pressure, substitute the following values:

$$T_2 = 273 + \theta_{amb(min)} \tag{13.3}$$

$\theta_{amb(min)}$ = minimum ambient temperature specified
V_2 = volume of air/gas space at minimum ambient temperature
The pressure limits are schematically shown in Figure 13.1.
Sealed-type construction can be any one of the following types:

- Rigid tank with pressed steel radiators with air/gas cushion
- Corrugated tank with air/gas cushion
- Completely filled with corrugated tank {no air/gas cushion}

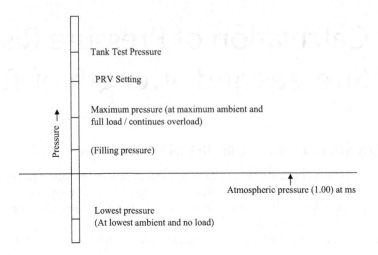

FIGURE 13.1   Pressure limits of tank.

13.1.1 TANK WITH PRESSED STEEL RADIATORS WITH AIR/GAS CUSHION

In this case, the tank top will have air/gas cushion and the expansion and contraction of the oil are absorbed by the corresponding pressure rise/pressure reduction of the air/gas.

The volume of air/gas space at the time of oil filling is decided by calculating the maximum and minimum pressure in the air/gas cushion during the specified loading and ambient conditions (including overloading).

The calculations are as follows:

V_o = total oil volume in transformer litres including radiator/fins
{Volume at the time of filling}
T_1 = temperature at the time of filling oil (°C)
T_2 = average oil temperature at the maximum loading conditions
{If overloading is specified, the oil temperature corresponding to the overload shall be considered}. Maximum ambient temperature specified shall be used for calculating T_2.

For distribution transformers up to 2500 kVA,

$$T_2 = T_{a(max)} + 0.8\,T_o \tag{13.4}$$

where

$T_{a(max)}$ = maximum ambient temperature specified (°C)
T_o = temperature rise of oil at the maximum loading specified (°C)
α = coefficient of volumetric expansion of transformer oil/fluid
= 0.000718 / °C for mineral oil

Oil expansion corresponding to temperature T_2

$$V_{eo} = V_o \propto (T_2 - T_1)L \tag{13.5}$$

V_1 = volume of air/gas space at temperature T_1
V_2 = volume of air/gas space at temperature T_2
= $V_1 - V_{eo}$

$= V_1 - V_o \, \alpha \, (T_2 - T_1)$

P_1 = air/gas pressure at the time of oil filling (kg/cm²)

P_2 = Max air/gas pressure at temperature T_2 (kg/cm²)

P_2 can be solved by using the equation

$$\frac{P_1 V_1}{(T_1 + 273)} = \frac{P_2 V_2}{(T_2 + 273)}$$

$$P_2 = \frac{P_1 V_1 (T_2 + 273)}{V_2 (T_1 + 273)} \tag{13.6}$$

The maximum pressure acceptable is decided after considering the following:

- The safe pressure withstand capacity of tank and radiator, which shall be less than the test pressure of the tank after considering sufficient factor of safety.
- The pressure relief device shall be set to operate at a pressure above P_2, but below the tank test pressure.

13.1.2 CORRUGATED TANK WITH AIR/GAS CUSHION

The calculations of pressure rise are the same as in Section 13.1.1 above.

The point to be noted is that the test pressure and maximum allowable pressure are generally lower than tanks with radiators.

The maximum pressure limit (P_2) for 1.2 mm corrugated fin tank is limited to 0.35 kg/cm². In order to get this pressure withstand capacity, the fins will have to be strengthened by embossing and spot welding, if the corrugated fin depth is more than 220 mm.

13.1.3 CORRUGATED TANK WITH COMPLETE FILLING OF OIL (NO AIR/GAS CUSHION)

In this case, the expansion and contractors of the oil will have to be absorbed by the elastic expansion and contractors of the fins. Hence, embossing and spot welding of fins cannot be used, and the fin depth will have to be limited to 220 mm for 1.2 mm thick corrugations.

If the fin depth is less than 180 mm with 1.2 mm thick corrugation, the elasticity fins may not be adequate to absorb the oil expansion.

13.2 CALCULATION OF PRESSURE AND STRESSES ON CORRUGATED FINS FOR COMPLETELY FILLED TRANSFORMER

a. Pressure rise in completely filled corrugated fins

$$\Delta P = \frac{30 \; E \; T^3 \; (\Delta V)}{NL \; w^5} \tag{13.7}$$

ΔP = pressure rise (pascals)

E = Young's modules = 200×10^9 N/m²

T = thickness of corrugated fin (m)

N = number of fins

L = height (length) of fin (m)

w = width/depth of fin (m)

ΔV = change in oil volume (m³)

$= 0.000718 \times Q \left[T_{av} - T_{fill} \right] \times 10^{-3}$ (m³)

Q = oil quantity (litres)

T_{av} = average oil temperature rise

T_{fill} = temperature at the time of filling

Calculate T_{av}, based on the following oil temperature

    a.   Top oil temperature at full load or

    b.   Top oil temperature at specified overloading condition or

    c.   Top oil temperature for calculation specified by customer

    d.   Top oil temperature as per standard overloading conditions as per applicable standard whichever is higher

  b. Ultimate pressure calculation

$$\Delta P_1 = \left(\Delta P + P_o \right) K \tag{13.8}$$

P_o = initial pressure at top of oil at the time of filling ~0.05 bar

K = factor of safety = 1.20

  c. Maximum stress at the centre of fin

$$\sigma_{max} = \frac{W^2 \times \Delta P_1 \times 10^{-6}}{4T^2} \tag{13.9}$$

$$\sigma_{max\,test} = \frac{W^2 \times P_t \times 10^{-6}}{4T^2}$$

where

σ_{max} = maximum stress at the center of fin (mpa)

σ_{max}(test) = maximum stress at the center of fin during test (mpa)

P_t = test pressure of transformer/tank (MPa)

Note

   i. The permissible maximum stress is 250 Mpa

  ii. w is the fin depth of the maximum unreinforced length; that is, w is full depth of fin if there is no embossing of the fin (generally, reinforcement is not done for completely filled.). However, embossing can be done, if the pressure test requirement is more than 0.15 kg/cm². In this case, the pressure rise is to be calculated for different widths based on the number of embossing positions)

  d. Calculation example for 1000 kVA 11/0.433 kV transformer

    Maximum ambient temperature = 50°C

    Oil filling temperature = 30°C

    Initial pressure at top of oil = 0.05 bar

    Test pressure = 0.35 kg/cm² = 350 Mpa

    Fin wall thickness = 1.5 mm

    Depth of fins = 350 mm 52 Nos + 370 mm 24 Nos

The fins are embossed, and the lengths between embossing are

350 mm = 220 + 130 mm

370 mm = 220 + 150 mm

Oil quantity = 9401

Top oil temperature rise = 45°C

Top oil temperature for calculation of pressure and stress on fin = 105°C
{specified by customer}

The maximum stress will be on 220 mm unembossed length of the fins.
The oil quantity in the fins is proportional to the number of fins and the depth of fins.
The total expansion of the oil quantity is shared between the fins.
Total oil quantity of 9401 will expand when temperature increases, and it is assumed that the expanded volume is shared by various sections in the following ratio:

$(54 + 22)$ 220:52×130:24×150
= 16,720:6760:3600
= 1672; 676:360
= 418:169:90

The maximum stress is on 220 mm fin section, and this is calculated as below:
  Δv = oil expansion for $(54 + 22)$ sections of 220 mm

$$\text{The proportional oil quantity} = \frac{940 \times 418}{677} = 580$$

Top oil temperature = 105°C (specified value for calculation)
Oil temperature rise = 105–50 = 55°C
Average oil temperature rise = 0.8×55 = 44°C
Average oil temperature = 50 + 44 = 94°C
Filling temperature = 30°C
∴ oil expansion corresponding to 94 – 30 = 64°C is to be calculated
ΔV = 580×64×0.000718 = 26.651
= 0.02665 m^3

$$\therefore \Delta P = \frac{30\,E\,T^3\,(\Delta)}{N\,L\,W^5} \times 10^{-5} \text{ bar}$$

$$= \frac{30 \times 200 \times 10^9 \times 0.0015^3 \times 0.02665 \times 10^{-5}}{76 \times 1 \times .22^5} = 0.1378 \text{ bar}$$

Ultimate pressure for stress calculation:
  $\Delta P_1 = (\Delta P + \text{Initial pressure}) \times 1.2$
  = (0.1378 + 0.05) 1.2 = 0.225 bar
  = 0.225×10^5 pa
Maximum stress at the centre of fin (220 mm section):

$$\sigma_{max} = \frac{w^2\,(\Delta P) \times 10^{-6}}{4\,T^2}$$

$$= \frac{0.22^2 \times \left(0.225 \times 10^5\right) \times 10^{-6}}{4 \times .0015^2} = 0.121 \times 10^5 \text{ Pa}$$

Tank test pressure is 0.35 kg/cm^2 (0.35×10^5 Pa)
Stress on 220 mm fin section at the test pressure of 0.35×10^5 Pa:

$$= \frac{0.22^2 \left[0.35 \times 10^5\right] \times 10^{-6}}{4 \times 0.0015^2}$$

=188.2 MPa

If fin with no embossing of depth 280 mm is used

Assuming

$N = 76$ (number of fins)

$W = 280$ mm

All other parameters remain unchanged.

$$\Delta P = \frac{30 \times 200 \times 10^9 \times 0.0015^3 \times (\Delta V) \times 10^{-5}}{76 \times 1 \times 0.28^5}$$

$\Delta V = 940 \times 64 \times 0.000718 = 43.2 \, \text{L}$

$\quad = 43.2 \times 10^{-3} \text{m}^2$

$$\therefore \Delta P = \frac{30 \times 200 \times 10^9 \times 0.0015^3 \times 43.2 \times 10^{-3} \times 10^{-5}}{76 \times 1 \times .28^5}$$

$$= 0.0655 \, \text{bar}$$

[When the fin depth is increased, i.e. the unembossed depth is 280 mm, the fin expands at a lower pressure, compared to fin of smaller depth]

Maximum stress assuming initial pressure of 0.05 bar and 20% extra margin

$\Delta P = 1.2 \, (.0655 + 0.05) = 0.1386 \, \text{bar}$

$$\text{Maximum Stress} = \frac{0.28^2 \times 0.1386 \times 10^5 \times 10^{-6}}{4 \times 0.0015^2}$$

$$= 120.7 \, \text{Mpa}$$

Hence, the maximum test pressure feasible on the fin to limit the stress within

$250 \, \text{Mpa} = \dfrac{0.1386 \times 250}{120.7} = 0.287 \, \text{bar}$

13.3 GAS PRESSURE CALCULATION OF SEALED TRANSFORMERS, CONSIDERING SOLUBILITY CHANGES OF GAS WITH TEMPERATURE AND PRESSURE

Sealed transformer with nitrogen gas filled above oil is considered for the calculations. The effect of the expansion or contraction of the tank as a result of the varying temperature is not considered. For tanks with corrugated fins, the pressure change on the fins would create small volume change. However, this is not appreciable, when the fins are embossed and spot welded to withstand higher pressure. Corrugated fins with sufficient elasticity to absorb the expansion and contraction of oil with temperature are used for completely filled transformers. In such cases, the pressure variation is low and the gas solubility variation can be ignored for practical calculation.

Symbols and terms used:

C_g = Concentration of gas component in gas space at temperature T and partial pressure P_i. If one gas only is present, C_g represents the concentration at the pressure P.

$C_{g,o} = C_g$ at $T = T_o$ and $P_i = P_{i,o}$

C_l = concentration of gas component I, in oil at temperature T and partial pressure P_i

(or P if one gas only is present)

$C_{l,o} = C_l$ at $T = o$ and $P_i = P_{i,o}$
ppm = parts per million
[Also denoted by microlitres/litre]
P_i = partial pressure of gas component I, at equilibrium condition at temperature T
$P_{i,o} = P_i$ at temperature T_o
P = total pressure at T
P_o = total pressure at T_o
T = final temperature (K)
T_o = initial temperature (K)
V_g = volume of gas space at temperature T and pressure P
V_l = volume of oil at T
$V_{l,o} = V_l$ at T_{og}

- Distribution Coefficient K_i
    This is the ratio of C_l/C_g and is numerically equal to the Ostwald solubility coefficient. This is the volume of gas component I, dissolved per unit volume of liquid at the specified temperature and partial pressure after equilibrium is reached (in the case of one gas, partial pressure is the total pressure).
    α = coefficient of volumetric expansion of oil
    = 0.000718/°C
    $\Delta T = T - T_o$
    This can be negative when transformer is cooling down.
Change of gas concentration in oil due to oil expansion/contraction

$$K_i = \frac{C_l}{C_g} \tag{13.10}$$

$$\frac{V_g}{V_l} = \left[\frac{1 + \dfrac{V_{g,o}}{V_{l,o}}}{1 + \propto \Delta T} - 1 \right] \tag{13.11}$$

The solubility of nitrogen increases with temperature. As the oil expands due to increase in temperature, more nitrogen from the gas space will dissolve in the oil. Also, the increase in pressure adds to the solubility of the gas.
    The K_i values of nitrogen at different temperatures can be calculated by the following formula:

$$\text{Log}_{10}\ K_i = -\frac{249.5}{T} - 0.176$$

The temperature T will have to be in $(273 + t)$ where t is the temperature in °C.
    The final equilibrium pressure P is calculated by the following formula:

$$P = P_o \left[\frac{T}{T_o} \right] \left(\frac{V_{l,o}}{V_l} \right) \left[\frac{K_{i,o} + \dfrac{V_{g,o}}{V_{l,o}}}{K_i + \dfrac{V_g}{V_l}} \right] \tag{13.12}$$

The observed pressure will be more than this, if the temperature change is faster than the solubility rate. The time required for the gas to dissolve and reach equilibrium is generally longer than the temperature variation due to load and ambient variation, and the observed pressure would be more than the calculated value assuming solubility of gas.

If the solubility of gas is ignored, the above equation is modified to

$$P = P_o \left[\frac{V_{g,o}}{V_g} \right] \left[\frac{T}{T_o} \right] \tag{13.13}$$

The observed pressure would be less than the pressure obtained by applying this formula because the gas gets dissolved in oil.

13.4   CALCULATIONS OF STRENGTH OF RECTANGULAR TANK WHEN PRESSURE TESTED

13.4.1   Tank Dimensions and Constants

Figure 13.2 shows the sketch of a rectangular tank with stiffeners.

H = tank height (cm)
W = tank length (cm)
D = centre-to-centre distance of stiffness (cm)
A = thickness of stiffeners (cm)
[The dimension of stiffener on tank plate]
T = thickness of tank plate (cm)
P = test pressure (kg/cm^2)
K = constant depending on the stiffener end structure and test pressure
Values of K are given in Table 13.1.
Z = section modulus (cm^3)
S = safe working stress of tank plate (kg/cm^2)
N = number of stiffness required
m = maximum allowable spacing of stiffness

$$m = \sqrt{\frac{ST^2 \, W^2}{0.25 \, P \, W^2 - ST^2}} \tag{13.14}$$

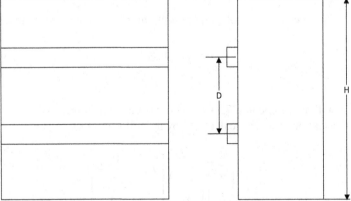

FIGURE 13.2   Rectangular tank with stiffeners.

TABLE 13.1

K Values and Tank Pressure test

Tank Test Pressure (kg/cm²)	K		
	Type of Stiffener		
	Stiffener Is Belted around the Tank	Tank Corner with Gussets	Stiffener with free Ends
0.35	0.00001	0.000013	0.000015
0.50	0.000014	0.000018	0.000021
0.70	0.000020	0.000025	0.000030
1.00	0.000030	0.000038	0.000045
1.40	0.000040	0.000050	0.000060

$$N = \frac{H - m}{A + m} \tag{13.15}$$

Round off to the next higher integer and recalculate the actual spacing N.

$$Z = KDW^2 \tag{13.16}$$

13.4.2 CALCULATION OF SECTION MODULUS REQUIRED FOR STIFFENERS

Calculate the section modulus of the stiffener. If the section modulus is not sufficient, select either a higher section or different type of stiffener {The stiffener can be flat, angle, channel, etc.}.

Example

Tank length = 210 cm (W)
Tank height = 190 cm (H)
Tank thickness = 0.5 cm (T)
Test pressure = 0.5 kg/cm² (P)
Stiffener size = 15×1 cm (flat)
Proposed stiffener type= free end

Safe working stress of Tank Plate, (S) = 2100 kg/cm²

m = Maximum allowable spacing of stiffner

$$m = \sqrt{\frac{ST^2\,W^2}{0.25PW^2 - ST^2}}$$

$$= \sqrt{\frac{2100 \times .5^2 \times 210^2}{0.25 \times 0.5 \times 210^2 - 2100 \times 0.5^2}}$$

$$= \sqrt{\frac{23{,}152{,}500}{5512.5 - 525}}$$

$$= \sqrt{\frac{23{,}152{,}500}{4987.5}}$$

$$= 68\,cm$$

Minimum number of stiffness

$$N = \frac{H - m}{A + m}$$

$$= \frac{190 - 68}{1.0 + 68}$$

$$= 1.77 \text{ Round off to } 2$$

Number if stiffners provided $= 2$

Actual distance between stiffners, $D = \dfrac{190}{(2+1)} = 63.3$

$$[D \text{ shall be less than } m]$$

Section modulus required $= KDW^2$

$$K = 0.000021 \text{ [Free end \& } 0.5 \text{ kg/cm}^2 \text{ (test pressure)}]$$

$$Z \text{ required} = 0.000021 \times 63.3 \times 210^2$$

$$= 58.622 \text{ cm}^3$$

Z of the stiffener provided

Z of the stiffner provided $= \dfrac{1 \times 15^2}{6} = 37.5 \text{ cm}^3$ The section modulus is not sufficient. Select a section with section modulus more than 58.622 cm³

The design alternatives can be one or more of the following:

- change the end structure of stiffener
- change the tank plate thickness
- change the size of stiffener
- use angle or channel as stiffener instead of flat, etc.

A flat stiffener of 1.6×15 cm gives a section modulus of $\dfrac{1.6 \times 15^2}{6} = 60 \text{ cm}^3$.

This size is adequate.

However, the stiffener height is high, and therefore, the radiator pipe length becomes longer and the overall dimension of transformer is increased.

In order to optimize the transformer dimensions, a stiffener with less height and more section modulus needs to be selected. A channel 80×80×8 mm has section modulus of 58.88 cm³ and can be used as the stiffener.

13.5 FASTENER SPACING AND TIGHTENING TORQUE OF GASKET JOINTS OF OIL-FILLED TRANSFORMERS

13.5.1 INTRODUCTION

Design of leak proof gasket joint of transformer is very important to ensure the safe long-term service of a transformer. Leakage of oil/insulation liquid through the joints in a transformer is a serious issue for the end user, the manufacturer and the environment. Several studies and continued development regarding the selection of gasket and spacing of fasteners are done by manufacturers in the past to ensure leak proof joints.

13.5.2 LEAKAGE RATE THROUGH GASKET JOINT

The leakage rate through a gasket joint is a function of the contact pressure, property of the gasket including load deflection characteristics, surface properties of the gasket seat, fastener spacing, etc. The relationship is approximate by the formula:

$$Q_p = f\left(\frac{P_c}{E}\right)$$

(13.17)

where

$$Q_p = \text{Leakage Rate}\left(\text{m}^3/\text{s}\right)$$

$$P_c = \text{Contact Pressure}\left(\text{kg/m}^2\right)$$

$$E = \text{Elastic modules of the gasket material}\left(\text{kg/m}^2\right)$$

$$f = \text{Functional Relationship Constant}$$

For molecular flow, the leak rate is given by the Knudsen equation:

$$Q_m = \left(\frac{2\pi RT}{m}\right)^{0.5}\left(\frac{D^3}{6L}\right)(P_1 - P_2)$$

(13.18)

where

Q_m = Volumetric leak rate of gas. [This equation is applicable for gas leak through gasket for sealed transformer with gas cushion]

R = universal gas constant
T = absolute temperature
M = molecular weight of gas
L = leak path length
D = leak path diameter
P_1 = Higher Pressure
P_2 = Lower Pressure

If the gas pressure inside the transformer becomes negative, the leak rate becomes negative showing flow of air from atmosphere into the transformer.

For the leakage of oil/liquid through gasket joint, the applicable equation is

$$Q_L = \pi \frac{\left(\dfrac{D}{4}\right)^4 (P_1 - P_2)}{8\mu L}$$

(13.19)

where

Q_L = Leak rate of oil/liquid
μ = Viscosity of oil/liquid

The design of a leak proof gasket joint will have to consider the following factors:

- properties of the gasket including load deflection curve
- correct fastener size and fastener spacing
- correct pre-stress on the gasket which is a function of the tightening torque
- surface finish of the metallic surface

To prevent leakage over a period of time, the pre-stress on the gasket will have to be maintained. This is achieved by the storage of elastic strain energy in the joint system. The elastic strain energy is received from the compression of the gasket, the flange deflection and the bolt strength in a joint.

13.5.3   Properties of Gasket Required for a Good Joint

- The internal structure of the gasket shall be elastic.
- The stress relaxation shall be low.
- The physical properties of the material shall not be affected by the fluid/oil/gas in the transformer. Also, the fluid/oil/gas shall not be affected by the gasket material.
- Property shall not be affected by temperature cycling
- At the lowest operating pressure, the gasket material shall not be porous
- The material shall conform to environmental regulations such as asbestos free, heavy metal free (i.e. free from lead, cadmium, mercury and chromium), etc.
- The material shall be capable of overcoming flange and surface imperfections such as distortion, waviness and roughness.

13.5.4   Types of Gaskets Used and Comparison of Properties

The common types of gaskets used for transformers including the applications are given in Table 13.2.

One of the most commonly used gaskets especially for power transformers is nitrile rubber conforming to Grade BO-70 as per BS 6996. The parameters of the material suitable for application of transformer are given in Table 13.3.

Nitrile rubber (NBR) and nitrile cork gaskets are most commonly used for oil-filled transformers. For distribution transformers, flat sealing gasket without grooves or stoppers is most commonly used where the bolt passes through gasket. In this case, nitrile cork gasket is most suitable.

TABLE 13.2
Comparison of Gaskets Used for Oil Filled Transformers

Sl. No.	Gasket Material	Temperature Range (°C)	Compression (%)	Typical Applications
1.	Neoprene	−40 to 90	30	Used where UV proof application is required. To be used with stoppers.
2.	Nitrile (NBR)	−40 to 140	25–50	Used with compression limiter or grooves. O-Rings flat and extruded/type available.
3.	Neoprene cork	−10 to 90	40	Can be used as flat-to-flat surface with no groove
4.	Nitrile rubber	−30 to 130	40	Can be used as flat-to-flat surface with no groove
5.	Viton	−25 to 200	30	For high-temperature O-Rings, flat and extruded gaskets. Used with grooves or recess or stoppers.

TABLE 13.3

Properties of Nitrile Rubber Gasket Material

Sl. No.	Property	Parameter	Test Method
1.	Type of polymer	Acrylonitrile butadiene	BS 4181
2.	Polymer content	45% (min) by weight	ASTM E 1131
3.	Ash content	7% max	ASTM E 1131
4.	Density	1.2 (±0.05) gm/cc	BS 903 part A1 method A
5.	Hardness	70 (0–4) shore A	BS 903 part A26 method N
6.	Tensile strength	12.5 N/mm² (min)	BS 903 part A2
7.	Elongation at break	250% (min)	BS 903 part A2
8.	Compression set	20% (max)	BS 903 part A6
9.	*Ageing in transformer oil*	±5 to 0%	BS 903 part A16
	Maximum change in volume	±5 Shore A	72 ± 2 hours at
	Change in hardness	±20%	100 ± 1°C in transformer Oil
	Change in tensile Strength	±15%	
	Change in elongation		

13.5.5 FASTENER SPACING

$$L = \left(\frac{480\ Y\ E\ t^3\ d}{13F_1 + 2F_2} \right)^{0.25} \times K \qquad (13.20)$$

$$F_o = \frac{(2F_1 + F_2)L}{3} \qquad (13.21)$$

L = distance between fasteners (m)
[For nitrile (NBR) and nitrile cork gaskets for transformer, the maximum spacing L is less than 110 mm]
Y = width of curb (m)
E = modules of elasticity of curb/cover plate (kg/m²)
t = thickness of cover plate (m)
d = deflection of gasket (m)
D = thickness of gasket (m)
F_1 = minimum force on gasket (kg/m)
F_2 = maximum (kg/m)
F_o = pre-stress to compress the gasket to 30%–40% of initial thickness (kg/m)
K = safety factor ≈ 0.8

It is assumed that the cover is parabolically loaded between two fasteners.

F_1 is the minimum force that will have to be maintained on the gasket to get a leak proof joint.

F_2 is the force required on the gasket to obtain a deflection "d" over and above the deflection obtained by applying force F_1.

A leak proof joint is realized when the gasket material flows into the imperfections of the contact surface. The amount of force to be applied on the contact surface for a good gasket flow is defined as the "Y" factor in ASME VIII. This is the yield force, and this is the minimum stress to be applied even for low-pressure application. When the internal pressure increases, the force on the gasket reduces. The ability of the gasket to maintain the seal at this reduced stress is defined as gasket factor "m" in ASME VIII.

13.5.6 BOLT TORQUE

The gaskets are usually purchased in rolls by the transformer manufacturer. The gasket manufacturers make butt joints in the roll and supply the material to the manufacturers; considering this in practice, the bolt hole shall be avoided at the gasket joint, if any. The correct pre-stress of the gasket is applied by torque. However, the relationship of torque and force depends on several factors and theoretical formula is not convenient for practical application. The relationship between force and torque can be represented as

$$T = \frac{F}{4}\left[\mu_H + (A+D) + \mu_t + (2d_e \sec\theta) + \frac{2\rho}{\pi}\right] \qquad (13.22)$$

where

T = torque (Nm)
F = axial load (kg)
A = across flat dimensions of the nut (m)
{Outside diameter of nut and washer in contact}
D = washer inside diameter (m)
μ_H = friction coefficient under the head of the nut
μ_t = friction coefficient of the thread
d_e = effective diameter of the bolt (m)
P = thread pitch
θ = half thread form angle

μ_H and μ_t change with various factors including lubrication. The above formula is simplified for practical application as

$$\text{Torque} = 0.2 \times \text{Load} \times \text{Diameter} \qquad (13.23)$$

When installing the gasket, initial tightening torque is necessary to give pre-stress to ensure leak proof joint at the lowest pressure of operation and also to ensure that the gasket is not overstressed at the maximum operating pressure. The pre-stress for the nitrile rubber and nitrile cork gaskets used in transformer is to limit the gasket compression to about 40%. The typical torque values are given in Table 13.4 based on the simplified formula.

13.5.7 JOINING/SPLICING OF THE GASKET

The distribution transformer manufacturers generally buy gasket in standard rolls. When this gasket is used for the top cover, joints at the four corners will be necessary and the common practice is to use cold splicing for the joint.

TABLE 13.4
Torque Values for Different Bolts

Size of the Bolt/Stud (mm)	Torque Required (Nm)
M10	20
M12	30
M16	70
M20	150
M24	200

Cold splicing uses quick setting glue, and the process does not require any special tools. However, the joint is a weak link in the system. The preferred practice is to use "hot splicing" where a rubber-based glue is used which has to be cured (vulcanized) in place. This creates a joint that retains the properties of the gasket at the joint, and the joint is not visible.

13.5.8 THICKNESS OF GASKETS FOR DISTRIBUTION TRANSFORMERS

The thickness of gasket for tank cover, cable box fixing, monoblock bushing fixing, cable box cover plate fixing on distribution transformers is dependent on the type of gasket used, the thickness of flange, fastener size and fastener spacing. The thickness given in Table 13.5 can be used as a general guidance.

TABLE 13.5
Gasket Thicknesses

Gasket Location	Flange Thickness (mm)	Typical Gasket Thickness (mm)
Tank cover	6–8	6
Tank cover	16	12
Cable box fixing	8	6
Cable box cover	3–4	3
Monoblock bushing fixing	6–8	6

14 Calculation of the Short Circuit Forces and Strength of Transformers

14.1 INTRODUCTION

Transformers may experience different types of short circuits in their lifetime. The short circuits can be 3-phase line to line, line to ground or double earth faults. The faults produce dynamic forces on the windings and components and also short-time temperature rise of the conductor bushings, bus bars, etc. In view of this, the design shall be capable of withstanding the thermal and dynamic effects of the short circuit. IEC 60076-5 specifies the requirements to demonstrate the ability of the transformers to withstand short circuits. The thermal ability and the dynamic ability shall be demonstrated considering the most onerous short circuit current.

14.2 CALCULATION OF THERMAL ABILITY TO WITHSTAND SHORT CIRCUITS

a. The thermal ability to withstand short circuits is demonstrated by calculation as per IEC 60076-5. Some customers can ask for demonstration of this by actual test.

b. Duration of short circuit current:
   The duration of short circuit current to be used for the calculation of thermal ability to withstand short circuits is 2 seconds unless otherwise specified by the purchaser.

c. Calculation of symmetrical short circuit current for transformers with two windings:

The rms value of the symmetrical short circuit current I (measured in kilo ampere) is calculated for 3-phase transformers as follows:

$$I = \frac{U}{(Z_t + Z_s)\sqrt{3}} \tag{14.1}$$

$Z_s = \dfrac{U_s^2}{S}$ (short circuit impedance of the system in ohms, per phase).

U_s = rated voltage of the system in kV.

S = short circuit apparent power of the system, in megavolt amperes.

U and Z_t are defined as follows:

- For rated tapping:
  U is the rated voltage U_n of the winding under consideration in kV
  Z_t is the short circuit impedance of the transformer referred to the winding under consideration

  $Z_t = \dfrac{U_z\, U_n^2}{100\, S_n}$

  U_z = impedance voltage at rated current and at the reference temperature as a percentage
  S_n = rated MVA of the transformer

- For tappings other than rated tapping:
    U = tapping voltage of the winding under consideration
    Z_t = short circuit impedance of the transformer referred to the winding and tapping under consideration (ohms per phase).

Note

- For transformers of category I, impedance of the system is neglected in the calculations if it is equal to or less than 5% of the short circuit impedance of the transformer.
- If the short circuit power of the system is not specified by the purchaser in the enquiry, its value can be taken from Table 14.1.

TABLE 14.1

Short Circuit Power of the System [a]

Highest System Voltage (kV)	Short Circuit Power (MVA)
7.2, 12, 17.5 and 24	500
36	1000
52 and 72.5	3000
100 and 123	6000
145 and 170	10,000
245	20,000
300	30,000
362	35,000
420	40,000
525	60,000
765	83,000

[a] Copyright permission from IEC 60076-5.

---

d. Maximum permissible value of the highest average temperature θ_2:
    On the basis of an initial winding temperature θ_o, calculated by adding the maximum permissible ambient temperature and the winding temperature rise measured (or if measured winding temperature rise is not available, the guaranteed winding temperature rise), the highest average temperature θ_1 of the winding after loading with a symmetrical short circuit current I calculated as in 14.2.C above and of duration "t" seconds shall not exceed the value of θ_2 in any tap position, as shown in Table 14.2.

e. Calculation of θ_1:

$$\theta_1 = \theta_0 + \frac{2(\theta_o + 235)}{\dfrac{106000}{J^2 t} - 1} \; (\text{Copper}) \tag{14.2}$$

$$\theta_1 = \theta_0 + \frac{2\,(\theta_o + 225)}{\dfrac{45700}{J^2 t} - 1} \; (\text{Aluminium}) \tag{14.3}$$

The calculated values of θ_1 of any winding and any tap position shall be less than θ_2 as per Table 14.2. When the impedance of the transformer (e.g. a special transformer or autotransformer) is low, the

TABLE 14.2

Maximum Permissible Value of Winding Temperature after Short Circuit

Transformer Type	Temperature Class of Insulation	Maximum Permissible Value of Average Temperature of Winding after Short Circuit θ_2	
		Copper (°C)	Aluminium (°C)
Oil-immersed	105 (A)	250	200
Dry	105 (A)	180	180
Dry	120 (E)	250	200
Dry	130 (B)	350	200
Dry	150 (F)	350	200
Dry	180 (H)	350	200
Dry	200	350	200
Dry	220	350	200

[a] Copyright permission from IEC 60076-5.

short circuit winding temperature rise can exceed permissible limits, especially when the design current density is comparatively high.

14.3 ABILITY TO WITHSTAND THE DYNAMIC EFFECT OF SHORT CIRCUITS

When a short circuit happens on a transformer, heavy short circuit currents flow thorough the windings. The peak value of the short circuit current depends on the time of short circuit on the alternating waveform and X/R ratio of the transformer. The forces experienced by the windings are proportional to the square of the peak fault current on the winding.

As per standard, the ability to withstand the dynamic effect of short circuit is demonstrated by the following:

a. Test on the transformers
b. Calculation, design and manufacturing considerations.

The ability to withstand the dynamic short circuit can be established by comparing the design of a transformer with the test results of similar transformers or by design review, if test reports of a similar transformer are not available. The calculation of the dynamic forces and stresses is given in this section.

a. Calculate peak value of asymmetrical short circuit current.

$$I_{SC} = K\sqrt{2}\ \frac{I_{ph}}{Z} \tag{14.4}$$

I_{SC} = peak value of asymmetrical short circuit current (A)
$K\sqrt{2}$ = asymmetry factor based on $\dfrac{X}{R}$ ratio
X = percentage reactance
R = percentage resistance

The value of "K" can be calculated by the following formula:

$$K = 1 + Sin\theta \left(e^{-\left(\phi + \frac{\pi}{2}\right)\frac{R}{X}} \right)$$

$$\phi = Tan^{-1} \frac{X}{R}$$

Z = per-unit impedance (for category 1 transformer, system impedance is neglected).

The values of $K\sqrt{2}$ for some X/R ratios are given in Table 14.3. Intermediate values can be extrapolated or calculated from the above formula.

b. Calculate I_{SC} for HV and LV

c. Calculate asymmetrical peak short circuit ampere turns = NI_{SC}

N = number of turns

I_{SC} = peak asymmetrical short circuit current (use the number of turns and short circuit values of the same winding)

TABLE 14.3
X/R Ratio and $k\sqrt{2}$

$\dfrac{X}{R}$	$K\sqrt{2}$
	2.828
1000	2.824
500	2.820
333	2.815
250	2.811
200	2.806
167	2.802
143	2.798
125	2.793
111	2.789
100	2.785
50	2.743
33.3	2.702
25.0	2.662
20.2	2.624
16.7	2.588
14.3	2.552
12.5	2.518
11.1	2.484
10.0	2.452
5.0	2.184
3.33	1.990
2.5	1.849
2.0	1.746
1.67	1.669
1.43	1.611
1.25	1.568
1.11	1.534
1.00	1.509

---

d. Calculate maximum hoop stress in winding

$$\sigma_m = \left[\frac{\gamma I_{ph}^2 R_{dc}}{h_w \, (ez)^2} \right] \text{kg/cm}^2 \tag{14.5}$$

σ_m = Hoop Stess kg/cm^2

$\gamma = 0.03 \left(\dfrac{K\sqrt{2}}{2.55} \right)^2$ for copper

R_{dc} = DC resistance/phase at 75°C
H_w = winding axial height in cm
This shall be less than 1250 kg/cm^2

e. Radial bursting force (outer winding)

$$F_r = \frac{2\pi \, \sigma_m \, I_{ph} \, N}{\delta} \, \text{kg} \tag{14.6}$$

F_r = Radial bursting force kg
σ_m = Hoop stress kg/cm^2
δ = Current density A/cm^2 (rated current density)
N = turns/phase in the circuit

f. Number of supports to be provided in inner winding with flat conductor

$$N_s = \frac{D_{mi}}{b_i} \sqrt{\frac{12\sigma_m}{E}} \tag{14.7}$$

$\in = 1.13 \times 10^6$ kg/cm^2 (copper)
D_{mi} = Mean diameter of inner winding (cm)
b_i = Thickness of inner winding conductor (cm)

g. Number of supports to be provided in inner winding with round conductor

$$N_s = 8 \frac{D_{mi}}{d_i} \sqrt{\frac{\sigma_m}{\pi \in}} \tag{14.8}$$

d_i = Diameter of inner winding conductor cm

h. Internal axial compression

$$F_c = \frac{-34 \, (\text{kVA})}{e_z h_w} \text{kg} \tag{14.9}$$

The $-$ sign indicates that force is compressive towards the axial centre of the winding
$\dfrac{1}{3} F_c$ is acting on the outer winding

$\dfrac{2}{3} F_c$ is acting on the inner winding

i. Axial unbalance force due to tapping

$$F_a = \frac{a}{2} \left(N\, I_{SC} \right)^2 \times 10^{-7}\, kg$$

a = per-unit turns out of circuit in the winding

If taps are at both ends of the winding, F_a is $\frac{1}{4}$ th of the above value.

If taps are from the middle of the winding and compensating gap is given on the untapped winding, F_a will be 50% of the value calculated by the above formula.

j. Calculation of maximum compressive pressure in the radial spacers.

$$P_i = \frac{F_a + \frac{2}{3} F_c}{A_i} \tag{14.10}$$

$$P_o = \frac{Fa + \frac{1}{3} Fc}{Ao} \tag{14.11}$$

A_i = total supported area of inner radial spacer in cm^2
P_i = compressive pressure on inner radial spacer in kg/cm^2
P_o = compressive pressure on outer radial spacer in kg/cm^2

k. Tensile stress on tie rod

$$P_t = \frac{F_a - \frac{1}{3} F_c}{N_t\, A_t} \tag{14.12}$$

N_t = number of tie rods
A_t = area of cross section of tie rod in cm2
P_t = tensile stress on tie rod in kg/cm^2

l. Resistance to collapse of disc winding with rectangular conductor

$$F_{crit} = \frac{1.5E \left(I_{ph} \right)^2 m}{b_o D_{mo} \delta^2 \times 10^8} + \frac{450\, A_v\, \delta b^3}{I_{ph}} \tag{14.13}$$

F_{crit} = strength of winding tons against collapse
E = modulus of elasticity of conductor in kg/cm^2
m = number of turns per disc x number of parallel conductors per turn of the disc
I_{ph} = rated current in A
b_o = thickness of the outer winding conductor in cm
D_{mo} = mean diameter of the outer winding in cm
δ = current density in A/cm^2
A_v = total supported area of outer radial spacer in cm^2

m. Most highly stressed coil in tapped winding

$$f_a = 0.733 \ Q \ F_r \log_{10} \left(2a \, N_c + 1\right) \tag{14.14}$$

f_a = radial bursting force on the most highly stressed coil in tapped winding (Tons)
Q = turns/coil adjacent to tapped coil expressed as a fraction of total turns in the limb
F_r = radial bursting force in Tons
a = per-unit number of turns out of circuit
N_c = number of discs/limb

$$W_1 \left(\text{Mechanical loading/cm of periphery}\right) = \frac{f_a \times 1000}{\pi D_m} \tag{14.15}$$

D_m = mean diameter of tapped winding in cm
Add 25% extra for force concentration
$\therefore W = 1.25 \ W_1$

$$\sigma_{\max} = \frac{(1.25 \ W_1) L^2 Y}{12 \ I_o} \tag{14.16}$$

$\sigma_{\max}$ = maximum bursting stress in tapped coil (kg/cm²)
$L = (\text{span}) = \dfrac{\pi D_m}{n_s} - b_s \ (\text{cm})$

n_s = number of spacers
b_s = width of the spacer in cm
Y = maximum distance from neutral axis for conductor in cm
I_o = moment of inertia of the coil $= \dfrac{bd^3}{I_2}$
b = radial depth of the coil (disc) in cm
d = axial height of the coil (disc) in cm
The maximum permissible value of $\sigma_{\max} = 1250 \, \text{kg/m}^2$

n. Bending stress on clamping ring

$$\sigma_{\max} = \frac{6\pi \ F \ D}{8 \ bt^3 \ n^2} \tag{14.17}$$

Total axial force $= F_a - \dfrac{1}{3} F_c$ in kg

D = diameter of the ring in cm
b = width of the ring in cm
t = thickness of the ring in cm
n = number of jacking points
The maximum value for permawood ring is 1100 kg/cm² (for circular ring)

o. Thrust force acting on the low-voltage winding lead exits
    When the low-voltage winding lead exit (i.e. the lead take out) is subject to the thrust force exerted by any winding (e.g. if the lead exit is below a clamp ring), the stress on the lead exit will have to be calculated.
    Thrust force acting on the low-voltage winding lead exit $= \sigma \times A$

where

σ = mean hoop compressive stress on LV winding

A = LV winding lead area under compression

If the LV winding leads are taken to the top directly from the outside and inside layers of a foil winding, the hoop compressive stress is not applied to the leads.

p. Compressive stress on conductor paper insulation and radial spacers

$$P_{LV} = \frac{F_a + K_l \times F_c}{A_{LV}} \tag{14.18}$$

$$P_{HV} = \frac{F_a + K_h \times F_c}{A_{HV}} \tag{14.19}$$

K_l = axial force sharing by LV = 2/3 of axial force

K_h = axial force sharing by HV = 1/3 of axial force

P_{LV} = compressive stress on LV radial spacer (kg/cm^2)

P_{HV} = compressive stress on LV radial spacer (kg/cm^2)

F_a = axial unbalance force due to tapping (as per Section 14.3.9)

F_c = internal axial compression (as per Section 14.3.8)

A_{LV} = support area of LV radial spacer

A_{HV} = support area of LV radial spacer

14.4 DESIGN REVIEW AND EVALUATION

Design review and evaluation when a comparison with the test results of a similar transformer is available is covered in Section 14.4.1, and Section 14.4.2 covers the situation where no type-tested transformer of similar design is available for comparison.

14.4.1 COMPARATIVE EVALUATION WITH A TYPE-TESTED TRANSFORMER OF SIMILAR DESIGN

A transformer is considered to be similar to another transformer if it has the characteristics as per the definition of similar transformer as per Annex B of IEC 60076-5.

A core-type transformer is considered to withstand the dynamic effect of short circuit if the calculated stresses are not exceeding the stresses of the type-tested design as per Table 14.4.

TABLE 14.4
Permissible Limits of Stress/Force

Type of Stress/Force	Permissible Stress/Force
Mean hoop compressive stress on disc, helical or layer winding	≤1.1 times the type-tested design
Equivalent mean hoop compressive stress on multilayer winding	≤1.1 times the type-tested design
Thrust force acting on the low-voltage winding lead exits	≤1.1 times the type-tested design
All other forces and stresses	≤1.2 times the type-tested design

14.4.2 EVALUATION BY CHECK AGAINST THE MANUFACTURER'S DESIGN RULES FOR SHORT CIRCUIT STRENGTH

When comparison with a type-tested transformer of similar design is not possible, evaluation by checking against manufacturer's design rules can be carried out by agreement between the purchaser and the manufacturer. In this case, the manufacturer should have a solid basis for establishing the design rules from tests done on several transformers, tests on models, and supportive evidence such as long-duration trouble-free operation of a number of transformers. The critical values of each stress established by the manufacturer will have to be compared with the calculated stresses and allowed stresses.

The allowed stresses or forces for a core-type transformer are given in Table 14.5.

$R_p = 0.2\%$ proof stress.

TABLE 14.5
Allowed Stresses or Forces for a Core-Type Transformer

Type of Stress/Force	Permissible Stress/Force
Mean hoop tensile stress for disc, helical or individual layer of multilayer winding	$\leq 0.9\, R_{p0.2}$
Mean hoop tensile stress for disc, helical and layer winding	
a. With CTC (non-bonded)	$\leq 0.35\, R_{p0.2}$
b. Resin-bonded CTC	$\leq 0.60\, R_{p0.2}$
Stress due to radial bending or axial bending on conductors between axial spacers	$\leq 0.9\, R_{p0.2}$
Compressive force on radial spacers	$800\ \text{kg/cm}^2$
a. with paper covered conductor	$1200\ \text{kg/cm}^2$
b. with enameled wire	
Compressive stress on conductor paper insulation with layer winding	$350\ \text{kg/cm}^2$
Compressive stress on wound-type pressboard end rings	$400\ \text{kg/cm}^2$
Compressive stress on stacked-type pressboard end rings	$800\ \text{kg/cm}^2$
Tensile stress on tie rods (steel rods)	$2500\ \text{kg/cm}^2$
Tensile stress on flitch plates (steel plate)	$2500\ \text{kg/cm}^2$

15 Rectifier Transformers

15.1 WINDING ARRANGEMENTS AND HARMONICS PRODUCED

When the secondary of a transformer is connected to rectifiers, harmonics are generated. For a rectifier system with "P" pulses, the harmonics generated are given by

$$m = nP + 1$$

where
m = harmonic number
n = 1, 2, 3,...
P = number of pulses of a rectifier.

For a 3-phase full-wave rectifier, the number of pulses is six, and therefore, the pulses are spaced at an angle of 60° between the adjacent ones. The primary supply to the transformer is normally delta connection for 6- and 12-pulse operations. However, zigzag connection, extended delta connection or polygon connection on the primary side can also be used. When a rectifier transformer for 18 pulses or more is required, several design options are available. The impedance between HV and each one of the secondary windings shall be equal, and therefore, a loosely coupled axi-symmetrical winding arrangement is required.

Table 15.1 gives the typical winding arrangement, harmonics produced and the angle between pulses of 6-, 12-, 18- and 24-pulse rectifier transformers.

TABLE 15.1
Winding Arrangement and Harmonics Present in Typical Rectifier Transformers

Number of Pulses Required	Number of Secondary Windings	Harmonics Produced	Angle between Pulses	Typical Winding Arrangement	
				Primary	Secondaries
6	1	5, 7, 11, 13,...	60	Delta	Star or delta
12	2	11, 13, 17, 19, 23,...	30	Delta	Star
					Delta
18	3	17, 19, 23, 25,...	20	Delta	Extended star-20
					Delta
					Extended star+20
24	4	23, 25, 29, 31,...	15	Delta	Star
					Extended star-15
					Delta
					Extended star+15
24 (two core and coil assemblies in one tank)	2 × 2 = 4	23, 25, 29, 31,...	15	Extended delta −7.5°	Star
					Delta
				Extended delta +7.5°	Star
					Delta

15.2 CALCULATION OF THE NUMBER OF TURNS WHEN EXTENDED STAR OR EXTENDED DELTA WINDING IS USED

By using extended winding, any desired phase shift can be realized. The calculation procedure of the number of turns of the main winding and the extended section winding is explained by a calculation (Figure 15.1).

Design parameters
Required phase angle shift = 7.5°
Volt/turn = 12.5
Phase volts = 250 V
OA = 250 V
Angle OPA = 120°
Angle POA = 7.5°
∴ Angle OAP = 180° − (120° + 7.5°) = 52.5°

$$\frac{OA}{Sin\ 120} = \frac{OP}{Sin\ 52.5} = \frac{AP}{Sin\ 7.5}$$

$$\frac{250}{\sqrt{3}/2} = \frac{OP}{0.7934} = \frac{AP}{0.1305}$$

Solving,
OP = 229.04 volts
AP = 37.67 volts

Volts/turn is known, and therefore, the number of turns required for the windings is

$$\text{Turns of OP} = \frac{229.04}{12.5} = 18\ \left(\text{Nearest Integer}\right)$$

$$\text{Turns of AP} = \frac{37.67}{12.5} = 3\ \left(\text{Nearest integer}\right)$$

For getting phase shift of delta connection, either a polygon connection or extended delta connection can be used. The winding arrangements of these connections are shown in Figures 15.2–15.5.

When using the extended delta connection, the losses and impedance calculations shall be done considering the fact that the line current flows through the extended winding.

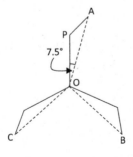

FIGURE 15.1   Extended winding for a 7.5° shift.

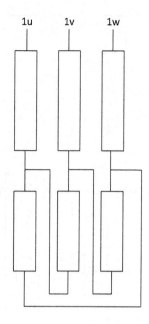

FIGURE 15.2   Extended delta connection.

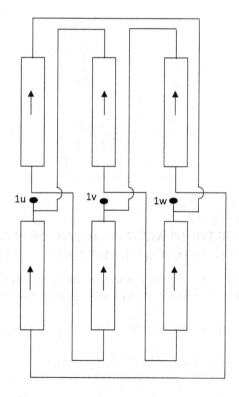

FIGURE 15.3   Polygon connection.

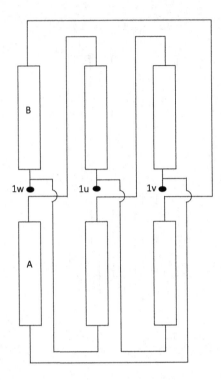

FIGURE 15.4 Polygon connection Pd0 +7.5.

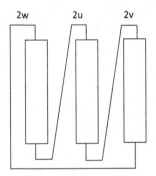

FIGURE 15.5 Delta connection.

15.3 CALCULATION OF THE NUMBER OF TURNS OF POLYGON DELTA CONNECTION WITH VECTOR GROUP PD0 +7.5 (7.5° LAG)

To realize a 7.5°C phase shift with polygon connection, the turns ratio of the polygon limbs will have to be in a defined ratio. The number of turns of limbs is shown as A and B, respectively, in Figure 15.4.

The following calculation shows that the ratio of the number of turns in A and B is 61:10.

$$A = \frac{\mathrm{Sin}\,(60 - 7.5)}{\mathrm{Sin}(120)} = 0.916$$

$$B = \frac{\mathrm{Sin}\,7.5}{\mathrm{Sin}\,120} = 0.151$$

$$\therefore \frac{A}{B} = \frac{0.916}{0.151} = 6.066 \cong 6.1$$

$$\therefore \text{ If } A = 1 \qquad B = \frac{1}{6.1} \text{ (Turns)}$$

$A = 61; B = 10.$

15.4 DETERMINATION OF THE PHASE DISPLACEMENT AND RATIO BY SINGLE-PHASE TURNS RATIO MEASUREMENT

The phase displacement and ratio can be determined by measuring the single-phase ratio by feeding in a cyclic permutation in HV phases with the other two phases short-circuited and measuring the ratio with respect to LV phases. The procedure is shown by an example.

Example of Single-Phase Ratio Measurement

Short V and W of the HV side.
  Apply single-phase voltage across U & V-W.
  Measure the ratio with u – v of LV winding.
  This ratio is termed K_b.
  With the HV side supply kept the same, measure the ratio with v – w of LV winding.
  This ratio is termed K_c.
  Repeat the same procedure by shorting WU of HV and then by shorting UV of HV in a cyclic manner.
  The ratios are tabulated in Table 15.2.
  The average of the three values of K_b and the average of three values of K_c are considered for the calculation of ratio and phase displacement:

$$\text{Ratio} = \frac{K_b}{1 + \left(\frac{K_b}{K_c}\right)^2 - \frac{K_b}{K_c}} \tag{15.1}$$

The phase difference θ is calculated from the following formula:

$$\text{Tan}\theta = \frac{\sqrt{3}\left(1 - \left(\frac{K_b}{K_c}\right)\right)}{\left(1 + \left(\frac{K_b}{K_c}\right)\right)} \tag{15.2}$$

The above formulas can be used for all winding connections such as star, delta, zigzag and polygon.

TABLE 15.2

Turns Ratio Measurement

K_b	K_c	K_b	K_c	K_b	K_c
U – VW	U – VW	V – WU	V – WU	W – UV	W – UV
u – v	v – w	v – w	v – u	w – u	w – v

15.5   CALCULATION OF RATIO AND PHASE ANGLE ERROR

When extended windings (e.g. star, delta) are used for the desired phase shift, the number of turns of the main winding and extended winding will have to be selected to get the minimum deviation of the ratio and phase shift. This becomes complex when the number of pulses increases and the primary winding requires taps.

The winding design of an 18-pulse transformer is given as an example.

Transformer details are as follows:

kVA	1000
No. of phases	3
Frequency	50 Hz
Cooling	ONAN
Primary voltage	11000 ± 5% in 2.5% steps V
Secondary voltage	1902 V
Secondary connection	$d^{5\circ}$, $d^{15\circ}$, $d^{25\circ}$

Calculations

The secondary windings are axially placed on the limb, and the primary is three axially placed parallel connected coils per phase. Extended delta-connected winding design is selected for the 5°, 15° and 25° phase shifts.

Volt per turn selected ≈12.85.

In this design, all the secondary windings are with extensions, and therefore, the turns of the primary windings are fixed initially.

$$\text{Number of turns for primary rated tap} = \frac{11000}{\sqrt{3} \times 12.85} = 494$$

The primary tap turns are now calculated and tabulated in Table 15.3.

Voltages and turns of secondary (LV) windings are as follows:

Secondary line voltage = 1902 V

Secondary 1 has 5° phase shift.

Then, $\dfrac{1902}{\text{Sin } 120} = \dfrac{V_1}{\text{Sin } 55} = \dfrac{V_2}{\text{Sin } 5}$

where

V_1 = voltage of main winding
V_2 = voltage of extended winding

TABLE 15.3
Primary Tap Turns and Voltage

Primary Tap Number	% Phase Voltage	Line Voltage	Turns per Phase
1	105.0	11550	519
2	102.5	11275	507
3	100.0	11000	494
4	97.5	10725	482
5	95.0	10450	470

TABLE 15.4

Secondary Turns and Voltages

Winding	Phase Shift Req.	LV Main Volts	LV Ext. Volts	LV Main Turns	LV Ext. Turns	LV Actual Volts	LV Nominal Volts	LV Actual Phase Shift
Secondary 1	5°	1607	191.41	125	15	1920.87	1902	5.03
Secondary 2	15°	984.6	586.4	77	44	1901.81	1902	14.921
Secondary 3	25°	331.5	928.2	26	72	1899.34	1902	24.954

TABLE 15.5

Ratio Errors Corresponding to the Five Primary Taps with Three Secondary Voltages

		Nominal Ratio and Ratio Error %					
		5°		15°		25°	
Tap Number	Voltage	Ratio	Ratio Error	Ratio	Ratio Error	Ratio	Ratio Error
1	11550	6.071	0.033	6.074	−0.016	6.082	−0.148
2	11275	5.931	−0.051	5.934	−0.101	5.942	−0.286
3	11000	5.779	0.069	5.782	0.017	5.789	−0.104
4	10725	5.638	0.018	5.641	−0.035	5.649	−0.177
5	10450	5.495	−0.073	5.501	−0.127	5.508	−0.255

Solving,

$V_1 = 1607.7$ V

$V_2 = 191.41$ V.

Similarly, the voltages of the second and third secondaries are calculated. The number of turns of the secondary, main and extended windings is calculated using the volt per turn and rounding off to the nearest integer as shown in Table 15.4.

The ratio errors corresponding to the five primary taps with three secondary voltages are calculated and given in Table 15.5.

The phase angle errors and the ratio errors are within tolerance limits.

15.6 TRANSFORMERS FOR VARIABLE-SPEED DRIVES (VSDs)

When a wide range of speed control is required for a motor, such as rolling mill drives, ship propulsion, and hoist drives, variable-speed drives are used. The transformer supplying to such variable speed drives are called VSD transformers. Some VSD transformers are equipped with an additional winding for connecting to a filter to reduce harmonic distortion and power factor improvement. The number of secondary windings depends on the requirement of pulses. Typical arrangement of a 12-pulse VSD transformer with filter winding is shown in Figure 15.6.

15.7 EFFECT OF WINDING GEOMETRY ON LOAD LOSSES

The ampere-turn balancing and eddy current losses of the windings of a rectifier transformer are affected by the winding geometry and design. The ampere turns are balanced in a 2-winding transformer (excluding the effect of no-load current). If harmonics are produced in the secondary

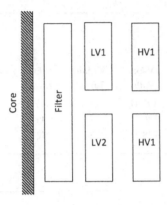

FIGURE 15.6   Winding arrangement of a 12-pulse VSD transformer.

TABLE 15.6

Common Configuration of a Rectifier Transformer with Two Secondary Windings and the Effects on Harmonics and Losses

Type of Coupling between Secondaries	Winding Configuration	Effect on Harmonics and Stray Losses
Close coupling	Interleaved secondary windings and one primary winding	Very low leakage flux between secondary windings
No coupling	a. Secondary windings are separated by an intermediate yoke b. Two core and coil assemblies	• Harmonics are balanced • Produces eddy current losses and stray losses
Loose coupling	a. Concentric arrangement of the secondary with LV–HV–LV arrangement b. Axially placed secondary windings	a. Harmonics in opposition produce eddy losses and stray looses b. The effect is almost similar to no coupling

winding, the ampere turns are balanced on the primary winding in equal magnitude. Due to this, the eddy losses on the primary and secondary have almost the same per unit magnitude. When the transformer has two or more secondary windings, the harmonic generation depends on the winding placement and coupling.

The common configuration of a rectifier transformer with 2 secondary windings and the effects on harmonics and losses are shown in Table 15.6.

When the rectifiers produce harmonics in loosely connected secondary windings, the radial component of leakage flux at the top and bottom ends of windings has a high magnitude. The eddy current losses produced by the winding conductors at the ends are therefore proportional to the square of the width of the conductor. If the secondary is made with foil conductor, the eddy current loss distribution needs detailed evaluation. When rectangular conductors are used, the end coils can be made with conductors of low width. Another practice is to make the end coils by "windings on the edge".

15.8   CALCULATION OF LOAD LOSS

The actual load loss in service of a rectifier transformer includes the additional eddy current losses from the harmonics and stray losses in metal parts. The actual losses after considering the effect of harmonics shall be the base for calculating and testing the winding temperature rise. The oil and winding temperature rises shall not exceed the values permitted by IEC 60076-2 for oil-filled transformers and IEC 60076-11 for dry-type transformers or the customer requirement as applicable.

The load loss at rated current is subdivided into the following four parts:

    i. I^2R = DC loss in winding and connection bus bars = A
    ii. Eddy current loss in windings from fundamental and harmonic currents = B
    iii. Eddy current loss in high current bus bars = C
    iv. Stray loss in structural parts = D

$$\text{Total loss} = A + B + C + D$$

A is calculated for each winding separately and added up.

Different windings of a rectifier transformer may have different proportions of I^2R losses and eddy losses. In extended delta connected winding, line current flows in the extended winding, and this aspect will have to be considered for the DC loss.

B is calculated by the eddy current losses of all windings arising from the fundamental power frequency and the harmonics. These losses are proportional to the square of the amplitude of the harmonic current and the square of harmonic number.

C is the eddy current loss from high current bus bars. When the bus bar or lead wire thickness is high, the eddy current losses from fundamental and harmonics are high.

D is the stray loss induced in structural steel parts including clamping structures and tank.

$$D \propto \sum_{1}^{n} \left(\frac{I_h}{I_1}\right)^2 h^{0.8}$$

where
I_h = current corresponding to the harmonic number, n
I_1 = rms load current of the transformer
h = harmonic order number

15.9 DUTY CYCLES FOR DIFFERENT APPLICATIONS

The rectifier transformers are used for several diverse applications, where the duty cycles are different. Examples of some applications and the duty cycles are given in Table 15.7.

TABLE 15.7
Duty Cycles and Examples of Loads

Duty Class	Typical Application	Rated Current
I	Electrochemical industry	1.00 p.u. continuously
II	Electrochemical industry	1.00 p.u. continuously + 1.5 p.u. for 1 minute
III	Light industry	1.00 p.u. continuously + 2.0 p.u. for 10 seconds
IV	Heavy-duty industry	1.00 p.u. continuously + 1.5 p.u. for 2 hours + 2.0 p.u. for 10 seconds
V	Mining, medium-traction substation	1.00 p.u. continuously + 1.5 p.u. for 2 hours + 2.0 p.u. for 1 minute
VI	Heavy-traction substation	1.00 p.u. continuously + 1.5 p.u for 2 hours + 3.0 p.u. for 1 minute

15.10   DESIGN EXAMPLE OF A RECTIFIER TRANSFORMER WITH THREE SECONDARIES

Transformer specifications
kVA – 3500 kVA
Frequency – 50 Hz
Cooling – ONAN
Vector group – $Dd^0d^{+20}d^{-20}$
Primary voltage – 6600 V (delta connection)
Secondary voltage – 2300 V
Percentage impedance – 7.25%
Winding material – Copper
Top oil temperature – 60°C
Average winding temperature rise – 65°C

Core Design
Core circle diameter – 290 mm
Volt/turn – 22.772
Working flux density – 1.68 T
Core window height – 845 mm
Core leg centre to centre distance – 525 mm

Winding configuration

- The three secondary windings are placed axi-symmetrical with winding with zero shift in the middle.
- The three secondary windings have several lead connections because of the extended delta windings of two of the secondaries. For convenience of these connections, all the three secondary windings are designed as the outer windings and 6.6 kV primary is the inner winding.

   For balancing of the ampere turns with the secondaries with phase shift, the primary winding is designed in three parallels of 3500/3 kVA each placed axially. The axial spacing between the coils of the primary and secondary is kept same to reduce the radial leakage flux and for magnetic balancing.

   The winding arrangement is shown in Figure 15.7.

   The winding design and calculated parameters are shown in Table 15.8 and I^2R loss calculations are shown in Table 15.9.

- Electrical clearances are shown in Table 15.10.
- I^2R loss in connection, bus bars + leads = 415 W
- Eddy current loss in windings from fundamental and harmonic currents = 2645 W
- Eddy current loss in high-current bus bars = 21.0 W
- Stray loss in structural parts = 1850 W
- Total load loss at full load = 34175 W

Note

- For designing the cooling surface area, the total sum of load losses including the losses from eddy currents and stray losses plus the no-load loss will have to be considered.
- When doing the temperature rise test, the fundamental power frequency current will have to be adjusted upwards to feed the total losses when the transformer is operating as rectifier transformer.

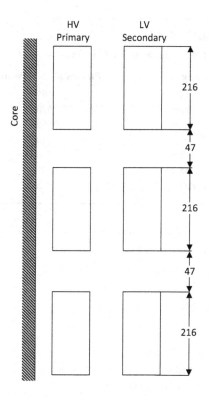

FIGURE 15.7   Primary and secondary winding arrangement.

TABLE 15.8
Winding Dimensions and Calculated Parameters

Particulars	HV (Primary)	LV1 (Secondary d+20°)	LV2 (Secondary d0)	LV3 (Secondary d-20°)
Line voltage (V)	6600	2300	2300	2300
Phase current (A)	176.8	293	293	293
Voltage/limb	6600	797 main	2300	797 main
		911 extended		911 extended
Inner diameter (mm)	305	416	416	416
Outer diameter (mm)	380	515	515	515
Radial thickness (mm)	37.5	49.5	49.5	49.5
Mean length of turn	1076	1462	1462	1462
Resistance/phase at 75°C (ohms)	0.14	0.0177 main	0.06655	0.0177 main
		0.0138 extended		0.0138 extended
I^2R loss (watts)	13185	1520+3555	5720	1520+3555

[a]   All dimensions are in millimetres.

Calculated Results

$\%X = 6.87$
$\%R = 0.97$
$\%Z = 6.94$

TABLE 15.9
Total I^2R Loss at 75°C in Watts

Winding	I^2R Loss at 75°C (Watts)
HV (three coils)	13185
LV1 – main	1520
LV1 – extended	3555
LV2	5720
LV3 – main	1520
LV3 – extended	3555
Total I^2R	29055

TABLE 15.10
Electrical Clearances

	Clearances
Core to HV (primary)	7 mm
HV to LV	18 mm
Phase to phase	10 mm
Axial distance between coils	47 mm
Clearance between main winding and extended winding	5.5 mm
Insulation between winding layers of LV	0.25 mm
Insulation between HV layers	0.9 mm
Insulation distance from winding to core yoke	51.5 × 2

16 Cast Resin Transformers

16.1 BASIC DESIGN PARAMETERS

16.1.1 MAXIMUM ELECTRICAL STRESSES

Maximum electrical stresses are given in Table 16.1.

16.1.2 INSULATION DESIGN

The minimum insulation clearances and winding arrangements for various voltage classes are given in Table 16.3.

16.1.2.1 Clearances to Enclosure and Live Parts

Clearances to enclosure and live parts are given in Table 16.2.

TABLE 16.1
Maximum Electrical Stress

Location	Maximum Stress Allowed (kV/mm at Test Voltage)	Remarks
Interlayer insulation/interturn	0.80–0.95	Mylar insulation. 0.95 can be used if mylar is in two or more layers
Voltage stress in air (creepage)	0.20	
Voltage stress in air gap (composite insulation)	0.80	For limiting PD, maximum stress will have to be less than 0.8 kV/mm
Voltage stress in resin	0.95	Preferred to limit this to 0.8 kV/mm

Note: Width of mylar insulation = foil width + 10 mm.

TABLE 16.2
Clearances to Enclosure and Live Parts

Rated Voltage	Impulse Level	Distance from Coil Outer Surface to Tank (Minimum)	Distance from Live Part Outside Coil to Tank (e.g. Tap Link Bolt)
12	75	80	125
17.5	95	110	170
24	95	110	170
24	125	145	225
36	145	175	270
36	170	220	325

Note: Minimum clearances to wire mesh /grid type walls from IP00 transformer are as follows:
12 kV – 230 mm up to 75 kV BIL
24 kV – 350 mm up to 125 kV BIL
36 kV – 450 mm up to 170 kV BIL.

TABLE 16.3

Minimum Clearances for Different Voltage Classes

Clearances	Voltage Class (kV)						
	<1	6.6	12	17.5	24	36	
Core to LV (use core shield for 6.6kV and above)	15	25 [from shield]	35 [from shield]	-	-	-	
LV–HV [LV voltage < 1 kV] [in air metal to metal]	-	25/30	35/40	38/40	58/50	93/105	*
LV–HV [LV voltage = 6.6 kV] [in air metal to metal]	-	-	35/46				*
LV–HV [LV voltage = 12 kV] [in air metal to metal]							*
Phase–phase in air			25	38	58	93	*
Phase–phase (metal–metal)			35	50	63	105	*
Cast coil – yoke			50	60	70	90	
Coil metal – yoke			70	90	110	150	
Resin on the outside diameter of winding [not applicable for LV with DMD]	-	5	6	6	6	6	
Resin on the inside diameter of winding [not applicable for LV DMD]	-	4	5	6	6	6	
Number of coils including tap coils	-	8	12	12	16	20	
Distance between coils (axial direction)	-	12	16	18	18	20	
Thickness of epoxy resin between coils	-	6	6	6	6	6	
Number of insulation cylinders between HV and LV [thickness of cylinder – 3 mm]	-	-	-	-	2	3	
Number of phase barriers [phase–phase] [thickness of phase barrier 3 mm]	-	-	-	-	2	3	

Notes:

1. The clearance of 24kV and 36kV classes (phase–phase) can be reduced by using phase barriers.
2. The number of insulation cylinders required for HV–LV gap can be increased for the 36kV class, and the clearances can be reduced.

16.1.2.2 Clearance between Windings, Core, Etc.

The minimum clearances for different voltage classes are given in Tables 16.2 and 16.3.

16.1.2.3 Clearance between Tap Links, Line Terminals to Tap Link, Etc.

The clearance through the surface for tap links, line terminal, etc., is decided by the creepage distance required for AC test voltage and impulse test voltage.

The minimum distance can be calculated based on the following:

a. Between tap terminals outside winding

$$\text{Minimum distance} = 1.1 \left[\frac{\text{Test voltage between Tap Terminals kV}}{0.2} \right] \quad (16.1)$$

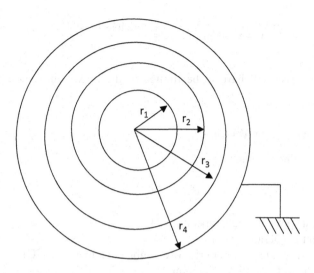

FIGURE 16.1 Schematic of dielectric stress in air and cylinder.

The maximum voltage appearing between any taps at induced over voltage test will have to be considered.

b Between line and tap terminals

$$\text{Minimum distance} = 1.1\left[\frac{\text{Test voltage between Line terminal \& Tap (kV)}}{0.15}\right] \qquad (16.2)$$

- Calculation of dielectric stress in air and cylinder, when insulation cylinders are used in the HV–LV gap (Figure 16.1)

V = applied test voltage
$r_1, r_2, r_3\ldots$ = inner radius of cylinders 1, 2, 3…
$\varepsilon_1, \varepsilon_2, \varepsilon_2$ = relative permittivity of dielectric
For air, $\varepsilon = 1$
For fibre glass cylinder, $\varepsilon = 4.5$
ε_i = stress on dielectric i
r_i = inner radius of dielectric i

$$\varepsilon_i = \frac{V}{r_i\varepsilon\left\{\dfrac{1}{\varepsilon_1}\log_e\dfrac{r_2}{r_1} + \dfrac{1}{\varepsilon_1}\log_e\dfrac{r_3}{r_2} + \dfrac{1}{\varepsilon_3}\log_e\dfrac{r_4}{r_3}\right\}}$$

16.2 CALCULATION OF TECHNICAL PARAMETERS

16.2.1 CORE TEMPERATURE RISE CALCULATION

$$\text{Core temperature rise} = 14\,[9.2w]^{0.7}$$

$$w = \text{watts/kg of core} = \frac{\text{core loss in watts}}{\text{core weight in kg}} \qquad (16.3)$$

The core temperature rise will have to be limited to the maximum winding temperature rise guaranteed.

16.2.2 Temperature Rise of Winding

- Hottest spot winding temperature

$$\theta_{HS} = \theta_a + \Delta\theta_{HS}$$

θ_{HS} = hottest spot winding temperature (°C)
θ_a = ambient temperature (°C)
$\Delta\theta_{HS}$ = hottest spot temperature rise over ambient temperature (°C)
- $\Delta\theta_{HS}$ as a function of load (for AN cooling)

$$\Delta\theta_{HS} = \Delta\theta_{Hsr}$$

$\Delta\theta_{Hsr}$ = hottest spot temperature rise at rated load (°C)
L = per-unit load
- $\Delta\theta_{HS}$ for fan-cooled operations

$$\Delta\theta_{HS} = \Delta\theta_{Hsr}$$

K_T = temperature correction for resistance change with temperature

$$K_T = \frac{T_K + \Delta\theta_{HS}}{T_K + \Delta\theta_{Hsr}}$$

T_K = 225 for aluminium
    = 235 for copper
$X = 1$ (different values can be used based on test data)
- Transient loading

$$\Delta\theta_t = (\Delta\theta_u - \Delta\theta_i)\left[1 - e^{\frac{-t}{T_r}}\right] + \Delta\theta_i$$

$\Delta\theta_t$ = hottest spot temperature rise at some time after t after the overload (°C)
$\Delta\theta_i$ = initial hottest spot rise at some period of load L_i (°C)
$\Delta\theta_u$ = ultimate hottest spot rise if the per unit overload L_u continued until stabilization (°C)
t = time (minutes)
T_r = time constant at rated load (minutes)
θ_{HS} = hottest spot temperature (°C)
$\Delta\theta_i$ = hottest spot temperature rise at some time t after the overload (°C)
θ_a = ambient temperature (°C)

16.2.3 Overload Capacity – For AN Cooling

$$L_u = \left(\frac{\Delta\theta_u}{\Delta\theta_{Hsr}} \right)^{\frac{1}{2m}} \qquad (16.4)$$

L_u = Overload (per unit)
$\Delta\theta_u$ = ultimate hottest spot rise if per unit overload L_u is continued till stabilization (°C)
$\Delta\theta_{Hsr}$ = rated hottest spot temperature rise over ambient temperature (°C)
$m = 0.8$

- Calculation of overload capacity with fan cooling

$$L_u = \sqrt{\frac{\left(\frac{\Delta\theta_u}{\Delta\theta_{Hsr}} \right)^{\frac{1}{x}}}{K_T}} \qquad (16.5)$$

x = empirical constant = 1 (or as per the test result)
K_T = temperature correction of resistance
- Loading capacity based on ambient temperature

$$L = \left[\frac{\theta_{HS} - \theta_a}{\Delta\theta_{Hsr}} \right]^{0.625} \qquad (16.6)$$

θ_{HS} = hottest spot winding temperature (°C)
θ_a = ambient temperature (°C)
$\Delta\theta_{Hsr}$ = rated hottest spot temperature rise over ambient temperature at 1.0 pu load (°C)
L = per-unit load

16.2.4 Altitude Correction Factor

De-rating factors for altitude greater than 1000 m at 30°C ambient are given in Table 16.4.

16.3 TYPICAL RESIN SYSTEM FOR CLASS F APPLICATIONS

Liquid resin with filler for different thermal classes is typically mixed as below. The resin components manufactured by Huntsman are given in the formulation.

TABLE 16.4

De-rating Factors for Altitude Corrections

Type of Cooling	De-rating Factor for Each 100 m above 1000 m (%)
Self-cooling	0.3
Forced air cooling	0.5

TABLE 16.5

Class F Resin Composition Ratio

Resin Component	Part No.	Part by Weight of Mixture	Remarks
Epoxy resin	Cy 205	100	
Hardener	Hy 905	100	
Plasticizer	Dy 40	10	
Filler	Si 02 (w12)	350	
Accelerator	Dy 061	0.3	
Colour paste		2.5	Colour shade as per requirement can be selected. Typical red colour shade in DW0133.

16.3.1 GENERAL CLASS F FILLED SYSTEM

Table 16.5 shows the general class F filled resin composition.

175 parts by weight of filler is mixed with resin and 175 parts by weight of filler is mixed with hardener.

- Hardener is a combination of anhydrides such as hexahydrophthalic anhydride, tetrahydrophthalic anhydride and phthalic anhydride.
- Plasticizer (flexibilizer) Dy 40 is polypropylene glycol.
- Accelerator Dy 061 is aminophenol.

Epoxy resin is unmodified epoxy based on bisphenol A (bisphenol A + epichlorohydrin). The chemical formula is as shown here.

- Resin system suitable to meet the requirement of $C_2E_3F_1$ (climatic environmental and fire withstand)

Typical combination is given in Table 16.6.

A fully premixed system for the above application is XB 5942 (100 parts by weight) and HB 5943 (100 parts by weight).

16.3.2 TYPICAL RESIN SYSTEM FOR CLASS H APPLICATIONS

Table 16.7 gives the resin mixture for class H applications.

TABLE 16.6

$C_2E_3F_1$ Class F Resin Composition Ratio

Component	Part No. of Huntsman	Parts by Weight
Resin	Cy 5538	100
Hardener	Hy 5571-1	100
Aluminium trihydrate		80
Silica	-	310

TABLE 16.7

Class H Resin Composition Ratio

Component	Part No. of Huntsman	Parts by Weight
Resin	Cy 5980	100
Hardener	Hy 5980	95
Plasticizer	Dy 040	15
Filler	Silica	410

16.4 ENCLOSURE FOR CAST RESIN TRANSFORMERS

16.4.1 THE AIR INLET AREA REQUIRED

$$S = \frac{0.188\,P}{\sqrt{H}}\,\text{m}^2 \qquad (16.7)$$

P = total losses (kW)
H = thermal head (M)
S_1 = air out let area = 1.1 S m^2

If the area is not adequate, forced cooling will be required, with airflow of 4.5 m^3/minute/kW.

16.4.2 STANDARD CLEARANCES TO ENCLOSURE

The minimum clearances required to uninsulated metal surface are given in Table 16.8.

Minimum air clearance required for a cast resin transformer with concrete wall for the prevention of disruptive discharges is shown in Figure 16.2. The air clearances required for cast resin transformer inside enclosure are given in Table 16.9.

TABLE 16.8

Minimum Clearance to Flat Uninsulated Metal Surfaces

Insulation Class	BIL	Clearance (mm)
7.2	60	90
12.0	75	120
17.5	95	200
24	125	220
36	170	320

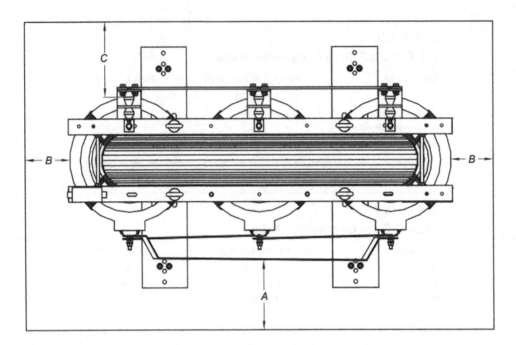

FIGURE 16.2 Minimum-clearance cast resin transformer.

TABLE 16.9
Air Clearances for CRT

Highest Voltage for Equipment U_M	Rated Lighting Impulse Withstand Voltage U_L		Dimensions (mm)		
kV	List 1	List 2	A	B	C
12	-	75	125	80	80
17.5	75	-	125	80	80
17.5	-	95	170	110	110
24	95	-	170	110	110
24	-	125	225	145	145
36	145	-	270	175	175
36	-	170	325	220	220

16.5 CALCULATION OF TIME CONSTANT OF DRY-TYPE TRANSFORMERS

16.5.1 TIME CONSTANT OF WINDING AT RATED LOAD

$$T_R = C \frac{\left(\Delta\theta_{\text{Hsr}} - \theta_e\right)}{P_r} \tag{16.8}$$

C = effective thermal capacity of the winding (watt-minutes per kelvin (W min/K))
  = 15 × mass of aluminium conductor in kg+24.5×mass of epoxy and other winding insulation material in kg
  = 6.42 × mass of copper conductor in kg+24.5×mass of epoxy and other winding insulation material in kg

P_r = winding total loss at rated load and rated temperature rise in watts (resistive loss+eddy loss)

$\Delta\theta_{Hsr}$ = winding hot spot temperature rise at rated load in kelvin

= core contribution to the winding hot spot temperature rise at no load

= 5 K for outer winding

= 25 K for inner winding (usually less than 1 kV class)

16.5.2 TIME CONSTANT AT ANY LOADING

$$T = T_R \frac{\left(\dfrac{\Delta\theta_u}{\Delta\theta_{Hsr}}\right) - \left(\dfrac{\Delta\theta_1}{\Delta\theta_{Hsr}}\right)}{\left(\dfrac{\Delta\theta_u}{\Delta\theta_{Hsr}}\right)^{\frac{1}{m}} - \left(\dfrac{\Delta\theta1}{\Delta\theta_{Hsr}}\right)^{\frac{1}{m}}} \tag{16.9}$$

$\Delta\theta_1$ = initial hot spot temperature rise at some prior load 'I'

$\Delta\theta_t$ = hot spot temperature rise at time t after changing load

$\Delta\theta_u$ = ultimate hot spot temperature rise if the per-unit overload I_u is continued until the hot spot temperature rise is stabilized.

t = time in minutes

T = time constant in minutes

θ_a = ambient temperature in °C

16.6 AGEING AND TRANSFORMER INSULATION LIFE EXPECTANCY

$L = a\,e^{\frac{b}{T}}$

L = expected life time

T = temperature in kelvin

a, b = constants (Arrhenius equation constants)

- Constants for life expectancy equation (Table 16.10)

TABLE 16.10
Life Expectancy Constants

Insulation System Temperature (Thermal Class)	Arrhenius Equation Constants		Rated Hot Spot Winding Temperature $\theta_{Hs,r}$	Maximum Hot Spot Winding Temperature
	a	b		
°C	h	k	°C	°C
105 (A)	3.1×10^{-14}	15900	95	130
120 (E)	5.48×10^{-15}	17212	110	145
130 (B)	1.72×10^{-15}	18115	120	155
155 (F)	9.6×10^{-17}	20475	145	180
180 (H)	5.35×10^{-18}	22975	170	205
200	5.31×10^{-19}	25086	190	225
220	5.26×10^{-20}	27285	210	245

16.7 DESIGN USING ROUND/RECTANGULAR CONDUCTOR

16.7.1 INTRODUCTION

The LV and HV windings can be manufactured using insulated rectangular/round conductors for the manufacture of cast resin transformers. This process does not require foil winding machine. The conductor insulation will have to be of relevant thermal class. If an enamelled conductor is used, the resin will have to be compatible with the type of epoxy resin used for casting.

16.7.2 CONDUCTOR INSULATION FOR CLASS F SYSTEM

Conductor insulation for a class F system is given in Table 16.11.

16.7.3 LAYER WINDING ARRANGEMENT

16.7.3.1 LV Winding

The LV is wound with preferably even number of layers so that line leads can be taken from the top of the coil. In order to limit the temperature rise limits, ducts in the winding can be provided by using tapered segments.

16.7.3.2 HV Winding

When ducts are not required, a typical winding arrangement is as shown in Figure 16.3.

TABLE 16.11
Conductor Insulation for Class F

Voltage Class of Winding	Round Wire	Rectangular Conductor
Less than 1 kV	Diameter between 1 and 3.15 mm. 0.3 mm glass fibre covering	Enamelled conductor with 0.3 mm glass fibre covering
3.6 kV	Diameter between 1 and 3.15 mm. 0.3 mm glass fibre covering	Enamelled conductor with 0.3 mm glass fibre covering
6.6 kV	Diameter between 1 and 3.15 mm. 0.4 mm glass fibre covering	Enamelled conductor with 0.4 mm glass fibre covering
11 kV	Diameter between 1 and 3.15 mm. 0.4 mm glass fibre covering	Enamelled conductor with 0.4 mm glass fibre covering
22 kV	Diameter between 1 and 3.15 mm. 0.5 mm glass fibre covering	Enamelled conductor with 0.4 mm glass fibre covering
33 kV	Diameter between 1 and 3.15 mm. 0.6 mm glass fibre covering	Enamelled conductor with 0.4 mm glass fibre covering

---

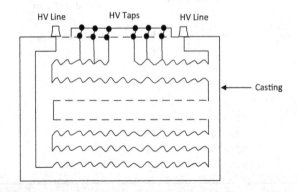

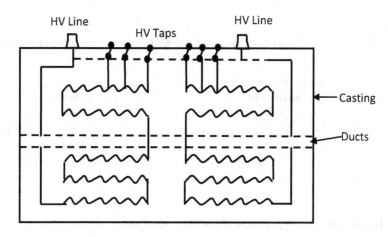

FIGURE 16.4 HV layer arrangements with ducts.

- HV layer winding with ducts
    A HV winding arrangement when ducts are provided is shown in Figure 16.4. It is preferable to have odd numbers of layers below the segments to enable line leads to be taken out from the ends.

16.8 CALCULATION OF COOLING FAN CAPACITY

The natural cooled (AN) rating of a cast resin transformer can be increased by forced air cooling. The fans are installed below the coils so that the maximum airflow is directed through the HV–LV gap and ducts.

The airflow required for the heat transfer is calculated by the following formula:

$$Q = \frac{0.05P}{T_2 - T_1} \tag{16.10}$$

where
  Q = airflow required in m³ per minute
  P = loss to be dissipated by airflow in watts
  $T_2 - T_1$ = temperature difference between the outlet and the inlet of air to the coil in °C (this difference is between 15°C and 20°C).

The airflow of the fan/fans shall be more than the above because some portion of the airflow from the fan flows outside the duct and away from the windings. The effective arrangement is to fix the fans below the coils to evenly cover the coils in all limbs. The airflow capacity of the cooling fans required is 3.3–4 times "Q" calculated as per the above formula. Table 16.12 gives the approximate airflow capacity of the cooling fans required for enhancing the loading capacity by 30%– 35% of the AN rating of typical ratings of cast resin transformers. The number of fans required shall be selected on the basis of the standard airflow capacity and the number of fans to be installed.

TABLE 16.12

Approximate Airflow Capacity of the Cooling Fans Required for Enhancing the Loading Capacity by 30%–35%

AN Rating of Transformer (kVA)	Airflow Rating of Fans (m³/minute)	Remarks
630–1000	16–20	• It is assumed that the fans are evenly distributed to give airflow uniformly to all coils.
1000–1600	20–25	to give airflow uniformly to all coils.
1600–2000	25–36	• This table is applicable for transformers
2000–2500	35–48	complying with the losses of EU (tier 1) losses

16.9 VENTILATION OF THE TRANSFORMER ROOM

Table 16.13 gives the airflow requirement, inlet area of air at the bottom of the room and outlet area of the air at the top of the room. The inlet and outlet areas given are the net values. If wire mesh protection is used, the effective area of the opening of the mesh will have to be considered (Figure 16.5).

S = air inlet area (m²)
S_1 = air outlet area = 1.15 S (m²)
H = centre to centre distance between the mid-point of transformer to the mid-point of air outlet

• Effective ventilation area and airflow requirements
    kW = total losses to be dissipated at reference temperature (e.g. 120°C for class F)
(No load loss + load loss at reference temperature in kW)

$$\Delta T = \text{Temperature between inlet air and outlet air}$$

TABLE 16.13

Effective Ventilation Area S at Inlet (m²) and Airflow Requirement (m³/minute)

	S = Inlet Ventilation Area (m²)								
	H (m)								
	1.5		2.0		2.5		3.0		Airflow
Total Losses (kW)	ΔT = 10°C	ΔT = 15°C	ΔT = 10°C	ΔT = 15°C	ΔT = 10°C	ΔT = 15°C	ΔT = 10°C	ΔT = 15°C	Required m³/minute
5	1.4	0.8	1.3	0.7	1.1	0.6	0.9	0.5	20
10	2.7	1.5	2.4	1.3	2.2	1.2	2.0	1.1	40
15	4.2	2.3	3.6	2.0	3.3	1.8	2.9	1.6	60
20	5.5	3.0	4.6	2.6	4.2	2.3	3.8	2.1	80
25	7.0	3.8	3.6	3.3	5.3	2.9	4.9	2.7	100
30	8.2	4.5	7.2	3.9	6.4	3.5	5.8	3.2	120

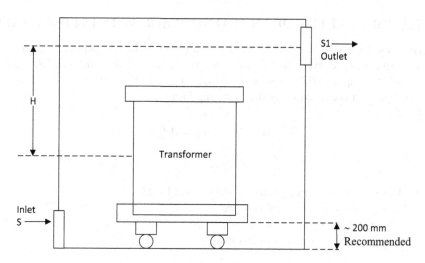

FIGURE 16.5 Schematic of a transformer-installed room.

16.10 EFFECT OF THE ENCLOSURE IP CLASS ON TEMPERATURE RISE

The no-load losses and full-load losses are dissipated by convection, conduction and radiation directly from the exposed surfaces in the case of IP00 execution (without enclosure). The average winding temperature rise is calculated by calculating the temperature gradient in the insulation.

When an enclosure is used for the transformer, the temperature rise of the winding increases, depending on the ingress protection class of the enclosure, the physical dimensions and colour. The temperature rise allowances to be considered for IP21, IP31 or IP41 enclosures for AN cooling are given in Table 16.14.

Example

IP21 enclosure for 11-kV class cast resin transformer, with class F insulation
Winding temperature rise allowed = 100°C
Allowance for IP21 enclosure = 10°C–12°C
Taking the higher value of 12°C
∴ The calculated value of winding temperature rise shall not exceed 100–12 = 88°C.
If a factor of safety of 3°C is considered, the designed winding temperature rise of IP00 shall be 85°C.

TABLE 16.14

Allowances for Temperature Rise for Enclosure IP Class

	Allowance for Temperature Rise (°C)		
	Voltage Class of Transformer (kV)		
IP Class of Enclosure	**11/13.8**	**22/24**	**33/36**
IP21 and IP23	10–12	10–12	12–15
IP31 and IP33	12–15	15–18	15–18
IP41 and IP44	15–18	18–20	18–20

16.11   TEMPERATURE RISE OF AN TRANSFORMER WITH IP54 ENCLOSURE

When the protection class of enclosure is IP54, there are no ventilation openings and the total losses need to be dissipated by convection and radiation from the enclosure surface. The air inside the enclosure gets heated up, and the temperature difference causes the heat transfer.

The loss dissipation is calculated by the following formula:

$$W = W_V + W_{\text{TC}} + W_F \tag{16.11}$$

where
W = total losses to be dissipated (watts) [no-load loss + load loss]
W_V = loss dissipated by vertical surfaces (watts)
W_{TC} = loss dissipated by top cover (watts)
W_F = loss dissipated by corrugated panels (watts)

$$W_V = A_V \left[2 \times \theta^{1.2} + 5.7 \times 10^{-8} \left\{ (273 + \theta_a + \theta)^4 - (273 + \theta_a)^4 \right\} \right]$$

θ = temperature rise of air inside the enclosure above ambient temperature
θ_a = ambient temperature
A_V = area of the vertical flat panels of the enclosure

When calculating the area, the vertical surface areas of the enclosure panels where cable boxes, bus duct flanges, etc., are fitted are excluded.

The vertical surface area where corrugated panels are fixed is excluded.

The vertical surface area of the enclosure below the level of the core bottom is excluded.

$$W_{\text{TC}} = A_{\text{TC}} \times 5.7 \times 10^{-8} \left[(273 + \theta_a + \theta)^4 - (273 + \theta_a)^4 \right]$$

A_{TC} = area of top cover of enclosure (m²)

$$W_F = 15 \times D^{-0.2} \times \theta^{1.2} \times n \times L \times 10^{-3}$$

D = depth of corrugation (mm)
L = height of corrugation (mm)
n = number of corrugations

The temperature rise of air inside the IP54 enclosure of a 1000-kVA cast resin transformer is calculated as an example here.

kVA = 1000
No-load loss = 1550 W
Load loss = 9000 W
Ambient temperature = 30°C
Area of top cover = $4\,\text{m}^2 = A_{\text{TC}}$
Area of flat vertical panels = $8\,\text{m}^2 = A_V$
Depth of corrugated fin = 300 mm = D
Height of corrugated fin = 1500 mm = L
Number of fins = 20 = n

Substituting the data in the equation,

$$W = W_V + W_{TC} + W_F$$

$W = 1550 + 9000 = 10550$

$$10550 = 8 \left[2 \times \theta^{1.2} + 5.7 \times 10^{-8} \left\{ (273 + 30 + \theta)^4 - (273 + 30)^4 \right\} \right]$$

$$+ 4 \times 5.7 \times 10^{-8} \left[(273 + 30 + \theta)^4 - (273 + 30)^4 \right]$$

$$+ 15 \times 300^{-0.2} \times \theta^{1.2} \times 20 \times 1500 \times 10^{-3}$$

Solving the equation,
$\theta \cong 27°C$
$\therefore$ The temperature of air at the top of enclosure is $27 + 30 = 57°C$

16.12   ANTI-VIBRATION PADS

The sources of vibration in a cast resin transformer are as follows:

- Vibration transferred from core due to magnetostriction
- Vibration of the winding transmitted through the structures
- Vibration of the tank (enclosure) by the transmitted vibration from core and winding

The vibration from the core depends on the working flux density, the material characteristics of the core, the clamping arrangement and the efficiency of the vibration clamping arrangements of the core structure. The vibration of the winding is not significant because it acts as a rigid solid body and resilient coupling is provided at the winding ends to the clamping structure. The vibration of winding is proportional to the square of the load current.

The vibration of the enclosure can go up, if any of the side plates or panels has resonant frequencies close to the integer multiples of 2f, where f is the rated frequency of the transformer. In order to reduce the vibration levels, anti-vibration pads are used at the coupling points to dampen the transmission of vibration.

The anti-vibration pads are applied between the core and bottom fixing plate of the core assembly and between the core and coil assembly and the bottom plate of the enclosure. Anti-vibration pads can be used to isolate the transformer from the floor also. The commonly used materials for vibration damping of the cast resin transformer are as follows:

a. Ribbed neoprene pad

Ribbed neoprene pads are available in different thicknesses. It is oil and weather resistant and is suitable for indoor applications.

The load capacity is in the range of 2.0–6.0 kg/cm². Maximum compression allowed is 30% at the loading of 6 kg/cm².

The calculation of the thickness of a ribbed neoprene pad is given by an example.
Pad between core and coil assembly and enclosure bottom plate
Weight of core and coil assembly = 3000 kg
Size of the core bottom flat = 50 × 500 (width × length) (mm)
Number of core bottom flats = 3

Total load bearing area of the pad $= \dfrac{3 \times 50 \times 500}{10 \times 10} = 750$ cm^2

Maximum load permissible on the pad $= 750 \times 6 = 4500$ kg (assuming maximum allowable pressure of 6 kg/cm^2)

Selecting a 10-mm-thick ribbed neoprene pad

Compression of the pad $= \dfrac{3000 \times 0.3 \times 10}{4500} = 2$ mm

Maximum compression permissible is 3 mm for a 10 mm-thick sheet, and the selected thickness is sufficient.

b. Core and ribbed neoprene pad

The material properties are as follows:

Loading capacity $= 2$–6 kg/cm^2

Maximum compression allowed $= 20\%$

Thickness of single sheet up to 25 mm is common. If the load exceeds, multiple pads can be used to build up thickness.

c. Cork and ribbed neoprene with steel

This is used when the load is high. The material properties are as follows;

Loading capacity $= 3$–7 kg/cm^2

Maximum compression $= 12\%$

d. Anti-vibration mount for transformer wheels

When the transformer is installed at site with rollers, anti-vibration mounts are required. The mount consists of an upper metal profile with two cylindrical axes coated with rubber to keep the transformer roller in place and a lower profile. The position of cylindrical axes can be adjusted to suit the wheel diameter.

The lower metal profile is fixed to the ground/foundation. An elastomeric element of about 25 mm is placed between the upper and lower metal profiles for vibration isolation.

16.13   RC SNUBBER FOR CAST RESIN TRANSFORMERS

Fast-rising transients are generated when the high-voltage side of a transformer is interrupted by a vacuum circuit breaker. A transient recovery voltage can develop when the primary breaker interrupts a fault on the transformer secondary causing fast transient oscillations between the capacitance and inductance of the transformer and connecting cable.

One method of reducing the effect of the fast-rising transient is to install RC snubbers very close to the high-voltage bushings of the transformer. For dry-type transformers, these can be installed inside the transformer enclosure to offer a better level of protection.

The RC snubber consists essentially of a surge capacitor and a damping resistor connected in series. A high-speed full-range current-limiting fuse also can be used in series. If the fuse is not used, the resistor melts when high current flows. The time constant of the RC snubber shall be within 1–10 µs.

Capacitors designed as surge capacitors only shall be used for the snubber as these are designed to withstand high-frequency transients. It is not recommended to connect surge capacitors in series or parallel to get a desired value.

The damping resistor shall be non-inductive and shall match with the surge impedance ($Z°$) of the incoming conductor. The continuous current rating of the resistor is selected as two times the capacitive current to factor for the harmonic content. The power dissipation capacity of the resistor is calculated from the current and resistance values.

- Design of a RC snubber for an 11 kV transformer.

Rated line-to-line voltage = 11 kV, frequency = 50 Hz
The standard surge capacitor of 0.25 microfarads, 11 kV is selected = C
The time constant of the RC snubber shall be less than 10 μs.

$RC = 10 \times 10^{-6}$, substituting the value of C, $R = \dfrac{10 \times 10^{-6}}{0.25 \times 10^{-6}} = 40\ \Omega$

In order to get a time constant less than 10 μs, a 20 Ω resistor is selected for this design.
The current through the surge capacitor assuming no harmonics is calculated as I_c;

$$I_c = \frac{2\pi \times f \times C \times V_{line}}{\sqrt{3}} \qquad (16.12)$$

where
  f = frequency (Hz)
  C = capacitance (farads)
  V_{Line} = line-to-line voltage

$$I_c = 2\pi \times 50 \times 0.25 \times 10^{-6} \times 11000/\sqrt{3} = 0.50\ A$$

The rated current of the capacitor is selected as $2 \times 0.5 = 1.0$ A.
If there is a short circuit in the capacitor, the resistor will have to carry the short-time fault current till the fuse operates.

Short-time current $I_{sc} = \dfrac{11000}{\sqrt{3} \times 20} = 317.5\ A$

Assuming that the fuse operates in 5 milliseconds, the energy absorbed by the resistor is calculated here:

$$W = I^2 \times R \times t \times 10^{-3}$$

where
  W = energy absorbed by the resistor (kilojoules)
  I = current
  R = resistance (ohms)
  t = time (seconds)

Substituting the values,

$$W = 375.5^2 \times 20 \times 5 \times 10^{-3} = 10.08\ kJ$$

The power and energy ratings of the resistor shall match the above.
The time constant of the RC snubber is 5 μs.

17 Earthing Transformers

17.1 INTRODUCTION

As per IEC 60076-6, earthing transformers (also termed as neutral couplers) are used to provide a neutral connection for earthing a 3-phase ungrounded network. Earthing transformer is termed grounding transformer as per ANSI C.57.12, and it is described as "a Transformer intended primarily to provide a neutral point for grounding purposes on a 3-phase ungrounded system to provide a return path for the fault current and to support a faulted phase above ground".

The neutral connection of an earthing transformer to the ground can be done directly, through a resister, through an arc suppression reactor or through a current limiting reactor.

The earthing transformer can be provided with an auxiliary winding for station auxiliary supply. The auxiliary winding is normally star-connected. An auxiliary winding can be provided for testing purpose.

The earthing transformer can be oil-filled type or dry type.

17.2 BASIS OF RATING

The rating of the transformer shall satisfy the following operating conditions:

- Rated short-time current and duration of the current
- Rated continuous current including harmonic content, if any
- Rated voltage
- Rated frequency
- Basic insulation level and test voltage
- Zero sequence impedance
- Auxiliary winding and its ratings including rated power, voltage, impedance and temperature rise
- Temperature rise limits
- Insulation class
- Taps required, if any
- Indoor or outdoor service
- Abnormal operating or service conditions

17.3 RATED SHORT-TIME NEUTRAL CURRENT DURATION AND CONTINUOUS CURRENT

Earthing transformers shall withstand short-time fault currents for a specified duration. The customer may specify the continuous neutral current, and in the absence of the same, the currents as per Table 17.1 can be used.

17.4 FAULT CURRENT FLOW THROUGH EARTHING TRANSFORMERS

When a line-to-ground fault occurs in a delta-connected system, it is detected as a normal current, if the magnitude is less than the load current or the unbalanced setting. It also affects the system stability. If a transformer with zigzag windings is connected to the delta system, the fault currents are distributed as shown in Figure 17.1.

TABLE 17.1

Continuous Neutral Current

Rated Time	Continuous Neural Current as % of Short-Time Neutral Current
10 seconds	3
1 minute	7
10 minutes	30
Extended time	30

Note: IEC 60076-6 is not specifying the continuous neutral current value. However, ANSI C.57.12 standard specifies it. In the case of extended time of operation the total time duration of operation considered over a period one year shall not exceed 90 days.

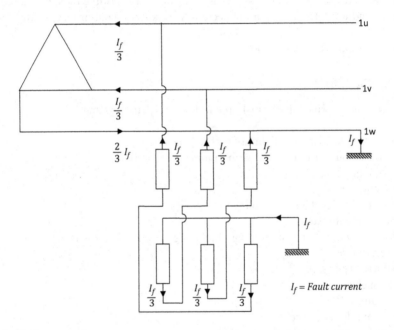

FIGURE 17.1   Fault current flow.

When a line to ground fault occurs, the zero phase sequence current flows through the neutral of the earthing transformer and returns to the system. From Figure 17.1, it can be observed that the current in each limb flows through the zig winding and zag winding in opposite directions. Thus, the net flux produced by the fault current in the earthing transformer is zero, which maintains the rated voltage as in the normal operation.

17.5   CALCULATION OF ZERO PHASE SEQUENCE IMPEDANCE (ZPS)

The zero phase sequence impedance required for the earthing transformer is calculated from the short time neutral current and the line voltage, if the neutral is directly earthed.

$$ZPS = \frac{3V_L}{\sqrt{3}I_f} \qquad (17.1)$$

where
ZPS = zero phase sequence impedance in ohms
V_L = line voltage
I_f = fault current at neutral

The ZPS is designed to limit the maximum fault current, and therefore, no-negative tolerance is allowed. The tolerance of ZPS as per standard is +20%.

If the neutral of the earthing transformer is earthed through a resistor, the ZPS required for the transformer can be calculated as follows:

$$\left(\frac{I_f}{3}ZPS\right)^2 + (RI_f)^2 = V_{ph}^2 \tag{17.2}$$

where
V_{ph} = phase voltage.

R = neutral grounding resistor in ohms
ZPS = zero phase impedance of transformer in ohms

The ZPS is the per phase value in ohms, and it is inductive reactance. The neutral current (I_f) flows through R, and $I_f/3$ flows through ZPS. The phase voltage is the vector sum of these voltage drops through R and ZPS.

17.6 DESIGN OF EARTHING TRANSFORMERS

17.6.1 CALCULATION OF MAXIMUM PERMISSIBLE CURRENT DENSITY

Earthing transformers are normally designed for short-time ratings, and the windings shall be capable of withstanding the short-time thermal requirement. The maximum allowable current density can be calculated as per IEC 60076-5.

$$\theta_1 = \theta_0 + \frac{2(\theta_0 + 235)}{\dfrac{106000}{J^2 t} - 1} \text{ for copper} \tag{17.3}$$

$$\theta_1 = \theta_0 + \frac{2(\theta_0 + 235)}{\dfrac{45700}{J^2 t} - 1} \text{ for Aluminium} \tag{17.4}$$

where
θ_0 = initial winding temperature in °C
J = current density when fault current flows in A/mm²
t = duration of fault current in seconds

The above equations can be used for short-time ratings up to 10 seconds as per IEC 60076-5. Though the above equations give indicative values of current density for short-time ratings above 10 seconds, these can be used as initial values for commencing a design for short-time ratings up to 60 seconds. For time durations exceeding the winding time constant, the effect of the heat dissipation from the winding can be considered while calculating the current density.

TABLE 17.2

Maximum Permissible Short-Time Current Densities

Type of Transformer	Insulation System Temperature (°C)	Fault Duration (Seconds)	Copper Winding Maximum Current Density (A/mm²)	Aluminium Winding Maximum Current Density (A/mm²)
Oil-filled	105	10	43.2	23.9
class A		60	17.6	9.8
Dry-type	155	10	55.6	15.0
class F		60	22.7	6.1

An important point to be considered is that the current density calculated on the basis of the short-time thermal withstand capability may need downward revision when calculating the dynamic force withstand and temperature rise requirements. The maximum permissible current densities for short-time thermal capability for copper and aluminium windings are given in Table 17.2. For practical designs, much lower current densities are used, especially for the 10 sec short time rating.

17.6.2 EQUIVALENT kVA FOR AN EARTHING TRANSFORMER

Earthing transformers are short-time-rated, and therefore, the concept of an equivalent kVA is considered for initial design of the core and winding. The empirical formula for determining the approximate equivalent rating is

$$kVA_{eq} = K \frac{V_{Line} I_{sc}}{\sqrt{3}}$$ (17.5)

where
  V_{Line} = line voltage (kV)
  I_{SC} = short-time current in neutral (A)
  K = as per Table 17.3

TABLE 17.3

Values of K for Calculating Equivalent Rating

Short-Time Rating (Seconds)	K
10	0.03–0.05
30	0.04–0.06
60	0.10–0.15
600 (10 minutes)	0.20–0.30

17.7 DESIGN OF AN ONAN EARTHING TRANSFORMER WITHOUT AUXILIARY WINDING

The basic input variables are as follows;

    a.   Line voltage
    b.   Short-time fault current in neutral
    c.   Duration of fault current.

When auxiliary winding is not specified, there will be limitations to carry out tests including induced overvoltage test.

By agreement between the manufacturer and the purchaser, an auxiliary winding for test purpose can be provided.

17.7.1 Sample Design of an Earthing Transformer without Auxiliary Winding

Specifications
Line voltage – 15000 V
Taps – not required
Short-time fault current – 200 A
Fault current duration – 30 S
Frequency – 50 Hz
Cooling – ONAN
Winding material – copper

Design Calculations

$$ZPS \text{ required} = \frac{15000 \times 3}{\sqrt{3} \times 200} = 129.9 \text{ ohms/phase}$$

$$\text{The equivalent kVA} = \frac{K \times 15 \times 200}{\sqrt{3}}$$

For 30 seconds short time duration, the value of K ranges from 0.04 to 0.06, from Table 17.3. Using $K = 0.04$,

$$\text{The equivalent kVA} = \frac{0.04 \times 15 \times 200}{\sqrt{3}} = 69.3$$

It can be assumed that the equivalent kVA is more than 69 as the lowest of K is used for the calculation. The lowest volt/turn can be estimated by using the equation $\frac{V}{T} = K\sqrt{kVA}$.

Substituting the lower value of K as 0.35, the lower range of volt/turn is calculated.

$$\frac{\text{Volt}}{\text{Turn}} = 0.35 \sqrt{069.3} = 2.91$$

Obviously, the V/T value more than 2.91 will give a more feasible design.
Alternatively, a core circle can be assumed and the design can be continued.

Core circle diameter = 109 mm
Net core area = 83.9 mm²

Flux density = 1.65 Tesla (assumed initial value)

$$\frac{\text{Volt}}{\text{Turn}} = 4.44 \times 50 \times 1.65 \times 83.9 \times 10^{-4} = 3.07$$

$$\text{Voltage of zig winding} = \frac{15000}{3} = 5000$$

$$\text{Voltage of zag winding} = 5000$$

$$\text{Turns of zig winging} = \frac{5000}{3.07} = 1628$$

17.7.2 CALCULATION OF THE SHORT-TIME CURRENT DENSITY

The maximum permissible current density for 30 seconds duration is calculated using the IEC 60076-5 formula. It is to be noted that the IEC formula is applicable for the fault duration of 10 seconds only. However, it gives a conservative initial design, because the temperature rise is calculated on the basis that the entire heat generated in the winding conductor is stored. Applying the formula for the maximum permissible current density,

$$\theta_1 = \theta_0 + \frac{2(\theta_0 + 235)}{\frac{106000}{J^2 t} - 1}$$

Substituting the parameters,

$$250 = 105 + \frac{2(105 + 235)}{\frac{106000}{J^2 \times 30} - 1}$$

Solving for J,
   Maximum current density = 24.89 A/mm²

$$\text{Current}/\text{Coil} = \frac{200}{3} = 66.66 \text{ A}$$

$$\text{Minimum area of copper conductor} = \frac{66.66}{24.89} = 2.68 \text{ mm}^2$$

If a round conductor with a bare diameter of 2 mm is considered, the actual current density works out to,

$$J = \frac{66.6}{\frac{\pi}{4} \times 2^2} = 21.19 \text{ A/mm}^2 < 24.89 \text{ A/mm}^2$$

17.7.3 Winding Design

Winding design particulars are given in Table 17.4.

ZPS calculation is done using the formula as per Section 17.6.

$$ZPS = 8\pi^2 f T^2 K\Delta \times 10^{-10}$$

$$f = 50$$

$$T = 1628$$

$$K = 1 - \frac{41+20+41}{\pi \times 246} = 0.87$$

$$\Delta = 145$$

$$ZPS = 8\pi^2 \times 50 \times 1628^2 \times 0.87 \times 145 \times 10^{-10}$$

$$= 132 \text{ Ohms}$$

The tolerance allowed for ZPS is +20%, and the calculated ZPS is within acceptable limit. The above calculations give an initial design, which requires iterations to get the optimum value.

TABLE 17.4

Winding Design Particulars

Particulars	Zig Winding	Zag Winding
Volts/coils	5,000	5,000
Current (A)	66.66	66.66
Number of turns	1628	1628
Number of layers	14	14
Turns/layer	117 × 13 + 107 × 1	117 × 13 + 107 × 1
Bare size of the conductor (mm)	2.0	2.0
Covered size of the conductor (mm)	2.1	2.1
Inside diameter of the coil	139	261
Outside diameter of the coil	221	343
Axial height (electrical)	246	246
Axial height including edge strip	296	296
Interlayer insulation	0.375	0.375
Number of ducts	2	2
Thickness of ducts	3	3
Radial thickness	41	41

All dimensions are in millimetres.

17.8 DESIGN OF AN ONAN EARTHING TRANSFORMER WITH AUXILIARY WINDING

In addition to the input requirement of the earthing transformer without auxiliary winding, the essential additional input details are as follows;

- Continuous rating of auxiliary winding
- Rated voltage and connection of auxiliary winding
- Whether taps are required or not
- Temperature rise of winding and oil
- Test voltage of auxiliary winding
- Winding material

Significant design considerations include the following;

- Thermal and dynamic short circuit withstand capability of the winding when fault occurs on the auxiliary winding
- Transferred surges to the auxiliary winding
- Relative placement of auxiliary, zig and zag windings
- If voltage variation is specified for auxiliary winding by providing taps on the zigzag winding, the same shall be provided on both zig and zag windings so that an equal number of turns are increased or decreased for every tap change from both the coils to ensure 120° phase shift on all tap positions.

Sample Design of an Earthing Transformer with Auxiliary Winding

1. Specifications

   Voltage ratio – 33/0.4 kV,

   Fault current – 800 A

   Fault current duration – 10 sec

   Frequency – 50 Hz

   Connection and vector group – ZNyn1

   Auxiliary winding – 100 kVA

   Rated voltage of auxiliary winding – 400 V

   Taps on the 33 kV side for ±5% variation in 2.5% steps

   Cooling – ONAN

   Impedance of auxiliary winding – 2%

   Winding material – copper

2. Design calculations

   The continuous rating of auxiliary winding is 100 kVA. For 10 second rating, the continuous neutral current is 3% of the fault current at the neutral.

$$\text{Continuous neutral current} = \frac{800 \times 3}{100} = 24\,\text{A}$$

The primary winding (33 kV) has therefore, 8 A per phase from the continuous neutral current and continuous current from the loading of the auxiliary winding.

$$\text{Rated load current of secondary} = \frac{100 \times 3}{\sqrt{3} \times 0.4} = 144.34\,\text{A}$$

Current in primary due to load current $= 1.75\,\text{A}$

The losses to be dissipated therefore will have to be calculated by considering the following currents in the windings:

$$\text{HV winding current} = 8+1.75 = 9.75\,\text{A}$$

$$\text{LV winding current} = 144.34\,\text{A}$$

The kVA corresponding to the continuous currents in winding is calculated as follows;
Rated kVA of auxiliary winding $= 100$ kVA

$$\text{Estimated kVA from continuous neutral current} = \frac{33 \times \sqrt{3} \times 8}{2} = 228$$

$$\text{Total kVA} = 100 + 228 = 328$$

$$\text{Short time kVA of the transformer} = \sqrt{3} \times 33 \times \frac{800}{3} = 15242\,\text{kVA}$$

Equivalent kVA using the factor K from Table 17.3 is

$$\text{Equivalent kVA} = 0.03 \times 15242 = 457.3$$

Core dimensions and volt/turn
Initial core diameter selected $= 185$ mm (based on equivalent kVA)
Core net area $= 0.0242\,\text{m}^2$
Assuming an initial flux density of 1.65 T,

$$V\!/_{T} = 4.44 \times 50 \times 1.65 \times 0.0242 = 8.864\,\text{V}$$

$$\text{Number of turns in LV} = \frac{400}{\sqrt{3} \times 8.864} = 26$$

The minimum area of cross section for the HV conductor is selected based on the permissible maximum current density for 10 second duration for copper from Table 17.2.

$$\text{Maximum permissible current density} = 43.2\ \text{A/m}^2$$

$$\text{Minimum conductor area required} = \frac{800}{3 \times 43.2} = 6.17\,\text{mm}^2$$

A higher cross-sectional area is required for the practical design to allow for a reasonable factor of safety, and a 3.85 mm diameter enamelled copper wire is selected for zig and zag windings. Winding dimensions are given in Table 17.5.

Zero sequence impedance is calculated using the formula as per Section 7.6, and the per cent reactance is calculated by the formula from Section 17.7.

Zero phase sequence impedance calculated $= 77.3\ \Omega/\text{phase}$

TABLE 17.5

Winding Dimensions and Clearances

Particulars	LV Winding	Zig Winding	Zag Winding
Volts/coil	230.94	11000	11000
Fault current	-	266.67	266.67
Turns/coil	26	1300 max tap	1300 max tap
Number of layers	2	10	10
Turns/layer	13	$132 \times 9 + 112 \times 1$	$132 \times 9 + 112 \times 1$
Bare size of the conductor	12×2.4	3.85	3.85
Covered size of the conductor	12.4×2.8	3.95	3.95
Number of conductors in parallel	3	1	1
Number in axial direction	3	-	-
Number in radial direction	1	-	-
Area of cross section of the conductor (mm²)	84.75	11.64	11.64
Current density at rated load current (A/mm²)	1.70	-	-
Current density at short-time current (A/mm²)	-	22.91	22.91
Inside diameter of the coil	194	258	410
Outside diameter of the coil	218	356	508
Radial depth of the coil	12	49	49
Axial height (electrical)	484	521	521
Axial height with edge strip	590	590	590
Interlayer insulation	-	0.625	0.625
Duct thickness	6	3	3
Number of ducts	1	1	1

$$\text{ZPS required} = \frac{33000 \times 3}{\sqrt{3} \times 800} = 71.4\,\Omega/\text{phase}$$

The tolerance allowed is +20%, and therefore, the calculated value is within permissible limit.

$$\%\text{Reactance calculated on } 100\,\text{kVA base for auxiliary winding} = 2\%$$

$$\% \text{ Resistance of auxiliary winding} = 0.45$$

$$\% \text{ Impedance of auxiliary winding} = \sqrt{2^2 + 0.45^2} = 2.05$$

This is within the tolerance limit of the impedance of the auxiliary winding.

17.9   RATED SHORT-TIME TEMPERATURE RISE

The short-time temperature rise of an earthing transformer of a 10 second rating shall be demonstrated by calculation.

In the case of short-time duration above 10 seconds but below 10 minutes, the temperature rise can be demonstrated by calculation by agreement between the manufacturer and the purchaser.

The short-time thermal capability calculation of an earthing transformer can be done using the following formula;

$$\theta_f = (T+\theta_s)\left[(1+K)m+0.6\,m^2\right] \tag{17.6}$$

where

θ_f = short-time winding temperature in t seconds (°C)

$$m = t\left(\frac{w_s}{C_{av}}\right)\left(\frac{1}{T+\theta_s}\right)$$

w_s = I^2R loss in watts/kg at the starting temperature θ_s

$$= \frac{I^2R}{W_{cu}}$$

I = Fault current of winding
W_{cu} = Weight of conductor in kg
C_{av} = Average specific thermal capacity at any temperature θx in Joules per degree celsius per kg
$C_{av} = 384 + 0.099\,\theta_x + 243\dfrac{A_i}{A_c}$ per kg of copper
A_i = cross sectional area of insulation
A_c = cross sectional area of conductor

$$T = \begin{array}{l} 234.5 \text{ for copper} \\ 225 \text{ for aluminium} \end{array}$$

K = per unit eddy current loss based on I^2R loss at referance temperature
θ_s = starting temperature

Reference ambient temperature + steady state hottest spot temperature rise above reference ambient temperature at continuous rating (for design calculation, the reference ambient temperature + guaranteed steady state hot spot temperature rise can be used).

C_{av} for aluminium is given by

$$C_{av} = 893 + 0.441\,\theta_x + 794\frac{A_i}{A_c} \text{ per kg of Aluminium}$$

18 Amorphous Core Transformers

18.1 INTRODUCTION

The word "amorphous" means lacking symmetrical structure or form. Amorphous metals are a group of alloys with metallic properties without the crystalline lattice of conventional metals. These metals have no atomic order and are consequently referred to as metallic glasses. The crystal and amorphous structures are shown in Figure 18.1.

Amorphous metals are formed into thin strips by rapidly cooling molten metals at very high cooling rates of 10^6 °C/second to prevent crystallization of the material during solidification. The development of glassy alloys began with gold and silicon in 1960 by Duwez and co-workers.

Of the iron-rich glassy materials, iron–boron alloy was the first to be considered for transformer applications. Further studies established that iron–boron–silicon alloys are better suited for transformer production. Allied Signal (now Hitachi Metals) introduced Fe78 B13 Si9, iron–boron–silicon alloy Metglas 2605S and later on introduced improved varieties. The schematic of the amorphous metal manufacturing process is given in Figure 18.2.

The comparison of properties of amorphous metal and CRGO steel are given in Table 18.1

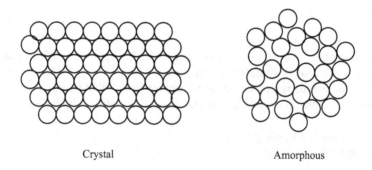

Crystal   Amorphous

FIGURE 18.1   Crystal and amorphous structure.

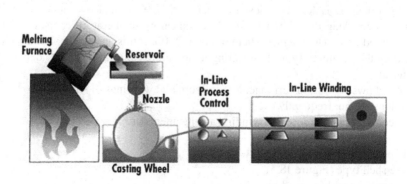

FIGURE 18.2   Amorphous metal manufacturing process.

TABLE 18.1

Comparison of Properties of Amorphous Metal and CRGO Steel

Properties	Amorphous Metal		CRGO Steel M5
	2605SA1	2605HB1M	
Density (g/cm³)	7.15	7.33	7.65
Specific resistance	130	120	45
Saturation flux density (tesla)	1.56	1.63	2.03
Typical working flux density (tesla)	1.35	1.40	1.70
Typical watts/kg at 1.35 tesla at 50 Hz	0.20 (built core)	0.17	0.70 (at 1.35 T) 1.29 (at 1.7 T) (Epstein value for sheets)
Thickness (mm)	0.025	0.025	0.3
Space factor	0.80~0.86	0.80~0.86	0.97
Brittleness	High	High	Low
Available form	Ribbon/foil Sizes: 106, 142.2, 170.2 and 213.4 mm	Ribbon/foil Sizes: 106, 142.2, 170.2 and 213.4 mm	Sheet/roll Coil width of 914, 1000 mm, etc.
Annealing temperature	350°C	310°C	810°C
Annealing atmosphere	Inert gas	Inert gas	Inert gas
Special requirement for annealing	DC magnetic field (approx. 10 oersted)	DC magnetic field (approx. 10 oersted)	–

The construction of a transformer using an amorphous magnetic material is fundamentally the same as that of a conventional wound core transformer. However, due to the special properties of the material, the manufacturing processes are different from CRGO core.

The core manufacturing process with amorphous metal consists of the following:

a. Core winding (ribbon winding)
b. Core cutting (ribbon cutting)
c. Core forming
d. Magnetic field annealing

The amorphous core material in as-cast state has high quenching stress.

Annealing of the core is done to relieve the quenching stress and also to induce a preferred axis of magnetostriction along the ribbon length. The annealing temperature is such that the glassy nature is preserved and a DC magnetic field of about 2400 A/m is applied to saturate the material at the annealing temperature. Typical annealing temperatures are 310°C–370°C depending on the type of material.

Figure 18.3 shows the various arrangements of core and windings of single-phase transformers. All these options are not frequently used.

The most commonly used core and winding arrangements are;

• Single phase shell type (Figure 18.4)
• 3-phase shell type (Figure 18.5)
• 3-phase core type (Figure 18.6)

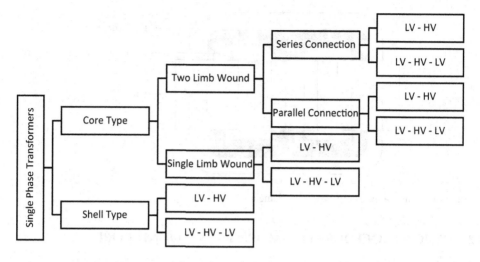

FIGURE 18.3 Core and winding arrangements.

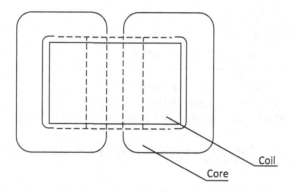

FIGURE 18.4 Single-phase shell-type arrangement.

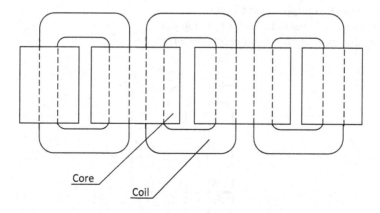

FIGURE 18.5 Three-phase shell-type arrangement.

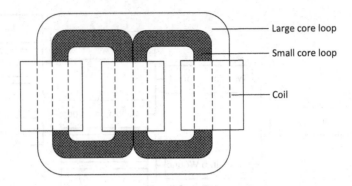

FIGURE 18.6   Three-phase core-type arrangement.

18.2   DESIGN PROCEDURE OF AMORPHOUS WOUND CORE

18.2.1   STRUCTURE OF CORE

Figure 18.7 shows the structure and dimensions of an amorphous step lap core.

D = ribbon width in mm
H = core window height in mm
W = core window width in mm
t = wound thickness of core yoke without joints in mm
t_1 = thickness of core limb in mm
t_2 = thickness of core yoke with step lap joints in mm
R_1 = inside corner radius of core at yoke without joints in mm (= 3 mm)
R_2 = inside corner radius of core at yoke with joints in mm (= 8 to 10mm)

$$\text{Volt}/_{\text{Turn}} = 4.44 \times B_m \times D \times T \times f \times 10^{-6}$$ (18.1)

where

f = frequency in Hz
B_m = working flux density in Tesla
D = ribbon width of amorphous core in mm
t_1 = core build-up (thickness of core build-up) in mm
T = effective thickness of core build-up in mm = 0.8 t

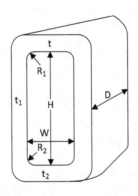

FIGURE 18.7   Amorphous step lap core.

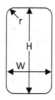

FIGURE 18.8 Winding mandrel size.

$$t_1 = \frac{T}{0.85}$$

Effective core cross sectional area $= D \times T$ in mm

$$\text{Core window height (Inside)} = H = \text{Coil Height} + A + B \qquad (18.2)$$

where

A = coil-to-top yoke clearance in mm
B = coil-to-bottom yoke clearance in mm
Radius of curvature of core $R_1 = 3$ mm
$R_2 = 10$ mm

For single-phase transformers and three-phase centre core, window width
W = coil winding thickness inside core window $\times 2 + C$
where C = phase to phase clearance

For 3-phase transformer side core
W = coil winding thickness inside core $+ P$
where P = winding to core clearance
t_2 = thickness of core at step lap point $= \left(\dfrac{n+1}{n}\right)t_1$
$n = \dfrac{W-15}{15}$ (rounded off to integer)
n = number of steps per block
Core weight $= D \times T \times 7.18 \times 10^{-6}$ kg
Core loss = core weight $\times w$
where
W = w/kg at the working flux density
Winding mandrel size is shown in Figure 18.8.
$H = D + r \times 2$
$W = t_1 + 9$ (for single-phase transformers)
$= t_1 \times 2 + 11$ (for three-phase transformers)
R = minimum bending radius of conductor in mm

18.3 DESIGN OF A SINGLE-PHASE AMORPHOUS CORE TRANSFORMER

Single-phase transformers offer advantages of reducing the line losses for power supply to remote areas, villages, etc., where the HV distribution system is employed. The construction of the core depends on the overall economy or specific requirements. One core with windings on both limbs

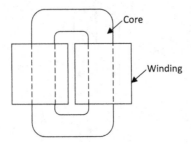

FIGURE 18.9   Core-type design (one core and two windings).

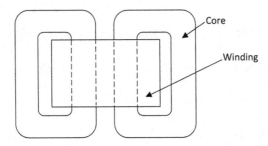

FIGURE 18.10   Shell-type design (one winding and two cores).

is one option, and one winding with two cores giving a shell type configuration is the other option. The normal design types required are as follows:

a. HV neutral earthed type, where the transformer primary is connected to one phase of the HV line, and the other end of the HV winding is grounded either inside the tank or through a neutral bushing. The HV line bushing is normally mounted on the top cover. LV can be single-voltage or dual-voltage type.

b. HV of the transformer connected across two incoming lines. In this case, the HV winding will necessarily have full insulation with respect to the LV and ground. The LV can be with single ratio or dual ratio.

c. Two modes of winding dispositions with respect to the core: core–LV–HV arrangement and core LV–HV–LV arrangement. In the former case, both LV and HV are full-capacity windings, and in the later case, 50% of the LV is wound after the core and the remaining 50% is wound above the HV. The LV–HV–LV arrangement offers low impedance, and it is normally used in the rural distribution system as per ANSI C57.12 standards.

d. Core and winding arrangements

A shell type or core type arrangement of the core is adopted. The selection depends on the design practices followed and the overall cost optimization requirements. Figures 18.9 and 18.10 show the different arrangements.

18.3.1   CORE DIMENSIONS (FIGURE 18.11)

A = window height
B = window width
C = core build-up (core limb thickness)
D = core width (ribbon width)
R = core window radius of core (3 mm typical)

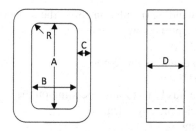

FIGURE 18.11   Core dimensions.

18.3.2   CORE COIL ASSEMBLY DIMENSIONS AND CLEARANCES

Figures 18.12 and 18.13 show the core and coil assembly dimensions and clearances.

E = core-to-winding clearance on the C side
F = core-to-winding clearance on the D side
G = clearance between cores (for shell-type construction)

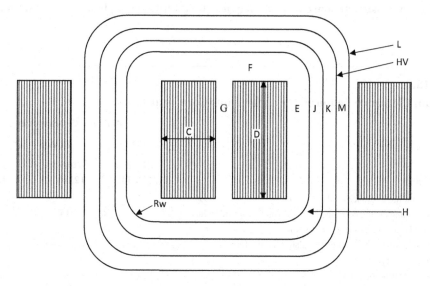

FIGURE 18.12   Core coil cross section and clearances.

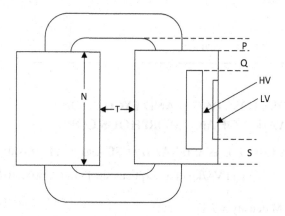

FIGURE 18.13   Core coil cross section and clearances.

H = inside insulation of LV winding (insulation above the former)
J = LV coil build-up (radial depth) below core
K = HV–LV gap (wrap and duct)
L = HV coil build-up (radial depth) below core
M = insulation wrap above HV coil
R_w = LV coil inside corner radius [3 to 5 mm depending on the conductor thickness]
N = winding axial height including edge strip
P = winding end-to-yoke distance
Q = LV edge strip
S = HV edge strip
T = phase-to-phase clearance (HV–HV clearance)

Applicable for core-type design (one core and two coils)

18.4 MINIMUM CLEARANCE REQUIRED FOR AMORPHOUS CORE TRANSFORMERS

Table 18.2 gives the minimum clearances required for oil-filled amorphous core transformers up to 33 kV class.

TABLE 18.2
Minimum Clearance Required for Amorphous Core Transformers

| | Minimum Clearances in mm | | |
| | 11 kV | 22 kV | 33 kV |
Description	28 kV AC, 75 kV BIL	50 kV AC, 125 kV BIL	70 KV AC, 170 kV BIL
E. Core-to-winding clearance on the "C" side	3	3	3
F. Core-to-winding clearance on the "D" side (including coil radius)	12	15	15
G. Clearance between cores (for shell type)	0	0	0
H. Inside insulation of LV winding (insulation on the former)	1.5	1.5	1.5
K. HV–LV gap (wrap and duct)	9	13	18
M. Insulation wrap above the HV coil	0.5	1.0	1.0
R_w – LV coil's inside corner radius	3–5 mm	3–5 mm	3–5 mm
P. Winding end-to-yoke clearance	12	15	15
Q. LV edge strip	15	15	20
S. HV edge strip	15–20	25	30
T. Phase–phase clearance [for core-type design]	10	15	20

18.5 TYPICAL CORE LOSS (W/kg) AND EXCITATION CURRENT (VA/kg) OF AN AMORPHOUS CORE

18.5.1 AMORPHOUS CORE LOSSES AND VA/KG AT 50 AND 60 Hz (GRADE 2605HB1M)

Table 18.3 gives the watts/kg and VA/kg of amorphous steel grade 2605HB1M at 25°C:

- Grade 2605HB1M density = 7.33 gm/cm³
- Saturation flux density = 1.63 T

TABLE 18.3
Watts/kg and VA/kg of Grade 2605HB1M

Flux Density (Tesla)	W/kg						VA/kg					
	Epstein Losses		Single-Phase Built Core		Three-Phase Built Core		Epstein Value		Single-Phase Built Core		Three-Phase Built Core	
	50 Hz	60 Hz	50 Hz	60 Hz	50 Hz	60 Hz	50 Hz	60 Hz	50 Hz	60 Hz	50 Hz	60 Hz
0.7	0.026	0.035	0.04	0.05	0.049	0.063	0.026	0.038	-	-	-	-
0.8	0.032	0.042	0.06	0.06	0.07	0.08	0.036	0.05	-	-	-	-
0.9	0.04	0.052	0.085	0.070	0.094	0.12	0.05	0.064	-	-	-	-
1.0	0.05	0.065	0.10	0.125	0.125	0.156	0.065	0.085	0.18	0.20	0.26	0.29
1.1	0.06	0.077	0.12	0.15	0.15	0.187	0.09	0.12	0.2	0.22	0.29	0.32
1.2	0.07	0.09	0.14	0.18	0.175	0.225	0.16	0.18	0.22	0.25	0.32	0.36
1.3	0.083	0.12	0.17	0.215	0.212	0.27	0.22	0.28	0.25	0.34	0.36	0.43
1.35	0.091	0.13	0.185	0.235	0.23	0.295	0.28	0.34	0.32	0.42	0.46	0.61
1.40	0.10	0.14	0.2	0.255	0.25	0.37	0.34	0.40	0.50	0.60	0.73	0.87
1.45	0.115	0.16	0.22	0.285	0.28	0.355	0.43	0.55	0.75	0.90	1.0	1.35
1.50	0.13	0.18	0.24	0.32	0.30	0.40	0.52	0.74	1.5	1.8	2.3	2.9

18.5.2 Amorphous Core – Losses and VA/kg at 50 and 60 Hz (Grade 2605SA1)

Table 18.4 gives the watts/kg and VA/kg of amorphous steel grade 2605SA1 at 25°C:

- Grade 2605SA1 density = 7.18 gm/cm³
- Saturation flux density = 1.56 Tesla

18.6 CALCULATION OF MEAN LENGTH OF WOUND CORE

This section gives the calculation of mean length of wound core. Figure 18.14 shows the dimension of the wound core.

TABLE 18.4
Watts/kg and VA/kg of Grade 2605SA1

Flux Density (Tesla)	W/kg						VA/kg					
	Epstein Value		Single-Phase Built Core		Three-Phase Built Core		Epstein Value		Single-Phase Built Core		Three-Phase Built Core	
	50 Hz	60 Hz	50 Hz	60 Hz	50 Hz	60 Hz	50 Hz	60 Hz	50 Hz	60 Hz	50 Hz	60 Hz
0.7	-	-	-	-	-	-	-	-	-	-	-	-
0.8	0.03	0.046	0.055	0.070	0.068	0.087	0.05	0.07	0.018	-	0.026	-
0.9	0.04	0.058	0.073	0.072	0.091	0.116	0.06	0.08	0.03	-	0.054	-
1.0	0.052	0.07	0.1	0.125	0.125	0.16	0.078	0.95	0.18	0.20	0.26	0.29
1.10	0.065	0.085	0.117	0.148	0.150	0.185	0.095	0.13	0.20	0.22	0.29	0.32
1.20	0.08	0.10	0.141	0.18	0.18	0.23	0.13	0.17	0.22	0.24	0.32	0.35
1.30	0.10	0.13	0.171	0.218	0.21	0.27	0.20	0.25	0.30	0.36	0.44	0.52
1.35	0.115	0.15	0.189	0.241	0.24	0.30	0.30	0.42	0.40	0.55	0.58	0.80
1.40	0.13	0.17	0.209	0.267	0.27	0.33	0.40	0.60	0.50	0.70	0.73	1.10
1.45	0.155	0.205	0.232	0.295	0.30	0.37	0.60	0.80	0.8	1.0	1.2	1.5

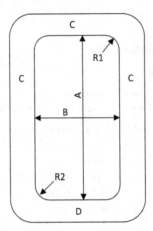

FIGURE 18.14 Dimensions of wound core.

L_m = mean length of core loop
A = window height (inside)
B = window width (inside)
C = stack thickness
R_1 = corner radius (corner without yoke staggering)
R_2 = corner radius (corner with yoke staggering)
D = stack thickness at the yoke with staggering

Mean length of core loop

$$L_m = \pi\left(R_1 + \frac{C}{2}\right) + \pi\left(\frac{R_2 + \frac{C}{2} + R_2 + \frac{D}{2}}{2}\right) + (A - R_1 - R_2)2 + (B - 2R_1) + (B - 2R_2)$$

When $C = D$, the above formula is reduced to

$$L_m = \pi\left(R_1 + \frac{C}{2}\right) + \pi\left(R_2 + \frac{C}{2}\right) + (A - R_1 - R_2)2 + (B - 2R_1) + (B - 2R_2) \qquad (18.3)$$

18.7 CALCULATION OF MEAN LENGTH OF WINDINGS WITH WOUND CORE

The cross section of core limb is rectangular, and therefore, the winding is done on a rectangular former. Figure 18.15 shows the dimensions of the core coil assembly.

H = winding mandrel long side (inside yoke)
W = winding mandrel short side (outside yoke)
W_1 = radial thickness of LV (inside yoke)
W_2 = radial thickness of LV (outside yoke)
W_g = HL gap
W_3 = radial thickness of HV (inside yoke)
W_4 = radial thickness of HV (outside yoke)
t = insulation above mandrel (insulation on ID of LV)
R = corner radius of mandrel

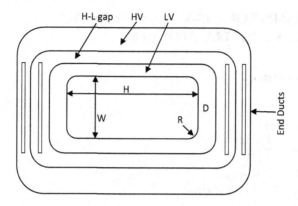

FIGURE 18.15 Dimensions of wound core coil assembly.

(Minimum bending radius of conductor + 1 mm)

$$LV_m = \text{mean length of LV} = (\text{H-2R}) + (\text{W} - 2\text{R}) + 2\pi\left[R + t + \frac{w_1 + w_2}{4}\right]$$

$$L_{gm} = \text{mean length of H-L gap} = (\text{H-2R}) + (\text{W-2R}) + 2\pi\left\{R + t + \frac{w_1 + w_2}{2} + \frac{w_g}{2}\right\}$$

$$HV_m = \text{mean length of HV} = (\text{H-2R}) + (\text{W} - 2\text{R}) + 2\pi\left[R + t + \frac{w_1 + w_2}{4} + W_g + \frac{w_3 + w_4}{4}\right]$$

18.8 REACTANCE CALCULATION

The winding is non-circular as it is done on a rectangular former. The reactance formula of Chapter 7 is modified as shown below for application to non-circular windings. The general formula remains unchanged.

$$X\% = 8\pi^2 f \frac{I}{V} T^2 KD \times 10^{-8} \qquad (18.4)$$

$$K = 1 - \frac{W_1 \times a + W_2 \times b + W_g + W_3 \times a + W_4 \times b}{\pi(h_L + h_h) \times 0.5}$$

where

$$a = \frac{H}{H + W} \text{ and } b = \frac{w}{H + W}$$

h_L = axial height of LV (electrical)
h_h = axial height of HV (electrical)

$$D = \left(\frac{W_1 \times a + W_2 \times b}{3h_L}\right) LV_m + \frac{W_g L_{gm}}{0.5(h_h + h_L)} + \left(\frac{W_3 \times a + W_4 \times b}{3h_h}\right) HV_m$$

Note: If W_g is not uniform throughout, the effective value of W_g is calculated as $aW_{g1} + bW_{g2}$, where

$W_{g1} = H_L$ gap inside yoke
$W_{g2} = H_L$ gap outside yoke

The other electrical parameters are calculated as per the same producers applicable for transformers with CRGO core.

18.9   DESIGN EXAMPLE OF A 15-KVA SINGLE-PHASE AMORPHOUS CORE TRANSFORMER

Transformer Specifications
- kVA – 15
- Phase – 1
- Frequency – 60 Hz
- Primary volts – $\dfrac{13,200}{\sqrt{3}}$
- Taps – ±5% @ 2.5% steps
- Primary neutral – solidly earthed
- Secondary volts – 240
- Winding material – copper
- Oil temperature rise – 60°C
- Winding temperature rise – 65°C
- No-load loss – 16 watts (max.)
- Load loss at 85°C – 205 watts (max.)
- Applicable standard – ANSI

Core Design
- Core cross section = 170 × 77
- Core net area = 107.464 cm²
- LV turns = 31
- Volt/turn = 3.871
- Working flux density = 1.35 T
- Core window height = 155
- Core leg centre to centre = 85

Winding Conductor Selection

$$\text{HV Current} = \frac{15 \times 1000}{13200 \big/ \sqrt{3}} = 1.97 \text{ A}$$

$$\text{LV Current} = \frac{15 \times 1000}{240} = 62.5 \text{ A}$$

Enamel-covered round copper wire is selected for HV winding, and enamel covered rectangular conductor is selected for LV winding. The design particulars are given in Table 18.5.

Insulation between windings, core and limbs
- Core to LV insulation = 2 mm
- HV–LV insulation = 4.5 mm
- Insulation between limbs = 6 mm
- Interlayer insulation of HV = 0.4 mm
- Interlayer insulation of HV = 0.25 mm

TABLE 18.5
Design Particulars for a Single-Phase Amorphous Core Transformer

Particulars	HV	LV
Number of limbs	2	2
Turns per limb	31	1033
Total turns	62	2066 (max. tap)
Number of layers/limb	2	13
Turns/layer	15.5	81
Bare size of the conductor	7.9 × 2.1	1.2
Covered size of the conductor	8 × 2.2	1.3
Number of conductors in parallel	2	1
Number of conductors in radial direction	2	1
Number of conductors in axial direction	1	1
Cross sectional area of the conductor	32.18	1.67
Current density	1.94	1.68
Winding mandrel size	182 × 86	–
Inside dimensions of winding	186 × 90	215 × 118
Outside dimensions of winding	206 × 109	260 × 162
Edge strip size	(5 × 13) × 2	16.5 × 2
Axial height of winding	142	142
Electrical height of winding	132	109
Radial thickness of winding	9	21
Mean length of winding	567	689
Bare weight of conductor	10	14.75
Weight of covered conductor	10.3	15.0
Interlayer insulation	0.25	0.40
Resistance at 85°C	0.0257	25.87
I^2R loss at 85°C	94	100
Eddy current loss	4	0.4
Insulation to yoke	10 (bottom)	21.5 (bottom)
	13 (top)	24.5 (top)
Winding gradient	9.8	7.8

No-Load Loss Calculation
Core weight = 57 kg
Flux density = 1.35 tesla
Watts/kg = 0.26
No-load loss = 57 × 0.26 = 14.8 watts (max. allowed as per requirement is 16 W)

Load Loss Calculation
LV I^2R loss = 94
LV eddy = 4
HV I^2R = 100
HV eddy current loss = 0.4
Miscellaneous loss = 2.0
Total losses = 200.4 (max. allowed as per requirement is 205 W)

Calculation of Impedance

$$\%R = \frac{200.4}{15 \times 10} = 1.34\%$$

– kVA	1000
– Phase	3
– Frequency	50 Hz
– Primary volts	11000 (Delta)
– Taps	±3 × 2.5%
– Secondary volts	400 (Star)
– Vector group	Dyn11
– Cooling	ONAN
– Winding material	Copper
– No-load loss	450 W (max.)
– Load loss at 75°C	5600 W (max.)
– 5 Leg core design	
– Stacking factor	0.86
– Net core area	643.2 cm²
– Core weight	1416 kg
– Width of core sheet	170 mm (width including coating = 174 mm; total core width = 348 mm)

$$\%X = 1.74\%$$

$$\%Z = \sqrt{1.34^2 + 1.74^2} = 2.2\%$$

Tank Size and Oil Quantity
Round tank design is selected.
Tank diameter = 425 mm
Tank height = 700 mm
Oil quantity = 70 l
Weight of tank = 26 kg

18.10   DESIGN EXAMPLE OF A 3-PHASE AMORPHOUS CORE TRANSFORMER

The core coil assembly details are schematically shown in Figures 18.16–18.18.

The core dimensions are shown in Table 18.6 (4 cores with 208 window and 4 cores with 113 window are used) and the design particulars are in Table 18.7.

- Winding former dimensions.

$$\text{Former length} = nw + g_{cc} + (2 \times g_w)$$

n = number of cores in parallel = 2
w = width of ribbon including coating = 174 mm
g_{cc} = gap between cores = 0 (when two core loops are used in parallel = 0)
g_w = core-to-winding clearance on width side (D side) including coil radius = 6 mm

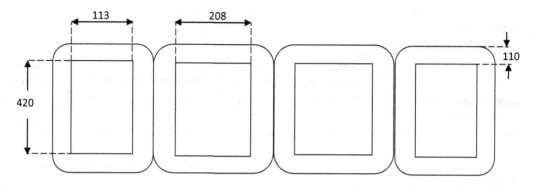

FIGURE 18.16 Schematic of core coil assembly.

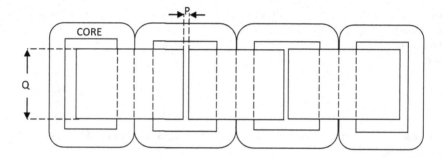

FIGURE 18.17 Schematic of core coil assembly.

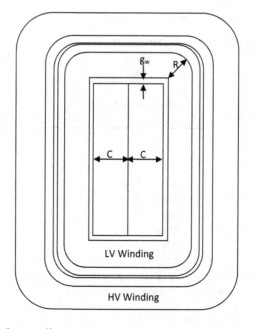

FIGURE 18.18 Top view of core coil assembly.

TABLE 18.6
Core Coil Assembly Dimensions

Core	Window Height (H)	Window Width (B)	Build Stack (C)	Width (D)	Number of Cores
Inside core (208 window)	420	208	110	170	4
Outside core (113 window)	420	113	110	170	4

TABLE 18.7
Design Particulars: 3-Phase Amorphous Core Design

	LV	HV
Turns	12	615-572-529
Conductor	$370 \times 1.27 \times 2$	2.5×9 (Enameled)
Conductor covering	–	0.1
Covered size of the conductor	–	2.6×9.1
Layers	12	16
Turns/layer	1	39/30
Layer insulation	0.125	0.25
Lmt	1310	1680
Coil radial depth (Calculated including manufacturing tolerance)	$31.85/34; (1.27 \times 2 \times 12 + 11 \times 0.125) = 31.85$	$45.35/48.5; (2.6 \times 16 + 15 \times 0.25) = 45.35$
Current density	1.54	1.39 (Rated tap)
		1.50 (Lowest tap)
I^2R @ 75°C	2195	2523
Eddy loss	35	195

Ducts

Winding	No. of ducts	Size of ducts	Position of ducts
LV	3	4×4	END (Outside window)
HV	3	4×4	END (Outside window)

HV–LV Gap

Full gap of 10 mm built of 2 ducts of 4mm and 2 mm wrap (Solid insulation)

Coil dimensions and clearances

Core to LV	:	4.5 mm
LV inside dimensions	:	363×229 mm
LV outside dimensions	:	455×297 mm
HV inside dimensions	:	475×317 mm
HV outside dimensions	:	620×418 mm
HV winding to yoke	:	25 mm
HV phase to phase	:	10 mm
Former to LV wrap	:	1.5 mm
LV radial thickness inside window	:	34 mm
HV radial thickness inside core window	:	48.5 mm

(Continued)

TABLE 18.7 (*Continued*)
Design Particulars: 3-Phase Amorphous Core Design

HV outer wrap	:	0.125×4 mm
Core to winding	:	10 mm
LV winding edge strip	:	12.5 mm
HV winding edge strip	:	15 mm
Core coil Length	:	1540 mm
Core coil Depth	:	620 mm
Core coil Height	:	700 mm
Core gross area		$= 110 \times 170 \times 2 = 37400$ mm²
Core Net area		$= 0.86 \times 37,400 = 32164$ mm²
Core Net area for flux density calculation		$= 321.64 \times 2 = 643.2$ cm²
No. of turns of LV		$= 12$
Volt/turn		$= \dfrac{400}{12 \times \sqrt{3}} = 19.245$
Flux density		$= \dfrac{19.245 \times 10^4}{4.44 \times 50 \times 643.2} = 1.348$ Tesla
Core weight		$= 1416$ kg

No Load Loss Calculation

Core loss = Core weight × w/kg × 3 phase factor
$$= 1416 \times 0.22 \times 1.33$$
$$= 414 \text{ watts}$$

Load Loss Calculation

LV I^2R loss	= 2195 watts
HV I^2R loss	= 2195 watts
LV Eddy current loss	= 35 watts
Stray loss in tank	= 335 watts
Other stray losses	= 185 watts
Load Loss	= 5468 watts

As per the design, the former length $= 2 \times 174 + 0 + 2 \times 6$
$= 348 + 12 = 360$ mm

$$\text{Former Width} = 2 \times C + 2d_w$$

C = core build-up (stack thickness) = 110 mm
d_w = core to coil clearance on C side = 3 mm
As per the design, former width $= 2 \times 110 + 2 \times 3$
$= 220 + 6 = 226$ mm
P = phase to phase clearance
Q = height of coil including edge strip
R = coil inside radius
C = core stack thickness

19 Design of Current-Limiting Reactors

19.1 AIR-CORE DRY-TYPE REACTORS

Series reactors are used in circuits to reduce the fault current and hence called current-limiting reactors. These can be dry type or oil (liquid) immersed. Aluminium or copper windings are used as per the customer requirements.

The above design options can be manufactured without magnetic core which are called air-core reactors. In this design, a magnetic material is used as a shield to limit the flux leakage.

When a magnetic core is used, the design is gapped-core type, where the gaps in core limbs are created by non-magnetic insulator inserts. The design procedures of a typical air-core dry-type current-limiting reactor with AN cooling are given in this chapter.

- Series reactor (current-limiting reactor)

$$S_{\text{rating}} = \sqrt{3} \ V_L \ I_L \tag{19.1}$$

$$V_{\text{drop}} = I_{\text{Line}} \ X_L \ \text{series}$$

$$V_{\text{drop}} = I_{\text{line}} \ 2\pi f \ L \ \text{series} \tag{19.2}$$

$$Q_{\text{series}} = \frac{3 \ V_{\text{drop}}^2}{X_L \ \text{series}} = 3 \ V_{\text{drop}} \ I_{\text{Line}}$$

$$= 3 \ I_{\text{Line}}^2 \ X_L \ \text{series}$$

$$Q_{\text{series}} = \left(6\pi f \ L_{\text{series}} \ \right) \left(I_{\text{line}} \right)^2 \tag{19.3}$$

where

S_{rating} = apparent power rating (kVA)
V_{drop} = voltage drop in the series reactor (choke voltage)
V_{Line} = line voltage (kV)
I_{Line} = line current (A)
L_{series} = series inductance (Henry)
f = frequency (Hz)
Q_{Series} = lagging reactive power (kVAr)

19.1.1 Magnetic Field Produced by a Cylindrical Winding

dB = magnetic flux density produced by a small current filament dL at a distance r (Figure 19.1)

$$\mu_o = 4\pi \times 10^{-7} \text{ H/m}$$

I = current
dL = small length of current-carrying filament
$\hat{r}$ = unit vector of r
$|r|$ = distance between the current filament and the point where flux density is to be calculated

$$dB = \frac{\mu_0}{4\pi} \frac{i \; dl \; \hat{r}}{|r|^2} \tag{19.4}$$

The field produced by a current-carrying loop of n turns at a distance of r from the coil centre is calculated here. Figure 19.2 shows the schematic of a current-carrying loop.

$$|B| = \text{magnitude of magnetic field}$$

$$\mu_o = 4\pi \times 10^{-7} \text{ H/m}$$

n = number of turns of the loop
I = current
D = mean winding diameter (loop diameter)
r = distance from the centre of the coil to the point where the flux is to be calculated

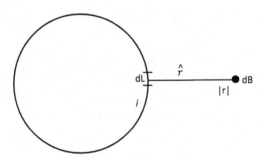

FIGURE 19.1   Magnetic field of a cylindrical winding.

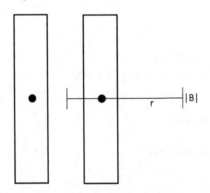

FIGURE 19.2   Magnetic field of a current-carrying loop of n turns at a distance of r from the coil centre.

$$|B| = \frac{\pi\, n\, I\, D^2}{4\, r^3} \times 10^{-7} \text{ tesla} \tag{19.5}$$

The flux density at a point which is shifted by an angle θ from the plane of the loop is calculated by introducing the parameter of θ.

$$|B| = \frac{(4\pi \times 10^{-7})n\, I\, D^2}{8\, r^3} \sqrt{\sin^2\theta + \frac{\cos^2\theta}{4}} \tag{19.6}$$

19.1.2 EDDY CURRENT LOSSES PRODUCED BY METALLIC PARTS

The eddy current losses produced by metallic parts (e.g. terminal, metallic spider) can be calculated once the flux density is known.

$$P = \frac{B^2\, t^3\, b}{6\left(4\pi \times 10^{-7}\right)^2 \delta^4\, k} \tag{19.7}$$

B = flux density (Tesla)
t = thickness of the terminal, spider, etc. (m)
b = width of the terminal, spider, etc. (m)
δ = penetration depth of the material (m)

$$= \sqrt{\frac{2}{wk\mu}}$$

$w = 2\pi f$ (f = frequency)
k = specific conductance of the material (S/m)
μ = permeability (H/m)
P = specific eddy current loss (W/m)
{The formula is applicable to non-magnetic conducting materials.}

Example

Losses produced in an aluminium spider of 10 mm thickness and 100 mm width are calculated at 75°C temperature:

B = 100 millitesla = 100×10^{-3} = 0.1 tesla
t = 0.01 m (thickness)
b = 0.1 m (width)
K = 28.6×10^6 s/m
δ = penetration depth

$$\delta = \sqrt{\frac{2}{2\pi fK\mu}}$$

$$= \sqrt{\frac{2}{2\pi \times 50 \times 28.6 \times 10^6 \times 4\pi \times 10^{-7}}}$$

$$= 0.0133 \text{ m}$$

Substituting the value of δ in the formula 19.7,

$$P = \frac{B^2\, t^3\, b}{6\left(4\pi \times 10^{-7}\right)^2 \delta^4 K}$$

$$= \frac{(0.1)^2 \times (0.01)^3 \times 0.1}{6 \times 16\pi^2 \times 10^{-14} \times 28.6 \times 10^6 \times (0.0133)^4}$$

$$= 118 \text{ W/m}$$

19.1.3 Clearances Required for Air-Core Reactors

Clearances are decided by considering the requirements of electrical clearance, ventilation clearance and magnetic clearance:

- Electrical clearance

  The outer surfaces of air-core reactors are live, and therefore, the minimum electrical clearances of live parts to ground as per standard substation clearances are to be provided.
- Magnetic clearance

  The magnetic clearances required are based on limiting the eddy currents induced on the conductive metallic parts in the vicinity which do not form a closed loop, which is shown as m_1 clearance in Figure 19.3.

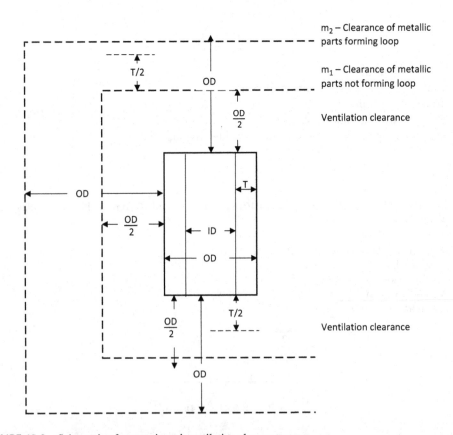

FIGURE 19.3 Schematic of magnetic and ventilation clearances.

There is a possibility of the flux linking a closed electrical loop in the substation such as fencing, structural member and reinforcing bars in the foundation/concrete. These clearances are shown as m_2 in Figure 19.3.

- Ventilation clearance

    The typical air-core reactor is mounted such that the cooling ducts are oriented to the vertical direction. Adequate clearances at the bottom and top of the coil are required. These clearances are normally less than the magnetic clearance requirement (m_1 and m_2) unless the reactor current rating is low.

The minimum magnetic and ventilation clearances are shown in Figure 19.3.

OD = outer diameter of the coil
ID = inner diameter of the coil
T = radial thickness of winding

19.1.4 DESIGN PROCEDURE OF DRY TYPE AIR CORE SERIES REACTORS

A design can be started by making an initial assumption of the internal diameter and axial height of the reactor coil.

D = internal diameter of the reactor coil (mm)
H = axial height of the reactor coil (mm) [electrical height]
L_r = inductance of the reactor (millihenry)

The required inductance is calculated from the specified rated current and kVAr.
T = covered thickness of winding conductor in the radial direction for one turn (mm)

The number of turns required is calculated by the formula:

$$N = \frac{1323}{D}\sqrt{H\ L_r} \qquad (19.8)$$

N is to be rounded off to the nearest integer + 0.5.

It is a normal practice to give a fractional number of turns of multiples of 0.5 so that the line leads come from the opposite sides of the reactor.

Most of the air-core reactors use aluminium as the winding conductor, and the insulation covering of the specified thermal class (e.g. class F, H etc.) is used, on the conductor.

The current density is selected by considering the cooling requirement and the short circuit thermal and dynamic requirements. The normal current density for AN-cooled aluminium winding ranges from 0.8 to 1.2 A/mm^2.

The typical insulation covering for various voltage classes is given in Table 19.1.

TABLE 19.1

Insulation Winding Covering of a Conductor of Air Core Reactor

Rated Voltage (kV)	AC Test Voltage (kV)	Insulation Thickness (mm) (Increase in Conductor Thickness)
11	28	0.4
22	50	0.5
33	70	0.6

When the rated current is high, parallel conductors are required.

Parallel conductors in radial direction, when required, are wound as separate concentric coils spaced by air ducts. This offers the advantage that transpositions are not required. The number of layers will be equal to the number of parallel conductors in the radial direction.

However, the reactance of the layers will be unequal, if the number of turns of all layers is kept the same. The ideal condition is to have equal impedance for all parallel layers. This is made nearly equal in the design stage by progressively reducing the number of turns of the outer layers.

19.1.5   SAMPLE CALCULATIONS OF A SINGLE PHASE AIR CORE DRY TYPE REACTOR COIL

Sample calculations of the dimensions and other parameters of a single phase air core dry type reactor coil are given here to make the calculation procedure clear.

- Specifications
- Design calculations

Application	-	Current limiting
Duty	-	Outdoor, air core
Frequency	-	50 Hz
Cooling	-	AN
Rated continuous current	-	55 A
Rated kVAr	-	4.8
Class of insulation	-	F
AC test voltage	-	70 kV
Impulse test voltage	-	170 kV
Winding temperature rise	-	90°C
Maximum losses at rated current (ref. temperature = 120°C)	-	500 watts
Winding material	-	Aluminium

$kVAr = I^2 X$
where $X = 2\pi f L_r$ = inductive reactance of coil (ohms)
L_r = inductance of the reactor coil
f = frequency

$$\therefore 4.8 \times 10^3 = 55^2 \times 2\pi \times 50 \times L_r$$

$$\therefore L_r = \frac{4.8 \times 10^3}{55^2 \times 2\pi \times 50}$$

$$= 5.05 \times 10^{-3} \text{ Henry}$$

$$= 5.05 \text{ mH}$$

Choke voltage $= I\, X = 55 \times 2\pi \times 50 \times 5.05 \times 10^{-3} = 87.2$ volts

To commence the initial design, the mean diameter and height of reactor coil are assumed as 480 and 490 mm, respectively.

Substituting in equation 19.8

$$N = \frac{1323}{D} \sqrt{H \, L_r}$$

$$= \frac{1323}{480} \sqrt{490 \times 5.05} = 137.10$$

This is rounded off to 137.5 turns.

The conductor width and thickness and the number of parallels are selected by considering the following;

- The wound coil height is kept nearly 490 mm when 137.5 turns are wound in one layer.
- The conductor width is within 4–8 mm (if the conductor is wound on the edge). If conductor is wound on the flat edge, width can be 5–15 mm.
- The number of parallel conductors is kept to optimum so that the number of winding layers is optimized to get sufficient cooling and nearly equal impedance and current sharing.

Considering an initial current density of 0.9 A/mm² in aluminium, the area of cross section of the conductor is calculated.

$$A = \frac{55}{0.9} = 61.1 \text{ mm}^2$$

The approx. covered size of the conductor in the axial direction $= \dfrac{\text{Winding Height}}{(N+1)}$

$$= \frac{490}{(137.5+1)} = 3.54$$

The test voltage is 70 kV, and therefore, 0.6-mm insulation covering is required.

The nearest standard size is 3.6 mm including covering.

It is obvious that the bare conductor size in the axial direction is 3 mm, and therefore, an edge-wound coil design is selected.

The conductor width (bare size) is selected as 5.0 mm, giving the covered size as 5.6 mm.

Area of cross section of 1 conductor $= 5 \times 3 - 0.85 = 14.15$

Number of parallel conductors $= 61.1 \div 14.15 = 4.3$

∴ Four parallel conductors are selected, and this will result in four-layer winding.

$$\text{Working current density} = \frac{55}{14.15 \times 4} = 0.971 \text{ A/mm}^2$$

The reactor rating is not high and therefore, the initial design is made without cooling ducts. The inner surface and the outer surface only are available for heat dissipation. If the temperature rise is within limit, cooling ducts are not required. (If cooling is not sufficient, axial ducts can be provided. In this design, four layers are used, and therefore, up to three cooling ducts can be provided between the layers in the axial direction.)

Insulation between layers is not required as the layers are connected in parallel and the voltage difference across the coil is only 87.2 (choke voltage).

19.1.5.1 Equalization of Current Sharing between Parallel Layers

The four layers of the winding are connected in parallel at the ends. The current distribution is made approximately equal by equalizing the reactances.

The innermost layer has the lowest diameter and the outermost has the highest diameter. The reactance of the layers can be adjusted by progressively reducing the turns of the outer layers. The electrical heights of the layers can be kept the same by providing suitable packing in the body of the outer layers. The overall heights are maintained equal by providing edge strips, of suitable sizes.

The following calculations assume that edge strips are provided on the outer layers when the turns are reduced.

The calculations are tabulated in Table 19.2.

Calculation of the number of turns of each layer
From Equation (19.8),

$$N = \frac{1323}{D}\sqrt{H\,L_r}$$

$$\therefore N^2 = \frac{1323^2\,H\,L_r}{D^2}$$

$$L_r = \frac{N^2\,D^2}{1323^2\,H} \tag{19.9}$$

Knowing that the mean diameter of each layer will be changing from the inner layer to the outer and the electrical heights of the layer change if the number of turns of a layer is changed, the inductance of each layer is equated in the following equation. L_{r1}, L_{r2}, L_{r3} and L_{r4} are the inductances of the first to fourth layers.

$$L_{r1} = \frac{137.5^2\,(485.6)^2}{1323^2\,(137.5 \times 3.6)} = 5.1456 \text{ m H}$$

TABLE 19.2
Design of a Four-Layer Reactor Coil

	Layer 1 (Inner Layer)	Layer 2	Layer 3	Layer 4 (Outer Layer)
Conductor size (covered)	3.6 × 5.6	3.6 × 5.6	3.6 × 5.6	3.6 × 5.6
Inside diameter	480	491.2	502.4	513.6
Radial thickness	5.6	5.6	5.6	5.6
Outside diameter	491.2	502.4	513.6	524.8
Number of turns[a]	137.5	132.5	126.5	121.5
Edge strip × 2	-	9 × 2	20 × 2	29 × 2
Electrical height	3.6 × 137.5 = 495	477	455.4	437.4
Wound height	3.6 × 138.5 = 498.6	480.6	459	441
Resistance at 120°C/layer (Ω)	0.5963	0.5879	0.5739	0.5634
Reactance at 120°C/layer (Ω)	5.4	5.19	5.18	5.19
Impedance at 120°C/layer (Ω)	5.174	5.223	5.212	5.220
Current sharing (amps)	13.79	13.67	13.78	13.76

[a] Rounded off to the nearest 0.5 turns.

$$L_{r2} = \frac{N_2^2 \ (491.2 + 5.6)^2}{1323^2 \ (N_2 + 1) \times 3.6} = 5.1456 \text{ mH}$$

Solving for N_2 and rounding off to the nearest 0.5 turns,
$N_2 = 132.5$
The turns of layers 3 and 4 are calculated by the same procedure.
Weight of bare conductor = 36 kg
$I^2 R$ loss at 120°C = 439 W

$$\text{Eddy current loss for aluminium winding} = 3.56 \times 10^{-8} \frac{I^2 \ T^2 \ t^2 \ f^2}{H^2} W$$

where
  I = rated current
  T = number of turns (consider the turns of first layer where turns are in parallel)
  t = conductor dimension orthogonal to the flux (mm) [for edge-wise winding, the conductor width is to be considered)
  f = frequency (Hz)
  H = electrical axial height of the winding (mm) [consider the height of the first layer, i.e. innermost layer]
  W = weight of aluminium (kg)
  Substituting the values,

$$\text{Eddy current loss} = \frac{3.56 \times 10^{-8} \times 55^2 \times 137.5^2 \times 5^2 \times 50^2}{495^2} \times 36 = 18.7 \text{ watts}$$

Miscellaneous losses due to connections, leads, etc., can be considered at 5% of $I^2 R$ loss

$$= 0.05 \times 493 = 22 \text{ watts}$$

Total losses = 439 + 18.7 + 22 = 479.7 W
  Round off to 480 W (less than 500 W guarantee)

19.1.5.2 Cooling Calculation
Winding temperature rise permissible = 90°C
  Total losses to be dissipated = 500 W (max.)
  Dissipation from outer surface of the coil = 7 W/m²/°C
  Dissipation from inner surface of the coil = 5 W/m²/°C
  Exposed cooling area of outside of the coil = $\pi \times 524.8 \times 441 \times 10^{-6} = 0.727$ m²
  Cooling area of inner coil surface

$$= \pi \times 480 \times 498.6 \times 10^{-6} = 0.75 \text{m}^2$$

Though 90°C is the allowable temperature rise, the calculation will have to be done by allowing a margin of 5°C.
  ∴ C calculations are done on the basis of 85°C rise.
  Heat dissipation from outer surface of coil = $7 \times 0.727 \times 85 = 432$ W
  Heat dissipation from inner coil surface = $5 \times 0.75 \times 85 = 318$ W
  The surfaces can dissipate more than the losses generated, and therefore, the design need not be revised by providing ducts.

19.2   OIL-FILLED AIR-CORE REACTORS

19.2.1   INTRODUCTION

The reactors can be single-phase or three-phase. When the overcurrent factor (i.e. the ratio of the fault current to the rated current) is typically more than 20, air-core reactors are commonly used. Where the overcurrent factors are low, gapped-core reactors are used.

$$\text{kVA of 3-phase reactor} = \sqrt{3}\ V\frac{Z}{100}I \qquad (19.10)$$

where
   V = line volts (kV)
   I = line current (amps)
   Z = impedance (%)
   In terms of the reactance, kVA for a three-phase reactor can be written as follows:

$$\text{kVA} = 3I^2 X_{\text{phase}} \times 10^{-3} \qquad (19.11)$$

where
   I = line current (A)
   X_{phase} = reactance/phase (Ω)
   $X_{\text{phase}} = 2\pi f L$

where
   f = frequency (Hz)
   L = inductance (Henry)

The electrical clearances of oil filled air core or gapped core reactors are the same as applicable for transformer of the same voltage class.

19.2.2   DESIGN PROCEDURE OF AIR CORE OIL FILLED REACTORS

The reactor coils are wound similar to transformer coils. Depending on the current and voltage class required, foil winding, disc winding, layer winding, etc., can be selected.

   The windings of the three phase coils are enclosed by a magnetic shield to prevent the flux from entering the tank or metal parts.

   The windings are clamped at top and bottom by clamp rings and tie rods.

   The shields are supported by structures similar to core clamps of the transformer. The losses in the windings and core and other stray losses are to be dissipated by the cooling system.

   The schematic arrangement of windings and shield is shown in Figure 19.4.

   Symbols used are as follows;

   D = mean coil diameter (m)
   h = leg height [coil height + top and bottom insulation and clamp]
   D_1 = inside diameter of winding (m)
   R = radial depth of winding (m)
   N = number of turns of the coil
   a = difference between limb height and coil electrical height

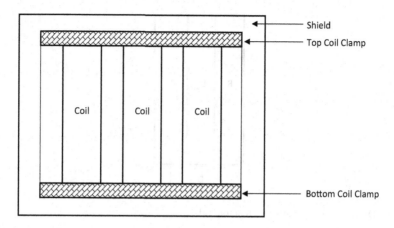

FIGURE 19.4 Schematic of air-core oil-filled reactors.

$$k = \text{function of} \left(\frac{h-a}{h} \right)$$

(19.12)

k can be calculated by Equation (19.12).

$$k = 0.3586667 + 1.139697 \left(\frac{h-a}{h} \right) - 0.5 \left(\frac{h-a}{h} \right)^2$$

(19.13)

L = inductance of the coil

$$L = \frac{4\pi \times 10^{-7} \, A_1}{h \, k} \text{ Henry}$$

(19.14)

$$A_1 = \frac{\pi}{4} \left(D_1 + \frac{2}{3} R \right)^2 + \frac{\pi}{18} R^2$$

$$B = \frac{4\pi \times 10^{-7} \, N \, I}{h \times k}$$

B = flux density in the coil (Tesla)
I = rated current (A)
ϕ = total flux in the coil
$= B \times A_2$ (Webers)

$$A_2 = \frac{\pi}{4} \left(D_1 + R \right)^2 + \frac{\pi}{12} R^2$$

A_1, A_2 = parameters related to radial depth and diameter of coil (m²)
To commence a design, the initial value of D_1 is assumed. Approximate core circle diameter of an oil-filled transformer of similar kVA rating can be used as a starting point.

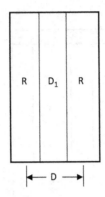

FIGURE 19.5 Dimensions of an air-core reactor coil.

If D is the mean diameter of coil and D_1 is the inside diameter, the relation between D, D_1 and R can be written as follows (Figure 19.5);

$D = D_1 + R$

Assuming that $R = 0.25\,D$,

$D = D_1 + 0.25\,D$

$\therefore D_1 = 0.75\,D$ or $D = \dfrac{D_1}{0.75}$

The leg height is now calculated as follows;

$h = 1.29\,D$

The wound height of coil is now calculated by subtracting the sum of end insulation heights from h.

19.2.3 SELECTION OF CONDUCTOR DIMENSIONS

The working current density is determined by considering the following parameters;

- Winding gradient and temperature rise
- Winding material
- Losses to be guaranteed
- Short-time withstand capability of the conductor [both thermal and dynamic]

The conductor insulation is decided based on the rated voltage and test voltage requirement.

Once the winding dimensions are calculated, the final values of resistance, reactance, flux, etc., are calculated.

Eddy current losses of copper winding are given by the formula:

$$W_{\text{eddy}} = 2.13 \times 10^{-8} \left(\frac{f}{50}\right)^2 \left(\frac{1.78\,N\,I}{h}\right)^2 \frac{b}{3} G \qquad (19.15)$$

where

W_{eddy} = eddy current loss (watts)

f = frequency (Hz)

N = number of turns

I = rated current per phase

h = axial electrical height of the coil

b = dimension of conductor orthogonal to the flux (for winding on edge, "b" is the width of the conductor) (m)

G = weight of copper (kg)

Other stray losses can be approximated as 7%–10% of the I^2R loss.

19.2.4 CALCULATION OF THE DIMENSIONS OF THE SHIELD

Stack thickness of shield = $D_1 + R$

CRGO steel can be used as the shield. The maximum flux density in the shield is kept low to reduce the noise level. The width of the shield is calculated by the following formula;

$$W_s = \frac{\phi}{(D_1 + R)B_s} \qquad (19.16)$$

where

W_s = width of the shield (m)

B_s = flux density in the shield

Losses in the shield = weight of shield × watts/kg × building factor

Watts/kg = the specific loss of the shield material

Building factor = core building factor for the type of construction of shield

The cooling system is to be designed to dissipate the total losses in the winding and shield.

19.3 DESIGN OF OIL-FILLED GAPPED-CORE REACTORS

Reactance of a gapped-core reactor

$$L = (k - k_1) \frac{\mu_o \pi N^2 d^2}{4 l} \qquad (19.17)$$

where

L = reactance in henry

$$k_1 = \frac{A_1 d_1}{A_{core} \times \text{Coil Height}}$$

$$k_1 = k (1+k)$$

$$\therefore k = \frac{-1 + \sqrt{1+4k}}{2}$$

l = axial length of the coil (m)

$$A_1 = \frac{\pi}{4} \left(D_1 + \frac{2}{3} R \right)^2 - A$$

D_1 = inside diameter of the coil (m)

R = radial depth of the coil (m)

A = area of cross section of core (m²)

$$d_1 = \text{air gap} = \frac{1.78\,NI}{B}$$

N = number of turns in the coil [this is calculated using the volt/turn formula]

B = flux density (Tesla) (typical value is 0.9 Tesla)

d_1 = approximate air gap in core (m)

$$d = \text{actual air gap in core} = \frac{d_1}{1+k}\ (\text{m})$$

The maximum air gap size is 9 mm. The number of core packets is decided by the total limb height.

$$\text{Actual number of turns} = \frac{N}{1+k}$$

$$\text{Flux density in yoke} = (1+k)\,B_{\text{leg}}$$

$\left(B_{\text{leg}} = \text{flux destiny in leg. It is assumed that the core leg and yoke have equal areas}\right)$

The $(k - k_1)$ factor depends on the ratio of $\dfrac{b}{d}$, i.e. the ratio of winding radial depth to the mean winding diameter and l/b ratio.

Table 19.3 gives the $(k - k_1)$ values for different b/d ratios corresponding to different l/b values.

TABLE 19.3

$(k - k_1)$ Factors

$\dfrac{b}{d}$ ratio	$\dfrac{l}{b}$ ratio $= \dfrac{\text{Axial length of coil}}{\text{Radial depth}}$			
	1	2	5	10
0.1	0.18	0.30	0.48	0.65
0.2	0.24	0.38	0.58	0.70
0.3	0.27	0.42	0.62	0.70
0.4	0.30	0.43	0.62	0.65
0.5	0.32	0.43	0.62	0.65
0.6	0.33	0.44	0.62	0.65
0.7	0.34	0.44	0.62	0.64
0.8	0.35	0.45	0.62	0.64
0.9	0.35	0.45	0.62	0.64

20 Scott-Connected Transformers

20.1 BASIC THEORY

Scott-connected transformers are used to connect a 3-phase system to a two-phase system and vice versa. In industries, it finds application where single-phase loads are connected to 3-phase transformers for balancing the power system loadings. The phasor diagram is shown in Figure 20.1.

The simple form of Scott-connected transformer consists of two single-phase transformers having primary turns in the ratio of $1 : \dfrac{\sqrt{3}}{2}$. The transformer with 86.6% turns is called "teaser". The main transformer is centre-tapped, and one end of the centre tap is connected to the teaser transformer. The 3-phase primary supply is given to the two terminals of the main transformer and the end of the teaser transformer as shown in Figure 20.2.

When the Scott connection was in common use in the past, it was often considered inconvenient that the main and teaser transformers were not interchangeable. As the cost difference of the main and teaser winding transformers was marginal, it was a common practice to make both transformers with 100% primary turns with taps at the centre and 86.6% of the turns.

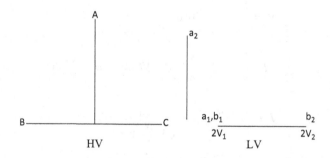

FIGURE 20.1 Scott connection: phasor diagram.

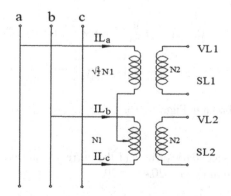

FIGURE 20.2 Connection diagram.

20.2   CONNECTION DIAGRAM AND CURRENT DISTRIBUTION

The connection diagram is shown in Figure 20.2. The current distribution of the primary and secondary windings is calculated as follows;

$$L_a = -\frac{\sqrt{3}}{2} \times \frac{N_2}{N_1} IL_1$$

$$IL_b = -\frac{IL_a}{2} - \frac{N_2}{N_1} IL_2 = \frac{N_2}{N_1}\left(\frac{1}{\sqrt{3}} IL_1 - IL_2\right)$$

$$IL_c = -\frac{IL_a}{2} - \frac{N_2}{N_1} IL_2 = \frac{N_2}{N_1}\left(\frac{1}{\sqrt{3}} IL_1 + IL_2\right)$$

$$VL_1 = -\frac{N_2}{N_1} V_{ab} \angle -30°$$

$$VL_2 = j\frac{N_2}{N_1} V_{ab} \angle -30°$$

$$V_a = \frac{V_{ab}}{\sqrt{3}} \angle -30°$$

$$V_b = \quad V_b = \frac{V_{ab}}{\sqrt{3}} \angle -30°a^2$$

$$V_c = \frac{V_{ab}}{\sqrt{3}} \angle -30°a$$

$$VL_1 = -\sqrt{3}\frac{N2}{N1} V_a , VL_2 = j\sqrt{3}\frac{N2}{N1} V_a$$

$$I_aL_1 = \frac{SL_1}{VL_1} = \frac{-SL_1}{\sqrt{3}\frac{N_2}{N_1}V_a}$$

$$I_aL_2 = \frac{SL_2}{VL_2} = \frac{-jSL_2}{\sqrt{3}\frac{N_2}{N_1}V_a}$$

Taking V_a as a reference,

$$V_{ca} = V_a + j\frac{V_{bc}}{2} = 0.866 + j\ 0.5V_{bc} \tag{20.1}$$

$$|V_{ca}| = |V_{bc}| = |V_{ab}|$$

This phase relationship is shown in Figure 20.3.

- Current distribution

    The primary and secondary turns and the current distribution of a Scott-connected transformer are shown in Figures 20.4 and 20.5.

$$I_m = \text{Secondary winding current of phase m}$$

$$I_T = \text{Secondary winding current of phase T}$$

$$I_{1m} = \text{Primary winding current of phase m}$$

$$I_{1T} = \text{Primary winding current of phase T}$$

$$I_{1m} = \frac{n}{N}\, I_m \tag{20.2}$$

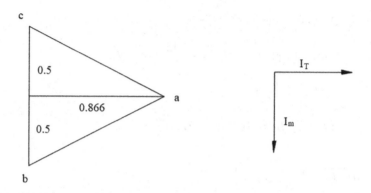

FIGURE 20.3   Phase relationships of a Scott-connected transformer.

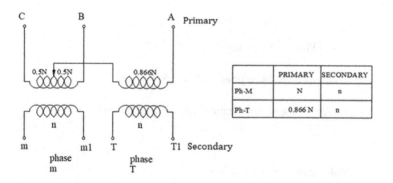

	PRIMARY	SECONDARY
Ph-M	N	n
Ph-T	0.866 N	n

FIGURE 20.4   Primary and secondary turns of a Scott-connected transformer.

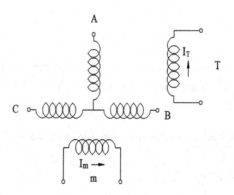

FIGURE 20.5   Current flow direction.

$$I_{1T} = \frac{2}{\sqrt{3}} \frac{n}{N} I_T \qquad (20.3)$$

When phase "m" and phase "T" have equal load,

Phase A current,

Current by phase $m = 0$

Current by phase $T = I_a = \dfrac{n}{0.866\,N} I_T = 1.154 \dfrac{n}{N} I_T$

Phase B current,

$$\text{Current by phase m} = I_{1m} = -j\,I_m\,\frac{n}{N}$$

$$\text{Current by phase } T = I_{1T} = \frac{-1.154}{2}\,\frac{n}{N}\,I_T$$

$$= -0.577\,\frac{n}{N}\,I_T$$

$$\text{If } I_m = I_T,\ I_b = \left(-0.577\,I_m - j\,I_m\,\right)\frac{n}{N}$$

$$\text{II}_b\ I = \frac{n}{N}\,I_m\,\sqrt{1+0.577^2} = 1.15\,I_m\,\frac{n}{N}$$

$$\text{Vector angle is } \left(90° + \tan^{-1} 0.577\right) = (90° + 30°) = 120°$$

Teaser,

$$E_3 = 3 \text{ Phase line Voltage}$$

$$\text{Voltage on the three phase side} = 0.866\,E_3 = E_2$$

$$\text{Current in the series winding} = I_3 = I_2 E_2 \times 0.866\,E_3$$

Main,

$$\text{Voltage on 3 phase side} = E_3$$

$$\text{Current in the series winding} = I_3$$

$$I_3 = \frac{I_2 E_2}{0.866 E_3}$$

$$\text{Current in the Common winding} = I_2\sqrt{1 - 2\frac{E_2}{E_2} + 1.33\left(\frac{E_2}{E_3}\right)^2} \qquad (20.4)$$

Assuming balanced 2 phase load, current in the main due to load across the main is

$$\text{Series winding} = I_2\frac{E_2}{E_3}$$

$$\text{Common winding} = I_2 \frac{(E_3 - E_2)}{E_3}$$

Teaser current due to teaser load

$$\text{Series winding} = I_2 \frac{E_2}{0.866\, E_3}$$

This current divides into two in the main hence adding the two currents at $90°$.

$$\text{Current in series winding} = I_2 \sqrt{\left(\frac{E_2}{E_3}\right)^2 + \left(\frac{E_2}{\sqrt{3}E_3}\right)^2}$$

$$\text{Current in series winding} = I_2 \frac{E_2}{0.866\, E_3} \tag{20.5}$$

Current in the common winding,

$$\text{Current in the common winding} = I_2 \sqrt{1 - 2\frac{E_2}{E_3} + 1.33\left(\frac{E_2}{E_3}\right)^2} \tag{20.6}$$

20.3   LE BLANC CONNECTION

An alternative connection to the Scott-connected transformers is the Le Blanc connection, which is less expensive. A core of a 3-limb 3-phase design is used for the construction of Le Blanc-connected transformers when compared to the requirement of two numbers of single-phase cores for a Scott connection. Since only a 3-phase core and hence a single tank are used for a Le Blanc transformer, it requires less active material and is cheaper, especially when it is compared with two single-phase transformers housed in separate tanks for a Scott connection.

20.4   APPLICATION OF SCOTT-CONNECTED TRANSFORMERS

- For linking a 3-phase power system with a two-phase power system and vice versa.
- To supply single-phase loads from a 3-phase system for better balancing of the primary power system.
- For single-phase or two-phase supply to electric trains.
- For electric furnaces to draw a balanced power from a 3-phase system.

20.5   EXAMPLE OF SCOTT-CONNECTED TRANSFORMER WINDING DESIGN

An example design of a 2500 kVA, 11 kV/400 V, 50-Hz Scott-connected transformer with percentage impedance of 10% and with copper as the winding material is given below. The voltages and currents of windings are tabulated in Table 20.1.

TABLE 20.1

Voltages and Currents of 2500-kVA Transformers

Particulars	Main Coil	Teaser Coil
Primary voltage	11000	$11000 \times \dfrac{\sqrt{3}}{2}$
Primary line current	131	131
Secondary voltage	400	400
Secondary current	3125	3125

Core circle diameter = 299 mm

Core net area = 625.5 cm²

Volt/turn = 22.22

Working flux density = 1.6 T

Number of turns of main coil = $\dfrac{11000}{22.22} = 495$

Number of turns of teaser coil = $\dfrac{11400}{\sqrt{3} \times 22.22} = 429$

Number of turns of LV = $\dfrac{400}{22.2} = 18$

The dimensions of the main coil and auxiliary coil are tabulated in Table 20.2.

TABLE 20.2

Sample Winding Design of Scott-Connected Transformer

Particulars	Main Coil		Teaser Coil	
	HV	LV	HV	LV
Voltage	11000	400	9526.3	400
Current	131	3125	131	3125
Bare conductor size	8.4 × 2.2	650 × 1.7 (foil)	8.4 × 2.2	650 × 1.7 (foil)
Covered size of conductor	8.8 × 2.6	-	8.8 × 2.6	-
Number of parallels	2	1	2	1
Arrangement of conductor	1 flat × 2 radial	-	1 flat × 2 radial	-
Area of cross section	17.98 × 2 = 35.96	1105	17.98 × 2 = 35.96	1105
Current density	3.64	2.83	3.64	2.83
Turns/coil	495	18	429	18
Turns/layers	73	1	73	1
Number of layers	73 × 6+57 × 1	18	73 × 5+64 × 1	18
Inside diameter	421	313	421	313
Outside diameter	519	399	507	399
Radial thickness	49	43	43	43
Number of ducts	2	2	2	2
Thickness of ducts	3	5	3	5
Interlayer insulation	1.0	0.25	1.0	0.25
Mean length of turn	1477	1118	1458	1118
Weight of conductor	234	198	201	198
Resistance at 75°C	0.427	0.382×10^{-3}	0.365	0.382×10^{-3}
I^2R loss at 75°C	7327	370	6264	370
Eddy current loss	141	48	124	48

21 Autotransformers

21.1 INTRODUCTION

An autotransformer is a transformer in which at least two windings have a common part. In the electric power system, it finds application in place of a two-winding transformer, especially when the interconnection or power transfer is required between two high-voltage systems. The major advantages of an autotransformer are as follows;

- It has higher efficiency when compared to a two-winding transformer of equal rating.
- It has lower material cost.
- It offers better voltage regulation.
- The excitation current is low.

The disadvantages of using an autotransformer in the power system include the following;

- There are high short circuit stresses.
- Since there is no galvanic isolation between the primary and the secondary, the stresses from surges can become high on the secondary side.

21.2 CURRENT DISTRIBUTION IN THE WINDINGS OF AN AUTOTRANSFORMER

Figure 21.1 shows the current distribution of an autotransformer without taps.

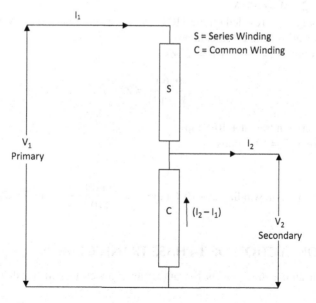

FIGURE 21.1   Current distribution of an autotransformer without taps.

The current through the common winding is the difference between the load current I_2 and the primary current I_1. When V_2 becomes close to V_1, I_2 becomes close to I_1 and the current through the common winding reduces. Thus, the winding material required for the secondary reduces, when the voltage ratio of transformation comes near unity.

The saving of the winding material when compared to a two-winding transformer is proportional to the auto ratio or "co-ratio", which is defined as follows:

$$\text{co-ratio} = \frac{V_1 - V_2}{V_1}$$

where

V_1 = primary voltage
V_2 = secondary voltage

Sample calculation of the current distribution

The winding currents of a 50 MVA, 132/110 kV 3-phase autotransformer are calculated as an example. It is assumed that taps are provided on the 110 kV side for + 10% variation.

$$\text{Current in series winding} = I_1 = \frac{50000}{\sqrt{3} \times 132} = 218.7 \text{ A}$$

Line current of the secondary at rated tap = I_2

$$\frac{50000}{\sqrt{3} \times 110} = 262.4 \text{ A}$$

Current in common winding at rated tap
$$I_2 - I_1 = 264.4 - 218.7 = 43.7 \text{ A}$$

$$r = \text{Co-ratio of the transformer} = \frac{132 - 110}{132} = 0.1667$$

It can be seen that the current in the common winding = $r \times I_2$
That is, $0.1667 \times 262.4 = 43.7$ A
Since the tap changer is provided on the 110 kV side for variation of 110 kV ± 10%, the currents at the extreme tap positions can be calculated.
At +10% tap position, $V_2 = 110 \times 1.1 = 121$ kV

$$I_2 = \frac{50000}{\sqrt{3} \times 121} = 238.6 \text{ A}$$

∴ Current in common winding at + 10% tap
$= I_2 - I_1 = 238.6 - 218.7 = 19.9$ amps
Similarly,

$$\text{Current in common winding at} - 10\% \text{ tap} = \frac{50000}{\sqrt{3} \times 110 \times .9} - 218.7 = 72.9 \text{ A}$$

21.3 AUTO CONNECTION OF 3-PHASE TRANSFORMERS

The most common auto connection is the auto star connection with earthed neutral shown in Figure 21.2.

Auto delta connection is rarely used, except in the case of phase-shifting transformers. The co-ratio of an auto delta connection is higher than that of an auto star connection.

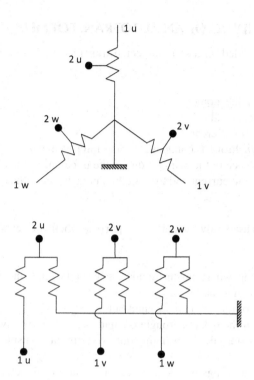

FIGURE 21.2 Auto star connection.

The other auto connections being used in specific applications are as follows;

• Auto open delta connection
• Extended delta connection
• Auto zigzag connection
• Auto T connection

When compared to auto star connection, the other connections are less economical. Table 21.1 gives the relative capacity required by other connections compared to an auto star connection.

TABLE 21.1

Relative Capacity Required When Compared to Auto Star Connection

	co-ratio				
	0.1	**0.2**	**0.3**	**0.4**	**0.5**
Auto star	1	1	1	1	1
Auto Scott T	1.072	1.072	1.072	1.072	1.072
Auto extended delta	1.133	1.18	1.10	1.085	1.072
Auto delta	1.22	1.30	1.4	-	-
Auto open delta	1.15	1.15	1.15	1.15	1.15
Auto zigzag	1.15	1.15	1.15	1.15	1.15

21.4   TERTIARY WINDING OF AN AUTOTRANSFORMER

A tertiary winding is provided on a star-connected transformer for one or more of the following reasons;

- To suppress the third harmonics
- To supply an auxiliary load
- To reduce telephone interference
- For connecting a synchronous condenser for the improvement of power factor
- To reduce the unbalance in the primary due to the unbalance of loads
- To redistribute the fault current and to limit the fault current in the case of a line to ground fault.

When tertiary winding is used only for stabilizing purpose, there are several options for connecting the terminals.

a. Buried delta winding where no terminals are brought out. One corner of the delta is grounded to the tank internally.
b. All three terminals of the delta are brought out.
c. Only one terminal of the delta is brought out and earthed externally.
d. The delta is closed externally by bringing out two terminals outside.

The customer specifies the ratings of a loaded tertiary winding. Generally, the rating specified is one-third of the transformer kVA. For an autotransformer, it is typically more than one-third of the transformer rating × co-ratio.

For a stabilizing tertiary winding, the designer can select the voltage if it is not specified by the customer. The winding is placed as the innermost winding, and the voltage can be typically 11 kV or below.

21.4.1   DESIGN FEATURES OF A TERTIARY WINDING

The rating of the tertiary winding is typically one-third of the transformer rating, and therefore, its impedance with the other windings will have to be sufficient to limit the fault current stresses. The impedance can become low especially in cases where the co-ratio is low. The transferred surges by electromagnetic and capacitive transfer from the primary or secondary can reach values above the designed withstand capability of the winding.

21.4.2   ELIMINATING THE TERTIARY WINDING

The tertiary winding is a weak link of the transformer, whether it is loaded type or unloaded type, because it is mechanically weak and it can fail from transferred surges, or short circuit forces. Increasing the mechanical and electrical strength of the tertiary winding is possible by increasing its rating, which increases the cost of the transformer. It is also to be noted that using a tertiary winding for auxiliary supply on a transformer of 100 MVA capacity can make the transformer vulnerable to any faults on the auxiliary power system.

21.5   LOCATION OF TAP CHANGER OF AUTOTRANSFORMERS

Most of the autotransformers are used in high-voltage power grid, and, therefore, on-load voltage regulation is normally specified. In a very few cases, an off-circuit tap changer is used. The on-load tap changer can be located at different electrical locations. Table 21.2 gives the locations and the types of voltage variations feasible.

TABLE 21.2

Location of the Tap Changer and the Voltage Variation

Location of the Tap Changer	Type of Voltage Variation
Neutral end of autotransformer	To regulate either HV or LV (variable flux voltage variation (VFVV))
Line end of LV winding	To regulate LV constant flux voltage variation (CFVV)
Fork of autotransformer connection	To regulate LV constant flux voltage variation (CFVV)
Separate tap winding connected to series winding (above autopoint)	To regulate HV (CFVV) To regulate LV (VFVV)
Main body of series winding	To regulate HV (CFVV) [used when the number of steps required is high for HV variation]

The optimum location of the placement of tap changer is determined by the following;

- Voltage to ground
- Voltage across tap winding
- Step voltage
- Current through the tap changer
- Number of turns of the tap winding

The geometrical location of the tap winding is selected on the following basis;

- Voltage between windings
- Impedance required and impedance variation over the tap range
- Consideration of parallel operation if required. In this case, the impedance characteristic is the deciding factor for the geometrical selection of the tap winding.

21.5.1 DIRECT VOLTAGE VARIATION AND INDIRECT VOLTAGE VARIATION

Direct voltage variation achieves the required voltage regulation by providing the taps or regulating winding on the main winding. Indirect voltage regulation uses a series regulating transformer. Indirect regulation is selected when there are limitations of the current rating or step voltage of the on-load tap changer used. Figures 21.3 and 21.4 show a typical winding arrangement of indirect voltage variation using a series transformer.

21.6 EFFECT OF GEOMETRICAL ARRANGEMENT OF TAP WINDING OF AUTOTRANSFORMERS

The voltage variation can be realized by constant flux voltage variation (CFVV) or variable flux voltage variation (VFVV). The desired voltage variation can be on the primary side or secondary side, and the selection of CFVV or VFVV depends on the cost, impedance variation over the range of tappings, voltage level, etc. Table 21.3 shows a comparison of the location of OLTC on the winding for CFVV and VFVV.

The location of the tap winding in relation to the core, common winding and series winding affects the design and the impedance variation, losses, etc. Some of the typical winding configurations and the corresponding impedance characteristics are shown in Table 21.4 for reference. The characteristics for voltage variation on the primary side and secondary side are included. A coarse fine winding arrangement is considered for all the cases.

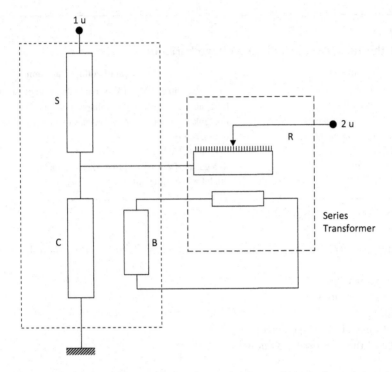

FIGURE 21.3   Voltage regulation with a series transformer (taps on series transformer).

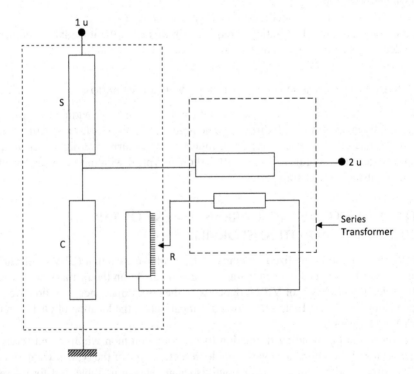

FIGURE 21.4   Voltage regulation with a series transformer.

TABLE 21.3
Comparison of VFVV and CFVV

Particulars	VFVV OLTC on Neutral End	CFVV OLTC on Line End
Weight of core and coil assembly	Heavy	Light
Load loss	High	Low
Core loss	Variable with taps	Constant
Per cent impedance	Wide variation within the tap range	Low variation within the tap range
Cost of OLTC	Less expensive	Costly

The symbols used are as follows;

T – tap coil
C– common winding
S – series winding
H – highest tap
L – lowest tap

It is assumed that the tap range is ±10%.

The per cent variation of reactance from the nominal value given is indicative and is for comparison only.

When the tap winding is placed between the common and series windings, the lowest impedance variation is realized when compared to the other arrangements. The per cent variations of impedance at the extreme taps shown in Table 21.4 are relative to the impedance at rated tap.

TABLE 21.4
Impedance Variation and Placement of Tap Winding

		Winding Configuration Relative to Core			
		T–C–S Impedance Variation	C–S–T Impedance Variation	C–T–S Impedance Variation	S–C–T Impedance Variation
Voltage Variation on	Location of Tap Winding	H-Tap–L-Tap	H-Tap–L-Tap	H-Tap–L-Tap	H-Tap–L-Tap
HV (primary)	Neutral of common winding	+25% −25%	−10% +10%	+10% −10%	+25% −25%
	Line end of common winding	+20% −20%	−20% +20%	+10% −10%	+20% −20%
LV (secondary)	Neutral of common winding	−10%+5%	+50%−30%	+15% −15%	−5% +5%
	Line end of common winding	−10%+10%	+50%−30%	+15% −15%	−5% +5%

21.7 DESIGN OF A 50–MVA, 132/66-KV AUTOTRANSFORMER

- Transformer specifications
  kVA – 30000/50000
  Phase 3
  Frequency – 50 Hz
  Cooling – ONAN/ONAF
  Primary volts – 132 kV
  Secondary volts – 66 kV
  Tertiary voltage (unloaded, stabilizing)
  Per cent impedance – 10, subject to IEC tolerance at rated tap
  On-load tap changer required for voltage variation of 66 kV from +5% to –15% in 1.25%
  steps
  Winding material – copper
  Load loss at rated tap at 75°C – 190 kW (max)
  No-load loss at rated voltage – 23 kW (max)
  Top oil temperature rise – 55°C
  Average winding temperature rise – 60°C
- Design of the transformer
  An initial design can be commenced by estimating volt/turn and then calculating a core
  diameter. For an autotransformer, the theoretical equivalent kVA is first calculated:

$$\text{Equivalent kVA} = 50000 \times \frac{132 - 66}{132} = 25000$$

(Co-ratio = 0.5)
Though the equivalent kVA is 25000, the overall material quantities would be more
than the requirements for a two-winding transformer of the same rating. Considering this,
the volt/turn can be calculated by assuming a slightly higher value of the constant in the
formula.

Initial volt/turn = $0.4\sqrt{25000} = 63.24$
Applying the emf equation and selecting 1.65 Tesla as the initial flux density,
$63.24 = 4.44 \times f \times B_m \times A$
$f = 50$ Hz, $B_m = 1.65$ Tesla
Solving for A = net core area = 0.1726 m^2
If d is the core circle diameter, the core net area $= \dfrac{\pi}{4} d^2 \times k_1 \; k_2$

d = core diameter (m)
k_1 = stacking factor of core = 0.97
k_2 = core filling factor ≈ 0.9

$$\therefore 0.1726 = \frac{\pi}{4} \times d^2 \times 0.97 \times 0.9$$

$\therefore d = 0.501$ m
Initially, a core circle of 0.5 m (500 mm) is selected.
For a 500 mm-diameter core, the net area using 11 steps is 0.18 m^2. This is used for
subsequent calculations.
- Calculation of voltages and currents of windings
  The design is done with a separate tap winding for +5% to –15% of the voltage variations.
20% of phase volt is $\dfrac{66000 \times 0.2}{\sqrt{3}} = 7621$

Common winding phase voltage $= \dfrac{66000 \times 0.85}{\sqrt{3}} = 32381$

Series winding phase voltage $= \dfrac{66000 \times 0.95}{\sqrt{3}} = 36200$

Tertiary winding

An unloaded tertiary winding is required, and therefore, the terminals are not taken outside. The voltage of the tertiary is arbitrarily selected as 11 kV. The current density of the tertiary considered for the design is low in order to strengthen the tertiary against short circuit forces.

- Calculation of the number of turns and working flux density

  The total tap range of 20% is divided into 16 steps. For medium-size transformers, one approach can be to try to keep integer number of turns per step.

  Total phase volts of tap winding = 7621

  Voltage/step = $7621 \div 16 = 476.31$

  Number of turns/step $= \dfrac{\text{volt/step}}{\text{volt/turn}} = \dfrac{476.31}{63.24}$

  Nearest integer values of 7 or 8 can be selected.

  $\therefore$ The volt/turn with turns/step of 7 $= \dfrac{476.31}{7}$

  The working flux density in core $= \dfrac{68.044}{4.44 \times 50 \times 0.180} = 1.702$ Tesla

  The number of turns of the tertiary, common, tap and series windings can be calculated from the volt/turn and the phase voltages of the respective winding.

- Winding dimensions, losses and impedance

  The arrangement of windings and clearances are shown in Figure 21.5.

  The tertiary winding is placed after the core. The series winding is designed as two axially placed parallel coils with the 132 kV line lead from the middle of the winding assembly.

  End insulation to core for series winding is $105 \times 2 = 210$ mm.

  Bunched conductors are used for the tertiary and common windings. The tap winding is designed as a multistart coil with one layer.

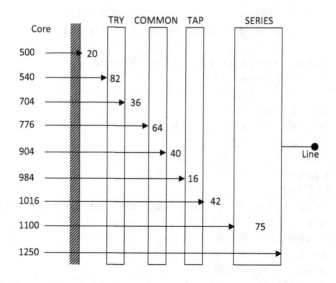

FIGURE 21.5   Winding arrangement and clearance.

TABLE 21.5

Design Particulars for Autotransformers

Parameters	Tertiary	Common	Tap	Series
Number of turns	161	476	112	532
Number of discs	81	100	7 × 16 (multistart)	46 × 2
Turns/disc	2	5	1	12
Bare conductor size (thickness × width)	4.5 × 10.85	2.3 × 7.9	3 × 7.25	2.4 × 8.2
Conductors/bunch	4	2	-	2 radial
Number of bunches	2	2	-	-
Covering per conductor	0.4	0.4	-	-
Overall covering	0.45	0.45	2	0.65
Overall conductor size (radial × axial)	20.5 × 11.7	12.5 × 8.75	9.25 × 16	6.1 × 8.85
Cross-sectional area	383.4	77	64.65	76.7
Current density	-	2.84	3.38	2.85
Inside diameter	540	776	984	1100
Radial thickness	82	64	16	75
Outside diameter	704	904	1016	1250
Average size of duct	3.65	3.40	-	4.2
Axial height of winding (electrical)	1240	1240	1175 (physical height)	1240
Size of compensation blocks	-	25	32 × 2	46
Weight of bare conductor (kg)	3340	2643	663	4129
Weight of covered conductor (kg)	3410	2745	738	4282
Resistance (ohms/phase)	0.0539	1.0577	a	1.6647

a   The tap winding resistance is calculated in two parts.

The duct sizes shown in the design are average values. Standard thicknesses of radial spacers are used, and the average of the spacer thickness is indicated.

Winding design (Table 21.5)

- Resistance of tap winding in series connection = 0.09399 Ω
- Resistance of tap winding in common winding = 0.28198 Ω

- Load loss calculation at rated tap at 75°C at 50 MVA base
    Common winding I^2R loss = 50.59
    Tap winding I^2R loss = 17.98
    Series winding I^2R loss = 79.62
    Common winding eddy current loss = 8.15
    Tap winding eddy current loss = 0.07
    Series winding eddy current loss = 7.46
    Total stray loss including loss in the tank = 20.40
    Total = 184.27 kW
    Guaranteed load loss is 190 kW, and the calculated value has about 3% margin.
- Calculation of no-load loss
    Core circle diameter = 500 mm
    Core window height = 1450 mm
    Core leg centre distance = 1320 mm
    Core weight – 14380 kg

The working flux density of the core is 1.7 tesla. The guaranteed core loss is 23 kW (maximum). The specific loss of M5 grade CRGO at 1.7 tesla is 1.3 w/kg at 50 Hz, and assuming a step lap core construction, a building factor of 1.15 is expected.

$\therefore$ No-load loss = 14,380 × 1.3 × 1.15

= 21498 W

= 21.498 kW

From the guaranteed maximum no-load loss, this gives 6% margin.

- Alternate calculations

The core weight of 14380 kg is divided into two parts, i.e. the core limb and yoke where the flux flow is along the grain oriented direction and the other part where the flux flow is almost across the grain orientation.

The building factor along grain orientation is assumed to be unity, and the building factor across the lamination is considered as a variable depending on the grade of material. For M5 grade of CRGO, the building factor across grain is assumed as 1.75.

Core weight along grain orientation = 10400 kg

Core weight across grain orientation = 3980 kg

$\therefore$ No-load loss = 10400 × 1.3 × 1 + 3980 × 1.3 × 1.75

= 13520 + 9054 = 22547 W

= 22.547 kW

The calculated loss has 2% margin from the guaranteed loss.

- Calculated parameters (Table 21.6)

Impedance at rated tap = 10.04% (common winding to series winding)

TABLE 21.6

Calculated Parameters for the Designed Autotransformer

Parameter	ONAN Rating	ONAF Rating
No-load loss (kW)	22.547	22.547
Load loss at 75°C (kW)	66.337	184.270
Total losses (kW)	88.884	206.817
Top oil temperature rise (°C)	46	52
Average winding temperature rise (°C)	51	57

22 Transformers for Special Applications or Special Designs

22.1 TRANSFORMERS FOR USE IN EXPLOSIVE ATMOSPHERES [ATEX-/IECEX-CERTIFIED TRANSFORMERS]

22.1.1 INTRODUCTION

This chapter covers the design aspects of the following types of transformers intended for special applications that are not covered in other chapters.

- Transformers for use in explosive atmospheres [ATEX-/IECEx-certified transformers]
- Furnace transformers
- Multiwinding transformers
- Dual-ratio transformers
- Liquid-filled transformers using high-temperature insulation materials
- Traction transformers
- Symmetrical core transformers

ATEX means "atmosphere explosible", and it deals with equipments and protective systems intended for use in potentially explosive atmospheres. The cause for an atmosphere becoming explosive can be from flammable gases, vapours or combustible dust.

A certification of an equipment intended for use in a potentially explosive atmosphere within the European Union (EU) is called ATEX certificate. Outside the EU, a similar certification system is called IECEx. Some customers outside the EU also specify ATEX certification. The relevant EU norm/IEC standard is IEC 60079 series.

The EU and IEC hazardous areas are classified into zones, whereas the US standards classify the areas as classes and divisions as shown in Table 22.1.

TABLE 22.1
Hazardous Area Zones/Divisions

Zone or Division	EU and IEC Classification	US Classification
An area in which an explosive mixture is continuously present or present for long periods	Zone 0 (gases)	Class I Division I (gases)
	Zone 20 (dust)	Class II Division I (dust)
An area in which an explosive mixture is likely to occur in normal operation	Zone 1 (gases)	Class I Division I (gases)
	Zone 21 (dust)	Class II Division I (dust)
An area in which an explosive mixture is not likely to occur in normal operation, and if it occurs, it will exist only for short time	Zone 2 (gases)	Class I Division II (gases)
	Zone 22 (dust)	Class II Division II (dust)

TABLE 22.2

Equipment Coding for Oil-Filled Transformers

Title	Mark	Application
Equipment group	I	Underground mines
	II	Use in all other areas
Category	1	Intended for use in areas where an explosive atmosphere is present continuously or for long periods.
	2	Intended for use in areas where an explosive atmosphere is likely to occur in normal operation and must have high level of protection
	3	Intended for use in areas where an explosive atmosphere is unlikely to occur in normal operation and must ensure normal protection
Gas/dust	G	Certified to use in flammable gas
	D	Certified to use where dust is present in the atmosphere
[a]Type of protection	d	Flameproof
	I_a	Intrinsically safe
	e	Increased safety
	n_a	Non-sparking
	O	Oil immersion
	P_d	Pressurization
	P	Electrical equipment type of protection "P"
	n	Electrical equipment with type of protection "n"
Gas group	I	Mines
	II	Surface above-ground industries
Gas subgroup	A	Less easily ignited gases
	B	Easily ignited gases
	C	Most easily ignited gases
Temperature classification [maximum surface temperature produced at an ambient of 40°C]	T_1	450°C
	T_2	300°C
	T_3	200°C
	T_4	135°C
	T_5	100°C
	T_6	85°C

[a] There are more types of protection. Typical protection types are indicated here.

22.1.2 EQUIPMENT CODING AS PER ATEX MARKING

The ATEX marking of a transformer as per IEC 60079 denotes the application, category, type of protections, etc. The relevant details applicable for an oil-filled transformer are given in Table 22.2.

22.1.3 TRANSFORMER DESIGN CONSIDERATION TO COMPLY WITH ATEX CERTIFICATION REQUIREMENT

a. Applicable standards

The design requirements and tolerances have an impact on the parameters such as surface temperature, components and cooling design.

b. Voltage rating of the equipment

The type of protection for cable box is dependent on the voltage class, which decides the requirement of pressurized protection. As per IEC 60079-2, cable boxes for application on 15 kV and above shall be pressurized above atmosphere.

TABLE 22.3

Special Features of Component for Mounting on the Transformer

Name of the Component	Special Features Required for Hazardous Zone
Oil temperature indicator	Low-energy contacts
	24V DC, 0.2A is a standard rating. Contact material is gold
Winding temperature indicator	Low-energy contacts
	24V DC, 0.2A is a standard rating. Contact material is gold
Magnetic oil level gauge	Low-energy contacts
	24V DC, 0.2A is a standard rating. Contact material is gold
Pressure oil level gauge	Low-energy contacts
	24V DC, 0.2A is a standard rating. Contact material is gold
Sudden pressure relay	Low-energy contacts
	24V DC, 0.2A is a standard rating. Contact material is gold
Buchholz relay	Low-energy contacts
	24V DC, 50mA contact
Enclosure gaskets	Thermal endurance to heat and cold tests as per IEC 60079-0
Marshalling box	For Ex-e increased safety box, ATEX-certified components can be used.
	For Ex-d flameproof box, ATEX/non-ATEX components can be used.

c. Type of hazardous gases present

The selection of gaskets, paint and cooling design is affected by the type of gas present in the atmosphere.

d. Hazardous zone

The selection of components mounted on the transformer externally depends on the hazardous zone. The special features required for the major components are given in Table 22.3.

22.2 FURNACE TRANSFORMERS

22.2.1 INTRODUCTION

Furnace transformers are used for various industrial applications such as steel melting, production of calcium carbide and manufacture of ferroalloys. The major applications are for electric arc furnace (EAF) and ladle furnace (LF) for steel production.

The EAFs have small capacities of few MVA to several hundreds of MVA with the secondary voltages ranging from 200 to 1000 V.

Many of the furnaces have load cycles of 3–8 hours. The melting part requires very high fluctuating currents, whereas the refining part requires smaller steady currents.

The furnace transformer may take unbalanced loads and produce harmonics. For steel making, the EAF furnace requires about 400–450 kWh of power/ton and the LF requires 20 kWh/ton of output. The transformer capacity is decided by the required output per hour, the efficiency of the system and the power factor. An approximate indication can be 1 MVA of transformer for 1 ton output per hour.

22.2.2 DESIGN FEATURES OF FURNACE TRANSFORMERS

The design principles are the same as those of a conventional transformer, except for the requirements of high current on the output side at low voltage. When the furnace cycle starts, a very high current flow is required for a short time, and this creates frequent mechanical impacts on the windings.

IEEE C57.17-2012 "IEEE standard requirements for Arc Furnace Transformer" is the applicable standard.

Tap Changer for Voltage Variation

The furnace requires voltage variation during the operation, and it is normally realized by using on-load tap changer (OLTC) on the high-voltage (input) side. The common methods used are direct regulation and booster-type regulation.

a. Direct regulation

OLTC or off-circuit tap changer can be used. The OLTC is connected on the HV side for voltage variations of the LV (output side), and therefore, it is variable flux voltage variation (VFVV). The OLTC is connected on the neutral side of the transformer HV (if it is star-connected winding). A typical winding arrangement is shown in Figure 22.1.

The step voltages of the HV are not constant, and the voltage variation is by changing the flux density in the core. The maximum working flux density shall be within limits allowed by the core material.

A sample calculation of the variable flux voltage variation is shown here for a clear understanding of the uneven step voltages and variable flux densities at different tap positions.

• Data of the transformer

Primary volts 33000, three phases; secondary volts, 240–480 V in steps of 40 volts

Frequency	50 Hz
Primary connection	Star
Secondary connection	Delta
kVA	10000

In this case, the maximum flux density is at 480 V as the volt per turn is the highest at this position.

Fixing the maximum flux density as 1.72 tesla at 480 V on LV, the calculations are shown in Table 22.4. It is to be noted that the number of turns on the LV is fixed and the incoming voltage on HV is constant. By changing the HV turns, the volt/turn is changed and the flux density also changes.

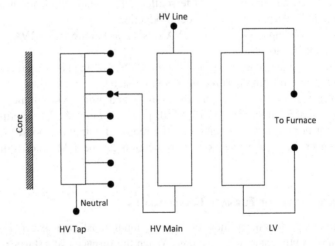

FIGURE 22.1 Direct regulation with VFVV.

TABLE 22.4

Number of Turns and Flux Density

Tap Number	HV Phase (V)	HV Turns	Volt/Turn	LV Turns	LV (V)	Flux Density (Tesla)
1	$\dfrac{33000}{\sqrt{3}}$	476	40	12	480	1.72
2	$\dfrac{33000}{\sqrt{3}}$	520	36.66	12	440	1.58
3	$\dfrac{33000}{\sqrt{3}}$	572	33.33	12	400	1.43
4	$\dfrac{33000}{\sqrt{3}}$	635	30	12	360	1.29
5	$\dfrac{33000}{\sqrt{3}}$	715	26.66	12	320	1.15
6	$\dfrac{33000}{\sqrt{3}}$	817	23.33	12	280	1.00
7	$\dfrac{33000}{\sqrt{3}}$	953	20.0	12	240	0.86

From the calculated results it is evident that the flux density has a wide variation and the turns per step in the High Voltage winding are not uniform.

b. Voltage regulation by booster

This design is done with two active parts assembled in a common tank. The regulating winding is independent of the HV winding.

The overall dimensions and material quantity requirement of this option are higher than those of a VFVV design where one active part only is used. This design is suitable for a wide range of secondary voltage (e.g. $LV_{max}/LV_{min} > 2.5$). Figure 22.2 shows the winding connections.

- Design features of furnace transformers

The LV winding has few turns, and it carries high current. The LV winding is designed as the outer winding for convenience of taking the high-current leads. It consists of a number of parallel disc sections arranged axially with the ends connected directly to vertical copper bus bars. For comparatively low currents, multiple parallel helical windings can be used. The uneven current distribution among winding parallels will have to be controlled by suitable design of winding geometry to obtain nearly equal impedance between parallels.

The high-current outer winding and bus bars will create severe leakage flux on tank non-magnetic structural parts is one solution.

Internally or externally closed delta connection of the LV can be used. The high-current leads are taken outside through non-magnetic fibre-reinforced plastic (FRP) plates or cast epoxy boards to reduce the stray losses.

It is necessary to analyse the critical parameters of leakage flux, hot spot, current distribution among parallel conductors, etc., by finite element analysis to ensure that the stray losses, temperature rise and hot spot gradients are within limits.

Typical cooling systems used are ONAN/ONAF for small ratings and OFWF for higher ratings. When OFWF cooling is used, cooling water flow of about 100 l/hour/kW of losses is required.

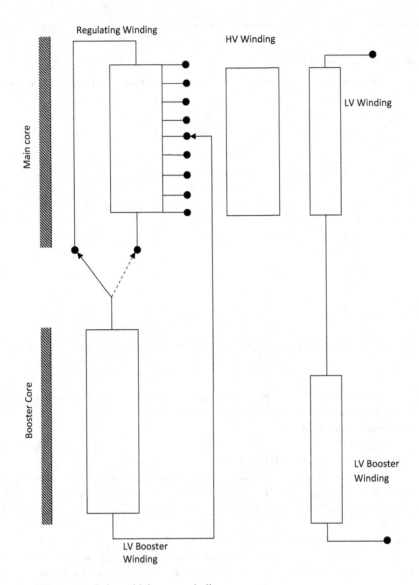

FIGURE 22.2   Voltage regulation with booster winding.

22.3   MULTIWINDING TRANSFORMERS

22.3.1   INTRODUCTION

Multiwinding transformers are used for various purposes where several windings are mutually coupled by a single magnetic core, to connect multiple circuits. For most of the applications, there will be one common input winding with several output windings of equal or unequal voltage and current ratings. In some applications, the output windings or input windings need parallel connection.

Three-winding transformers are commonly used for circuit interconnection. Multiwinding transformers find extensive application in solar photovoltaic generation where a number of inverter-fed windings are connected to a common output winding that is connected to the grid. The following sections give the different winding configurations to get equal impedance between the output windings and the equivalent circuits. Loosely coupled output windings are assumed for all the options.

22.3.2 THREE-WINDING TRANSFORMERS

Figure 22.3 shows the winding configuration for equal impedance between output windings 2 and 3.
The equivalent circuit of a three-winding transformer is shown in Figure 22.4.
Other winding configurations are as follows:

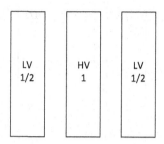

FIGURE 22.3 Configuration for a 3-winding transformer.

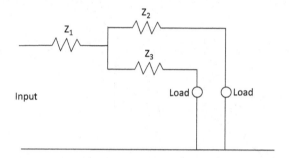

FIGURE 22.4 Equivalent circuit.

a. Axi-symmetric output windings (loosely coupled)
b. Axi-symmetric closely coupled winding where the impedance between the two output windings 2 and 3 is low

22.3.3 FOUR-WINDING TRANSFORMERS (1 INPUT AND 3 OUTPUT WINDINGS)

For equal impedance of output windings, two commonly used winding configurations are given in Figures 22.5 and 22.6.

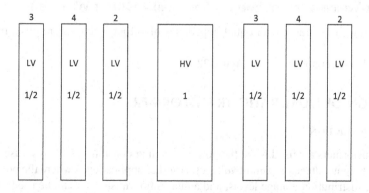

FIGURE 22.5 Four-winding transformer (arrangement 1).

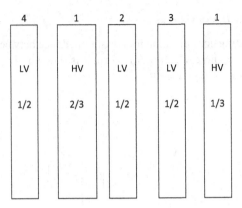

FIGURE 22.6 Four-winding transformer (arrangement 2).

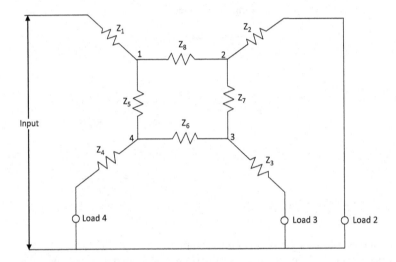

FIGURE 22.7 Equivalent circuit of a 4-winding transformer.

The equivalent circuit of a four-winding transformer is shown in Figure 22.7. The branch impedance values are the combinations of windings in pairs, Z_{12}, Z_{13}, Z_{14}, Z_{23}, Z_{24} and Z_{34} (Z_{12} is impedance between windings 1 and 2 so on).

22.3.4 FIVE-WINDING TRANSFORMERS (1 INPUT AND 4 OUTPUT WINDINGS)

The two winding configurations for equal impedance of output windings are given in Figures 22.8 and 22.9.

The equivalent circuit is shown in Figure 22.10.

22.4 DESIGN OF DUAL-RATIO TRANSFORMERS

22.4.1 INTRODUCTION

Dual-ratio transformers are used when the power system or consumer is likely to use it for connecting to different input voltages or output voltages. Another application is where the power system has more than one distribution voltage levels, and a dual-ratio transformer can be used on either of the voltages. Typical examples are 11 and 22 kV systems and 11 and 6.6 kV systems.

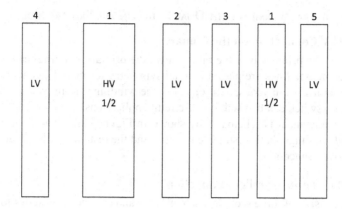

FIGURE 22.8 Five-winding transformer (arrangement 1).

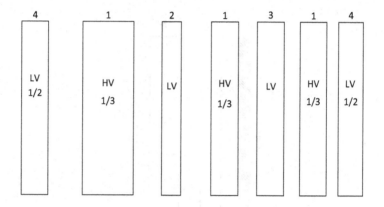

FIGURE 22.9 Five-winding transformers (arrangement 2).

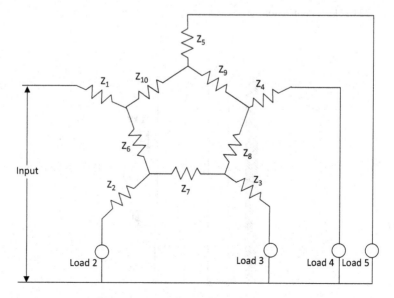

FIGURE 22.10 Equivalent circuit of a 5-winding transformer.

22.4.2   Design of Transformers with Dual Ratio on the Primary Side

22.4.2.1   22–11 kV Connection on the Primary

Since the voltage on the primary is in the ratio of two, the primary windings can be made into two sections of 11 kV. The windings are placed in axial symmetry for impedance matching. For 22 kV, the windings are connected in series, and for 11 kV, the windings are in parallel. A double-decker tap switch cum ratio switch can be used for off-circuit applications.

The winding arrangement is schematically shown in Figure 22.11. The clearance between the two axially placed windings shall be suitable to withstand the test voltage levels at parallel connection as well as series connection.

22.4.2.2   33–11 kV Series-Parallel Connection

The HV winding consists of three sections each of 11 kV and connected in series for 33 kV applications. For 11 kV, the three coils are connected in parallel. The schematic of the winding configuration is shown in Figure 22.12.

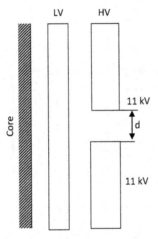

FIGURE 22.11   22 to 11 kV connection.

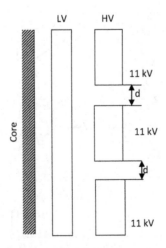

FIGURE 22.12   33 to 11 kV connection.

22.4.2.3   11–6.6 kV Connection

For 11–6.6 kV connection, the ratio is not suitable for connecting complete windings in parallel in one configuration and then making it in series for the full voltage. One solution is to make a part of the winding as common series connected for both 11 and 6.6 kV connections. The voltages of the different coils are calculated as follows;

$$2x + y = 11000$$
$$x + y = 6600$$

where

x = voltage of coil for series-parallel connection
y = voltage of common coil
Solving,
$x = 4400$
$y = 2200$

The two coils of 4400 volts each will have to be placed axi-symmetric to avoid circulating current in parallel connection.

Winding design of an 11–6.6 /0.4 kV, 1500 kVA, 50 Hz, 3-phase ONAN transformer with off-circuit taps on HV for + 7.5% HV variation in 2.5% steps is given as an example.

- Transformer specifications

kVA	1500
Phases	3
Frequency	50 Hz
Cooling	ONAN
Primary volts	11000–6600 delta
Secondary volts	400 star
Vector group	Dyn11
Taps on HV	+ 7.5% variation in 2.5% steps for 11 kV
Winding material	Copper

- Design calculations

Series-parallel-connected winding section – 4400 V
Common winding section – 2200 V
The 4400 V winding sections are placed axially for series-parallel connection, and the 2200-volt section is placed concentrically as common section.
Core circle = 250 mm
Volt/turn = 16.5
The winding design particulars are shown in Table 22.5.
Calculated parameters
LV $I^2 R$ loss = 4995 watts
HV $I R$ loss = 6138 watts
Total eddy current and stray loss = 2047 watts
Total load loss at 75°C = 13180 watts
% Reactance = 6.3
% Resistance = 0.878
% Impedance = 6.36%

TABLE 22.5

Design Particulars

Particulars	LV	HV Series-Parallel	HV Common
Phase volt	230.9	4400×2	2200 rated
			3025 max. tap
Turns/phase	14	267×2	183
Number of layers	14	11	4
Turns/layer	1	24.5	52
Conductor size (bare)	500×1.75	8.9×2.1	8.9×2.1
Conductor size (covered)	-	9.3×2.5	9.3×2.5
Inside diameter	259	354	432
Radial depth	37.5	35	11
Outside diameter	334	434	454
Axial height (electrical)	500	479	480
Axial height (wound)	500	$488\ (244 \times 2)$	490
Axial height including edge strip	530	530	530
Interlayer insulation	0.25	0.25	0.25
Number of ducts	3	2	
Size of the duct	3	3	-
Weight of conductor (kg)	307	335	131

22.5 LIQUID-FILLED TRANSFORMERS USING HIGH-TEMPERATURE INSULATION MATERIALS

22.5.1 INTRODUCTION

The conventional mineral oil-filled transformers use cellulose-based insulation materials for producing the transformers. The thermal class of almost all the insulation materials used is 105°C with a hot spot temperature limit of 98°C. The mineral oil used is hydrocarbon base with low flash point and high calorific value. With a view to improve the fire withstand capacity, synthetic fluids (e.g. silicone liquids), synthetic esters and natural esters came into use. These fluids have higher flash points and lower calorific values when compared to the conventional mineral oil.

Initially, the high-temperature fluids were used just as a substitute for mineral oil, and the other insulation systems remained unchanged.

Though the fluids were able to withstand higher temperature without impairing the life, the permissible temperature rise of the transformer remained unchanged primarily due to the limitations of the temperature withstand capability of class A insulation.

In order to utilize the optimum thermal capacity of the fluid, new designs using high-temperature insulation systems came into effect. This section gives the salient features of the design of liquid-filled transformers using high-temperature insulation.

22.5.2 THERMAL CLASS OF INSULATION MATERIALS

The thermal class of insulation materials is determined by ageing tests, and the concept is that the insulation is capable of working at the "hot spot" temperature assigned to the class of insulation for a defined lifetime. Table 22.6 gives the thermal class and corresponding hot spot temperatures for different insulation materials.

TABLE 22.6

Thermal Class and Hot Spot temperature of Insulation

Thermal Class	Hot Spot (°C)	Typical Materials
105	98	Cellulose-based materials
120	110	Thermally upgraded paper
130	120	Resin-bonded paper
155	145	Polyphenylene sulphide
180	170	Polyester glass
200	190	Polyimide
220	210	Aramid

22.5.3 CONCEPT OF HIGH-TEMPERATURE INSULATION

The basic concept of high-temperature insulation design is that all insulation materials used including the insulation liquid are capable of working at a higher than conventional top liquid temperature, average winding temperature and winding hot spots. It is permitted to use insulation components of lower class where high temperatures are not expected. Obviously, the winding carries the current and the insulation of the winding conductor and the materials in contact with it need to be of the high-temperature type.

IEC 60076-14 2013, Liquid-immersed Power Transformers using High-Temperature Insulation Materials, has classified the systems of insulation applicable for the design into different categories and specified the temperature rise limitations of each category. The salient points are as follows;

a. High-temperature insulation system

In this system, the insulation liquid, the conductor insulation, the spacers, interlayer insulation, edge strips, insulation barriers, etc., shall be high-temperature materials.

The temperature rises allowed as per IEC for conventional mineral oil insulation and the most commonly used insulation liquids are given in Table 22.7.

If the thermal class of solid insulation used is not 180°C class as shown in Table 22.7, the permissible temperature rise limits change correspondingly. For example, if the temperature class of solid insulation used is 155°C, the permissible temperature rises change as shown in Table 22.8.

b. Hybrid insulation system

In a hybrid insulation system, the insulation liquid can be conventional or high-temperature type and some of the solid insulation materials are either conventional or a combination with high-temperature type. Three categories of hybrid designs are identified in this category.

TABLE 22.7

Temperature Rise Comparison of Insulation Fluids

Particulars	Mineral Oil	Ester Liquid	Silicone Liquid
Thermal class of solid insulation	105	180	180
Top liquid temperature rise (K)	60	90	115
Average winding temperature rise (K)	65	125	125
Hot spot temperature rise (K)	78	150	150

TABLE 22.8

Temperature Rise Limits with Class 155 Solid Insulation

Particulars	Ester Liquid	Silicone Liquid
Thermal class of solid insulation	155	155
Top liquid temperature rise (K)	90	115
Average winding temperature rise (K)	105	105
Hot spot temperature rise (K)	125	125

- Full-hybrid winding system

    Insulation liquid – conventional or high-temperature

    Insulation of conductor – high-temperature

    Spacer and strips – high-temperature

    Other solid insulations – conventional

    i. Since the liquid insulation is either conventional or high-temperature, the top liquid temperature rise allowed is limited to the conventional only

    ii. The winding conductor is insulated with high-temperature insulation, and therefore, the average winding temperature rise and hot spot temperature rise allowed correspond to the temperature index of the conductor insulation.

    For example, if 155° class material is used for insulating the conductor, the average winding rise and hot spot rise are 105°C and 125°C, respectively. The top liquid temperature rise is limited to 60°C, because the insulation liquid can be either conventional or high-temperature.

- Mixed-hybrid insulation system

    Insulation liquid – conventional or high-temperature

    Insulation of conductor – conventional and high-temperature combination

    Spacer/strips – conventional and high-temperature combination

    Other solid insulations – conventional

    In this case, the top liquid temperature rise and average winding temperature rise correspond to the conventional class.

- Semi-hybrid winding insulation system

    Insulation liquid – conventional or high-temperature

    Conductor insulation – high-temperature

    All other solid insulations are high-temperature. The top oil temperature rise allowed is "conventional". The average winding rise and hot spot temperature rise correspond to the high-temperature class of the conductor insulation.

22.5.4 DESIGN PARAMETERS TO BE CONSIDERED

When high-temperature insulation liquids are used, the following design parameters need consideration:

- Heat dissipation from tank and cooling fins/radiators

    The viscosity of high-temperature fluids is high when compared to mineral oil, and it affects the heat transfer coefficient. When the transformer operates at low ambient temperatures, the viscosity of silicone liquid increases and the temperature rise of winding is comparatively higher in comparison with the operation at a higher ambient temperature. The approximate increase in cooling area required, when silicone liquid or ester liquid is used, is given in Table 22.9.

TABLE 22.9

Increase in Cooling Area Required for Silicone Liquid and Ester Liquid

Insulation Liquid	Increase in Cooling Area When Compared to Mineral Oil (%)
Silicone liquid	+15
Ester liquid	+12

22.5.5 THERMAL CLASS AND PARAMETERS OF HIGH-TEMPERATURE INSULATION MATERIALS

22.5.5.1 Typical Enamel Insulation for Winding Conductor

The following enamels are suitable for copper or aluminium, round or rectangular conductors and can be used with silicone liquid, synthetic ester and natural ester.

- Polyamide-imide as per IEC 60317 part 26
  Temperature class of this enamel is 200
- Polyimide as per IEC 60317 part 7, 30
  Temperature class of this enamel is 220

22.5.5.2 Insulation Liquids

The thermal classes and applicable standards of some of the high-temperature insulation liquids are given in Table 22.10.

- Compatibility of materials with silicone liquid and synthetic ester
  Some of the materials being used regularly for the mineral oil-filled transformers are not compatible with silicone liquid and synthetic ester. Special attention is necessary to exclude non-compatible materials.
- Non-compatible materials with synthetic ester
  Ethylene propylene rubber [self-amalgamating tape]
  Neoprene rubber
  Natural rubber
  Polychloroprene
  Polystyrene
  Low-modulus silicone sealant
- Non-compatible materials with silicone liquid
  Silicone rubber gaskets or sealing materials shall not be used when silicone liquid is used in the transformer. The fluid is readily absorbed by silicone rubber and causes swelling and loss of physical properties. Silicone rubber parts are likely to be found in bushing seals, tap changers, instrumentation, cover gaskets, etc. These are to be replaced with compatible materials.

TABLE 22.10

Characteristics of Insulation Liquids

Name of the Insulation Liquid	Applicable Standard	Thermal Class	Flash Point (°C)	Fire Point (°C)
Synthetic ester	IEC 61099	130	275	316
Silicone liquid (dimethylsiloxane)	IEC 60836	155	310	360
Natural ester	-	130	330	360

22.6 TRACTION TRANSFORMERS

22.6.1 INTRODUCTION

Driving forward of a vehicle is called traction. For supplying power to high-speed trains, electric multiple units (EMUs), electric locomotives, etc., a transformer is mounted on the vehicle that gets power from the catenary. These are called on-board traction transformers that operate at 50 Hz, 60 Hz, 25 Hz or 16.7 Hz AC systems. Typical voltages of the system are 15 kV and 25 kV. For EMUs and locos, transformer capacities up to 10–12 MVA are required. The transformer is the single heaviest component of the vehicle, and therefore, the design will have to be compact and lightweight.

22.6.2 DESIGN FEATURES OF ON-BOARD TRACTION TRANSFORMERS

22.6.2.1 Specifications and General Requirements

The applicable standard is IEC 60310: 2016 – Railway applications "Traction Transformers and inductors on board rolling stock".

The design of the transformer depends on various factors including the mounting location, the voltage class, the rating required and temperature rise limitations.

The location of the transformer in the vehicle can be classified into three groups as shown in Table 22.11.

The overall efficiency of the train system is influenced by the weight, and the optimization of the transformer weight should not affect the efficiency and reliability.

Due to the train movement, the transformer is subject to shock and vibration. As per the IEC standard, shock and vibration test is a type test, and the design of transformer and components shall be suitable to withstand this test.

The electromagnetic field generated by the transformer in operations needs to be within specified limits. A typical requirement is 75 μ Tesla at a distance of 100 mm from the transformer.

22.6.2.2 Harmonics

The transformer is subject to a high amount of harmonics, and the winding and cooling designs need to consider this.

22.6.3 DESIGN OF TRANSFORMERS

22.6.3.1 Core

Low-loss CRGO core material is used to reduce the losses.

For single-phase core type, 2-limb core construction is used. The windings are placed on both limbs for series or parallel connection.

The core configuration is shown in Figure 22.13.

TABLE 22.11

Mounting Location in the Vehicle

Mounting Location	Design Features
Under floor	This is usually used in electric locomotives. The transformer has a minimum height.
Machine room	Transformer has limitations in length and width.
Roof assembly	The transformer is located in the roof of the train. These are generally used in low-floor trains. The height and width are limitations, and the weight needs to be low.

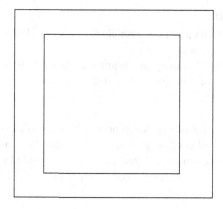

FIGURE 22.13 Single-phase core.

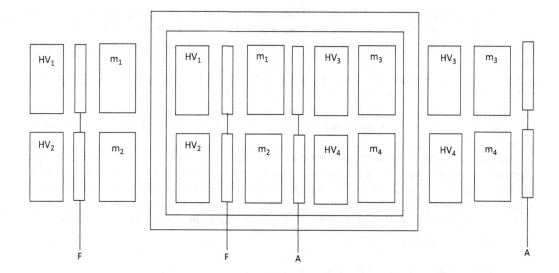

FIGURE 22.14 Winding arrangement of a traction transformer.

22.6.3.2 Windings

The on-board traction transformer has multiple windings. Copper is the preferred winding material to realize higher current density and to withstand a higher short circuit thermal and mechanical stress.

The windings in a typical on-board traction transformer are as follows;

Four traction windings
One auxiliary winding
One filter winding

The winding arrangement relative to the 2-limb core is shown in Figure 22.14.
Symbols used are as follows:
m_1, m_2, m_3, m_4 – traction windings
HV_1, HV_2, HV_3, HV_4 – HV windings (parallel connections)
F – filter winding
A – auxiliary winding

22.6.3.3 Insulation Design

Design of dry-type traction transformers is with class F or class H insulation. 220° class insulation can be used for reducing the size and weight.

For liquid-filled design, high-temperature liquids including silicone liquid, synthetic ester or natural ester with compatible solid insulation materials are used.

22.6.3.4 Cooling System

The traction transformer operates at high current densities to reduce size and consequently have higher losses. The requirement of cooling goes up further from the contribution of harmonic losses. Force-directed liquid flow with pumps and fans is used for higher MVA capacity. For lower MVA ratings, ONAF cooling or unit coolers can be used. The gradient calculation shall consider the effect of harmonics and the type of liquid flow.

22.7 SYMMETRICAL CORE TRANSFORMERS (TRIDIMENSIONAL CORE)

The 3-phase stacked core or wound core construction has "planar" symmetry, and therefore, the flux paths of the outer limbs are longer than the middle limb. This creates flux asymmetry when the core is excited by a 3-phase balanced power supply that is the cause for harmonics and higher excitation current.

Unlike the transformer with planner symmetry, the rotating electrical machines such as motors and generators have the poles arranged in circular symmetry.

The symmetrical core or 3D core transformer has complete magnetic symmetry of the three limbs of a three-phase transformer. The first symmetrical core on commercial scale was produced by Hexaformer, Sweden.

The three limbs of the core are produced by winding the magnetic core laminations that are diagonally slit. The core limb is continuously wound with progressive displacement to get a 30° inclined semicircular cross section. When such 3-core limbs are placed in a 3-dimensional structure, the combined core limb becomes circular. Figure 22.15 shows the cross section of the core.

The core manufacturing process is different from the conventional stacked core or wound core construction. The processes involved are as follows;

- Diagonal slitting of the laminations
- Core winding
- Annealing of wound core

[For CRGO material, the annealing is at vacuum and the temperature is about 820°C.]

The benefits of the symmetrical core construction when compared to three-phase stacked core construction are as follows;

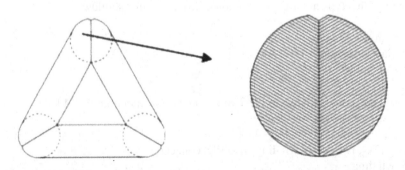

FIGURE 22.15 Cross section of the core.

- Reduced core loss and excitation current
- The core limbs are wound by diagonally slit material to form a circular cross section, and therefore, the process loss of material is practically negligible. For stacked core, the material wastage is high.
- For a given cross-sectional area of core, the core circle for symmetrical core is low when compared to stacked core, where a discrete number of steps are used, and therefore, the core circle is higher.
- Due to the symmetry of the flux paths, the harmonics generated are low.
- The noise level is comparatively low.
- In 3-phase stacked and wound core construction, the cooling efficiency of the middle phase is comparatively low when compared to the outer limbs. In symmetrical core construction, all three coils have the same cooling efficiency.

22.7.1 DESIGN OF SYMMETRICAL CORE TRANSFORMERS

The basic design procedures and concepts of a three-phase transformer are applicable for symmetrical core design except for the core design, winding construction and structural design in particular.

Two types of symmetrical core designs are used: one is the core without joints, and the other type is with removable yoke.

The comparison of the two types of core design is given in Table 22.12.

22.7.1.1 Design of Transformers

a. Core design

$$\text{Core gross area} = \frac{\pi}{4} d^2$$

where d is the core diameter.

This assumes that the core circle is completely filled.

The volt/turn, flux density and other parameters are calculated using the same procedure of the 3-phase stacked core transformer.

Core to LV clearance and winding-to-yoke clearance parameters remain unchanged.

b. Winding design

The winding design is essentially unchanged for the core with joints in the top yoke.

For the core without joints, the limitations of the toroidal winding machines are taken into consideration.

c. Core and coil assembly

The clamping structure of the core–coil assembly is different from the conventional stacked core assembly. The structure is of triangular formation.

TABLE 22.12

Comparison of the Two Types of Core Design

Particulars	Core Without Joints	Core with Removable Top Yoke
Core size and weight	Same	Same
Core loss	1 PU	1.05–1.07 PU
No-load current	1 PU	1.1–1.12 PU
Winding construction	Toroidal winding machine is required	Normal winding machine is sufficient
Manufacturing time for winding	Slow	Fast
Other details	The core is loaded on the winding machine, and LV and HV are wound directly on the core	-

d. Tank

In order to match with the triangular formation of core and coil assembly, irregular hexagonal-shaped tank can be used to optimize the oil quantity. However, rectangular-shaped tank configuration similar to the stacked core design can be used where the oil quantity required is higher by 25%–35%. In this case, the width of the tank is increased and the tank length is reduced.

23 Transformers for Renewable Energy Applications

23.1 INTRODUCTION

The future additions to the power generation capacity in the world will be predominantly solar photovoltaic (PV) generation and wind power generation. The special operating and performance requirements of these transformers call for different design and analysis techniques. This chapter covers these design requirements. The emerging scenario of the development of smart grid and the application of these transformers to the smart grid also is briefly covered.

23.2 TRANSFORMERS FOR DISTRIBUTED PHOTOVOLTAIC GENERATION

Electric power is generated by converting solar energy to direct current (DC) by using PV cells. The DC generated is converted to alternating current (AC) by inverters, and the AC is connected to the power grid by a step-up transformer.

The international standard applicable for the transformers for distributed photovoltaic (DPV) generation is IEEE C57.159 2016 "IEEE Guide on Transformers for Application in Distributed Photovoltaic (DPV) Power Generation System."

At present, there are limitations on the power rating and voltage level of the inverter system, and therefore, one or more inverters are connected to an equal number of secondaries of the step-up transformers. Though the most common configuration is 2–3 secondaries, at present, transformers with 6 secondaries are also manufactured.

Inverter manufacturers are now developing inverters with higher power and voltage ratings, and this would increase the transformer MVA ratings in future and reduce the number of secondary windings required.

The following paragraphs give the design requirements of the transformers for the inverter-fed working.

23.2.1 Special Design Features Required to Meet the Service Conditions

A. Nonsymmetrical load and voltage

The inverter voltage and load current to the three phases of the transformer can be unbalanced. If the transformer is fed by more than one inverter, there is a possibility of one of the inverters getting inactive, which can create unbalanced loading of the winding. The unbalanced voltage and current can create excessive leakage flux, stray loss, and overheating of winding and tank.

The design with vertically stacked loosely coupled low-voltage (LV) windings and an equal number of split high-voltage (HV) windings is preferred to reduce the effect of unbalances. The impedance characteristics will be defined based on the inverter system and the number of inverters connected to a transformer. A typical winding arrangement with two secondaries is shown in Figure 23.1

Winding arrangements commonly used along with advantages and disadvantages are shown in Table 23.1.

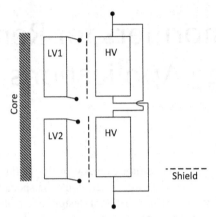

FIGURE 23.1   Winding arrangement with two secondaries.

Source: Copyright permission from Transformer for Distributed Photovoltaic (DPV) *Generation—Conference Proceedings 2018 International Conference on Electrical, Electronics, Computer and Optimization Techniques* (ICEECCOT) 14–15 December 2018, Mysore, India.

TABLE 23.1

Winding Arrangements Commonly Used Along with Advantages and Disadvantages

Winding Arrangement	Advantages	Disadvantages
Concentric LV$_1$–HV–LV$_2$	LV$_1$–LV$_2$ impedance is high	It is difficult to get LV$_1$–HV and LV$_2$–HV impedances nearly equal LV$_1$–LV$_2$ impedance is relatively low
Concentric with vertically interleaved LV (type A) LV$_1$–LV$_2$–HV	LV$_1$–HV and LV$_2$–HV impedances are approximately equal	
Concentric with vertically interleaved (type B)	LV$_1$–HV and LV$_2$–HV impedances are approximately equal	LV$_1$–LV$_2$ impedance is relatively low More expensive compared with (type A)
LV$_1$–LV$_2$–HV Vertically stacked LV winding	LV$_1$–HV and LV$_2$–HV impedances are almost equal	Less compact design LV$_1$–LV$_2$ impedance is high

Source: Copyright permission from Transformer for Distributed Photovoltaic (DPV) Generation- Conference Proceedings 2018 International Conference on Electrical, Electronics, Computer and Optimization Techniques (ICEECCOT) 14–15 December 2018, Mysore, India.

B. Presence of DC in winding

There is a possibility that a DC can come to the inverter-fed winding which can increase the core magnetizing current and inrush current peak.

C. Wave shapes of inverter output

The wave shapes of two or more inverters connected to one transformer may not be synchronized. This can cause change in wave shape and harmonics as well as disturbances of the flux.

D. Fast rising pulsed waveform on LV winding

The inverter produces a pulsed output to ground, and the pulse can reach a rate of rise (*dv/dt*) level of 500 V/microsecond.

The LV winding insulation will have to be designed to withstand the rapid rising voltage for the design life of the transformer. Figure 23.2 shows the typical voltage waveform from

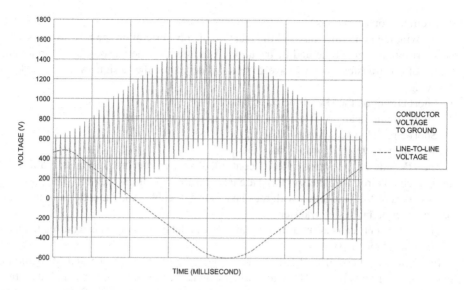

FIGURE 23.2 Voltage waveform from the inverter.

Source: Copyright permission from, IEEE C 57-159, IEEE Guide on Transformer for Application in Distributed Photovoltaic Power (DPV) Generation Systems.

the inverter. An electrostatic shield is provided between the LV and HV winding to isolate the HV winding from the effect of the fast rising voltage on the LV. The shield acts as an additional *dv/dt* filter and filters the voltage gradient of the pulsed inverter output. It also reduces the transfer of the transients from the HV winding to the LV winding.

Copper or aluminum can be used as shielding. Copper shields will produce less eddy current loss when compared with aluminum shield.

In the context, it is noted that "sunny tripower" inverters manufactured by SMA Solar do not require electrostatic shielding between primary and secondary windings of the transformer.

Accelerated aging test of prototype LV winding insulation can be carried out to check the effect of the fast rising transients on the insulation life. The effect is different for dry-type transformer insulation and fluid-filled transformer insulation.

E. Losses and efficiency of the transformer

The DPV transformer is designed with comparatively low no-load loss because the transformer draws exciting power from the system when there is no generation (i.e., at night). The efficiency at a specified load cycle is optimized to get overall economy of operation. If the power system is designed with a power storage facility like battery system, the transformer will be operating under load continuously and the efficiency level can be fixed on this basis.

F. Inrush current considerations

The LV winding of the DPV transformer is normally near the core, and therefore, the air core reactance of this winding is low. The inrush current when LV is switched on is comparatively high.

G. Thermal design

The transformer cooling system is designed to consider the effect of the following;
- Ambient temperature variation at the site
- Load curve
- Effect of harmonics
- Effect of reactive load, if any

H. Short-circuit considerations

   The winding configurations and the location of short circuit on the transformers affect the short-circuit magnitude and distribution. The effect of the various short-circuit conditions of the transformer will have to be considered for the design, which include the following;
   - Short circuit on the HV of the transformer
   - Short circuit on any one or more of the LVs of the transformer
   - Short circuit between any two LVs of the transformer

I. High-frequency switching transients

   The HV side of the transformer is controlled by a circuit breaker, and in almost all cases, vacuum circuit breakers (VCBs) are used for the circuit interruption. The prestrikes and restrikes of VCBs together with the capacitance of cables and inductance of the transformer generate fast rising transients.

   These transients can cause insulation failure, and detailed analysis is required to ensure adequate insulation design.

   IEEE Std C57.142-2010 "IEEE Guide to Describe the Occurrence and Mitigation of Switching Transients Induced by Transformers, Switching Device, and System Interaction" deals with this subject. Switching of an unloaded transformer with VCB can cause chopping overvoltage, multiple reignitions overvoltage and overvoltage due to virtual current chopping. Simulation study of transformer model for different frequency intervals (up to 2 MHz) using electrical parameters of cable used and transformer design data can be carried out to get the calculated values of overvoltages during VCB switching.

   Some of the methods adopted to protect the transformer from fast rising transients include the following;
   - Installation of surge arrester on the HV side of the transformer. However, it is not sufficiently effective to protect the transformer against very fast transients.
   - RC snubbers on the HV side of the transformer against very fast rising transients. Typical value of resistor is 25–50 Ω, and capacitor is 0.15–0.35 μF.
   - Insertion resistors on the ring main unit (RMU). This is an expensive solution, as RMU design needs modification. The protection method will have to be designed and implemented in construction with RMU manufacturer.

J. Environmental and climatic requirements

K. Special installation/operation practices
   - The inverters are connected to the star-connected LV windings of the transformer, and thus, neutral is kept floating. The neutral shall not be earthed/grounded. It is a safe design practice to keep the neutral isolated inside the transformer.
   - The electrostatic shields shall have single point earthing only.

   Power generation from PV system does not produce carbon emission. However, if mineral oil is used for the DPV step-up transformer, it is not environmental friendly. The options used now are as follows:
   i. Biodegradable oil–filled transformer where vegetable-based oil is used
   ii. Dry-type cast resin transformer where no insulation fluids are used

The design requirements of both types of transformers to meet the environmental and climatic requirements are mainly the following;

For biodegradable oil–filled transformer:

- The viscosity of biodegradable oil is high when compared with mineral oil, and therefore, about 15% more cooling surface is required to get the same heat dissipation. (The viscosity of biodegradable oil at 40°C is about 35 mm²/second against 12 mm²/second for mineral oil when measured as per ISO 3104.)

- Transformers using this fluid can be designed as per IEC 60076-14-2013 "Power transformers – Part 14: Liquid-immersed power transformers using high-temperature insulation materials" and can work at temperatures above conventional limits, subject to customer acceptance.
- Surface treatment and corrosion protection of all external parts such as tank and radiators will have to be as per ISO 12944, C4 for normal outdoor environmental and C5-M for highly corrosive environments such as sea coast and off-shore applications.

For dry-type cast resin transformer:

- Class F (155°C) and class H (180°C) insulations are generally used. The temperature rise limit will depend on the ambient temperature conditions at the installation location.
- The dry-type transformer requires suitable ingress protection. IP23D and IP31 are most commonly used classes of ingress protection. The ventilation requirement and air flow requirement, if installed inside a room, will have to be calculated to ensure that the temperature rise of the winding is within limits.
- The cast resin transformer will have to meet the climatic, environmental, and fire withstand requirements (C2E2F1 or C2E3F1) as per IEC 60076-11 "Power transformers – Part 11: Dry-type transformers".
- The design of the epoxy casting system shall meet the thermal expansion and contraction of the winding material (copper/aluminum). This is done by changing the filler loading to suit the temperature coefficient of expansion of respective metals.

23.3 TRANSFORMERS FOR WIND TURBINE

Humankind was using wind energy in the past for various applications. The first wind mill for electricity generation was installed in 1887, in Scotland for battery charging. By the beginning of the 1900s, there were several wind-driven electric generators. Utility-connected wind turbines were installed in the United Kingdom in 1951. There is a trend for using 66 kV for off-shore, which requires 66 kV for the transformer. However, the development of wind turbines for electric power generation has been progressing in the recent past to improve the efficiency and reliability. The type of wind turbines used can be classified as follows:

- Permanent magnet synchronous generations (these were early-stage wind power generators)
- Field-excited synchronous generators
- Doubly fed induction generators

The most commonly used wind turbine generator today is the doubly fed induction generator that provides variable voltage control and 100% control over slip. The power from the wind turbine generator is stepped up by a transformer for feeding to the grid. By virtue of its nature of application, the design of the transformer will have to consider the requirements explained in the following paragraphs.

The applicable standard is IEC 60076-16 "Power transformers – Part 16: Transformers for wind turbine applications." Most of the design considerations discussed under Section 23.2 are applicable to transformers for wind turbines. This section therefore discusses the requirements that are not covered in Section 23.2.

A. Loading cycle of transformer

Due to the variation of wind speed, the load on the transformer varies frequently, which may include abrupt changes or even shutdown. The sudden variations of load several times can induce severe mechanical stress on the winding. In the case of liquid-filled transformers, this may create bubbling in oil.

B. Temperature correction for transformers installed inside the tower or nacelle

The airflow and ventilation of the transformer installed inside the tower or nacelle will have to be evaluated. Suitable derating is necessary depending on the requirement.

C. Level of vibration

The transformer installed inside the nacelle will experience vibrations of varying magnitude, and the effect on the transformer's active part and components will have to be calculated. Possibilities of resonance of some components such as radiators need careful evaluation and structural design. Particular attention is required for connections which can get loosened or broken.

D. Humidity and salinity

Exposure to humidity and salinity especially when the installation is near coastal area or off-shore can lead to severe corrosion and eventual failure of cast resin transformers in ventilated enclosure. The core assemblies including clamping structure will have to be protected by suitable epoxy coatings in the case. Exposed line or tap terminals will have to be avoided. Also the enclosure needs C5M surface treatment and painting process for oil-filled transformers

23.4 EMERGING TRENDS IN THE DEVELOPMENT OF TRANSFORMERS FOR RENEWABLE ENERGY

The complexity of electrical grid is increasing rapidly due to the use of renewable energy distributed generating systems, usage of large number of nonlinear loads, electric vehicle charging, and so on. Along with this, the need for making the grid versatile and "intelligent" has prompted the development of "smart grid" concept. It is inevitable that the grid will get phenomenal transformation sooner or later and this will require "smart intelligent transformers" in future. The transformers for renewable energy are no exception, and the next generation of transformers for renewable energy will have to integrate with the demands of the smart grid Figure 23.3 shows the schematic of smart grid with renewable energy generation.

Technical challenges for commercialization of the transformers for smart grid

There are several technical issues to be solved and product development work required for the commercialization of the solid-state transformer (SST) for renewable energy such as follows;

* Introduction of smart grid by the utilities, which is a slow process now due to investment constraints and legacy issues.
* Commercial availability of HV (e.g., 11 kV, 13.2 kV, etc.) insulated gate bipolar transistor or silicon carbide components for converter/inverter applications. At present, cascade connection is employed to get the working voltage level.
* Protection of HV power electronic circuit from surges/impulses and system faults.
* Availability of low loss magnetic material for the high-frequency transformer core.
* Overall system efficiency needs improvement.

    The conventional transformer has high efficiency (more than 99% typically), whereas the overall efficiency of the SST is considerably low.
* Special winding material is required for HF application. It is expected that carbon nanotube may offer low weight and low loss solution in future.

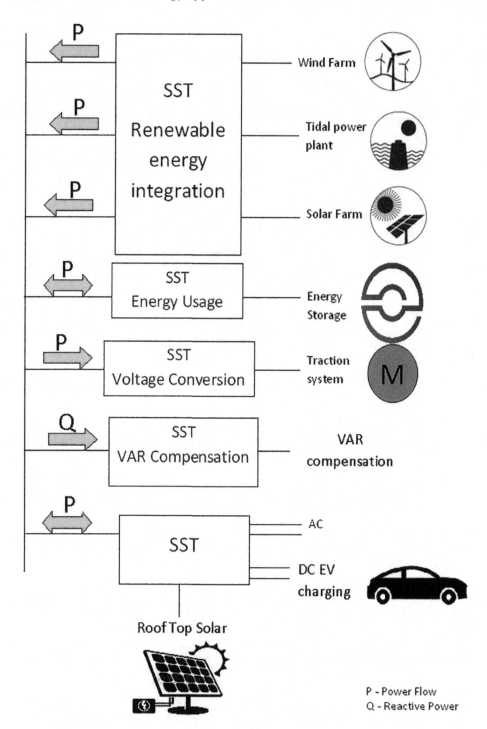

FIGURE 23.3 Schematic of smart grid.

24 Condition Monitoring of Oil-Filled Transformers

24.1 ONLINE AND OFF-LINE DIAGNOSTIC METHODS

The modern power system is mandated to provide reliable uninterrupted power supply, and therefore, it is important to ensure the health and longevity of the transformer. There are several effective online and off-line diagnostic tools available today for the diagnosis of the condition of the transformer which can identify incipient faults, estimate the remaining life, ascertain maintenance strategy, or schedule the retirement and replacement of the transformer. This chapter lists out the diagnostic/monitoring methods and reviews some of the commonly used techniques.

The online monitoring/diagnostic methods measure the inputs continuously without affecting the operation of the transformer, whereas the off-line methods need the planned shutdown of the transformer.

24.2 ONLINE MONITORING METHODS

Table 24.1 shows the commonly used online data monitoring methods. The measurement applicable for a particular transformer is selected based on the design, application, rating, etc.

24.3 OFF-LINE DIAGNOSTIC METHODS

Off-line measurements are taken when a transformer is disconnected from service. The major techniques are given in the following. Most of the online monitoring parameters given in Section 24.2 can be measured and analyzed off-line also, depending on the criticality of the installation and the type of diagnostic system envisaged.

- Insulation resistance and polarization index
- Capacitance and tan delta measurement
- Oil breakdown voltage
- Polarization and depolarization current
- Recovery voltage method
- Optical spectroscopy
- Sweep frequency response analysis (SFRA)
- Dielectric frequency response analysis
- Impulse frequency response analysis
- Transformer impedance circle characteristics
- Frequency domain spectroscopy

24.4 DISSOLVED GAS ANALYSIS

Dissolved gas analysis (DGA) detects incipient fault conditions, leading to future faults of transformer. Oil is a hydrocarbon molecule containing hydrogen and carbon atoms forming chemical bonds. Decomposition of electrical insulation materials in transformer (paper and oil) generates gases. Table 24.2 shows the various gases developed in the transformer oil.

TABLE 24.1

Online Monitoring Methods

S. No.	Transformer Subsystem	On Line Data Monitored
1.	Core and coil assembly	Load current of transformer
		Terminal voltage of transformer
		Winding temperature
		Top oil temperature
		Peak voltage of surge (if any)
		Short-circuit current/surge current (if any)
		Search coil–based monitoring of internal faults
		Dissolved gases in oil
		Core to ground current
		Tank to ground current
		Moisture in oil
		Operating noise
		Partial discharge measurement
2.	Oil preservation system	Oil level in tank/conservator
		Nitrogen/gas pressure level
		Moisture in oil
3.	Cooling system	Ambient temperature
		Cooling system supply
		Cooling fan status (on/off)
		Fan motor current
		Status of oil pump (on/off)
		Oil pump motor current
		Oil flow through the pump
4.	Bushings	Capacitance and tan delta
		Leakage current
		Bushing voltage through capacitive coupler
5.	On-load tap changer (OLTC)	OLTC oil level
		Dissolved gas in OLTC oil
		Moisture content in OLTC oil
		Current through OLTC (winding current)
		Oil temperature of OLTC diverter switch

Gases dissolved in oil are analysed by gas chromatography which separates each gas from other and directly measures their concentrations individually. The interpretation is carried out by the guidelines given in IEC 60599-2007-5 "Mineral oil–impregnated electrical equipment in service: Guide to the interpretation of dissolved and free gas analysis" or IEEE C57.104.2008 "IEEE Guide for the Interpretation of Gases Generated in Oil-Immersed Transformer" or CIGRE guidelines or Duval Triangle (DT) or a combination.

The concentration of various gases and the ratios of the concentration give conclusion regarding the type of fault/condition existing in the transformer (Table 24.3).

When the values of $CO > 500$ PPM and $CO_2 > 5000$ PPM, CO_2/CO ratio of <3 indicates thermal cellulosic degradation. (However, this ratio is not reliable for sealed-type transformers.) A ratio 3 to 11 is considered healthy cellulose insulation.

- Interpretation as per CIGRE SC15

    The guidelines for interpretation of DGA in oil-filled transformers as per CIGRE SC-15 are given in Table 24.4

TABLE 24.2
Various Gases Dissolved in the Transformer Oil

Gas	Formula	Temperature at Which Gas Forms	Source of Gases
Hydrogen	H_2	<150°C – corona in oil >250°C – thermal and electrical faults	Partial discharge thermal faults Power discharges
Methane	CH_4	<150°C–300°C	Corona, partial discharge, low- and medium-temperature faults
Ethane	C_2H_6	200°C–400°C	Low- and medium-temperature faults
Ethylene	C_2H_4	300°C–700°C	High-temperature thermal faults
Acetylene	C_2H_2	>700°C	High hot spot. Low energy discharge
Carbon monoxide	CO	105°C–300°C >300°C complete decomposition	Thermal faults involving paper press board, wood, etc.
Carbon dioxide	CO_2	100°C–300°C	Normal ageing, thermal fault involving cellulose
Oxygen	O_2	Vacuum when temperature drops	Exposure to air, leaky gasket, air breathing through conservator

TABLE 24.3
IEC 60 599-2007-5 Method

C_2H_2/C_2H_4	CH_4/H_2	C_2H_4/C_2H_6	Fault Type
Not significant	<0.1	<0.2	Partial discharge
>1.0	0.1–0.5	>1.0	Discharge of low energy (D_1)
0.6–2.5	0.1–1.0	>2.0	Discharge of high energy (D_2)
Not significant	>1.0	<1.0	Thermal fault<300°C
<0.1	>1.0	1.0–4.0	Thermal fault <300°C to 200°C (T_2)
<0.2	>1.0	>4.0	Thermal fault>700°C (T_s)

TABLE 24.4
Interpretation of DGA in Oil-Filled Transformer as per CIGRE SC 15

Name	Ratio	Significant Value	Interpretation
Key ratio 1	C_2H_2/C_2H_6	>1	Discharge
Key ratio 2	H_2/CH_4	>10	Partial discharge
Key ratio 3	C_2H_4/C_2H_6	>1	Thermal fault in oil
Key ratio 4	CO_2/CO	>10 indicates overheating of cellulose <3 indicates degradation of cellulose by electrical arcing	Cellulose degradation
Key ratio 5	C_2H_2/H_2	>2 (>30 PPM) diffusion from OLTC or through common conservator	Applicable for tank OLTC

DGA, dissolved gas analysis; OLTC, on-load tap changer.

Evaluation of Faults by Rogers Ratio Method (IEEE C57.104.2008)
Table 24.5 gives the six different cases as per IEEE C57.104.2008.

- Duval Triangle

DT method plots the relative percentage of CH_4, C_2H_4 and C_2H_2 on each side of a triangle from 0% to 100%. Six zones of faults are indicated in the triangle and a zone indicating mixture of thermal and electrical faults (DT zone). Several software packages are available for DGA interpretations using triangle method.

24.5 PARTIAL DISCHARGE MEASUREMENT

In an oil-filled transformer, the dielectric including cellulose and oil is continuously under electrical stress. The minor voids or imperfections, conductive particles and so on create localized stresses, leading to partial breakdown of the dielectric. These discharges are very fast electrical processes and radiate electromagnetic waves of high frequency and ultrahigh frequency ranges. If such partial discharges (PDs) are continuously increasing, it is a sure sign of a weak insulation, high stress concentration and so on. These are signs of incipient faults.

A healthy oil-filled transformer can have PD magnitude of 10–50 PC or lower at operating voltage. The increasing PD levels signify the following (Table 24.6).

The different methods of PD detection are shown in Table 24.7.

TABLE 24.5
Different Cases as Per IEEE C57.104.2008

Case	R_1 CH_4/H_2	R_2 C_2H_2/C_2H_4	R_3 C_2H_4/C_2H_6	Diagnosis
0	>0.1–1.0	<0.1	<1.0	Normal unit
1	<0.1	<0.1	<1.0	Low energy arcing – partial discharge
2	0.1–1.0	0.1–3.0	>3	Arcing – high energy discharge
3	>0.1–<1.0	<0.1	1.0–3.0	Low temperature thermal
4	>1.0	<0.1	1.0–3.0	Thermal <700°C
5	>1.0	<0.1	>3.0	Thermal <700°C

TABLE 24.6
Significance of Varying Partial Discharge Levels

100–300 pC	Presence of particles with small air bubbles. (This is typical after filling oil.)
Up to 500 pC	This level in an oil barrier space may be considered as normal deterioration. (This does not affect the dielectric withstand level.)
>1000 pC	This can be caused by different degradation process:

a. 2000–4000 pC level generally signifies increasing moisture level.
   (PD inception level is reduced in the case.)
b. 1000–2000 pC – indicates poor impregnation.
c. >2500 pC in cellulose – will cause destructive ionization if prolonged.
d. >10000 pC in oil – will cause destructive ionization if prolonged.

TABLE 24.7

Methods of Detection of PD

S. No.	Type of Sensor	Advantages	Disadvantages
1.	Electric Direct connection to the test tap of bushing or through high-frequency CT on ground wires	Good sensitivity	Sensor installation is possible only when transformer is deenergized
2.	Electromagnetic antenna	Easy to use	High disturbance of ambient
3.	Acoustic accelerometer placed on tank	• Easy to use • Can detect acoustic emission magnitude • Approximate location of PD can be identified	Low sensitivity (can detect PD > 10000 pC)
4.	UHF PD probes (UHF couplers)	Can be installed in drain valve or special flanges installed at factory on tank.	Less affected by ambient PD as the measurement is at UHF

PD, partial discharge; UHF, ultrahigh frequency.

24.6 FURAN ANALYSIS AND DEGREE OF POLYMERIZATION OF TRANSFORMERS

When the transformer operates the insulation ages (becomes weak), and it would ultimately lead to a breakdown causing the failure of the transformer. The number of glucose rings in the cellulose is reduced when the kraft paper ages, and this is called degree of polymerization (DP).

New cellulose will have 1000–1200 glucose rings when a transformer is manufactured. When it is processed under drying some of the rings are broken down and hence the DP value of new transformer is about 950.

A DP value of 250 is considered as the end point of insulation life, and at this point, the insulation becomes highly brittle, and the tensile strength of the cellulose becomes very low. (If left without disturbance, some transformers may give extended useful service after reaching this point.)

Measuring the DP value of the insulation paper can be done by taking sample from the transformer. However, this has several drawbacks like

a. The transformer will have to be switched off and opened to take a sample.
b. The sample collected may not be from the area of maximum ageing, i.e., the hot spot.

The disintegration of the cellulose inside the transformer generates furan compounds, which are dissolved in the transformer oil. Hence, it is convenient to analyse the transformer oil sample for furan content to determine the ageing (DP) of the insulation. This can be done without switching off the transformer. The furan compounds generated by various types of reactions are different. The concentration of furaldehyde (2-FAL) in the oil is due to the overheating of the paper, and the maximum aging is at the location of hot spot. The drawback of this method is that the measurement gives an average of the furan concentration, and therefore, the same at the hot spot may be much higher. The concentrations of furans are measured by high-pressure liquid chromatography. The concentrations of various furans indicate different types of stresses as shown in Table 24.8.

The concentration of 2-FAL value is used by different authors to estimate the DP value, and the results of DP values corresponding to the concentration of 2-FAL values as per "chengdong" are given in Table 24.9.

TABLE 24.8

Concentration of Furan which Indicates Different Types of Stresses

S. No.	Type of Furan	Symbol	Nature of Stress
1	5-Hydroxymethyl 2-furfuraldehyde	5-H_2F	Oxidation
2	Furfural alcohol	2-FOL	High moisture
3	2-Furaldehyde	2-FAL	Overheating
4	2-Furyl methyl ketone	2-ACF	Lightning
5	5-Methyl 2-furaldehyde	2-M_2F	Local severe overheating

TABLE 24.9

DP Values Corresponding to the Concentration of 2-FAL Values as per "chengdong"

S. No.	Furan (2-FAL) PPM	Estimate DP Value
1	0.01	1003
2	0.10	717
3	0.50	517
4	1.0	431
5	2.5	318
6	5	232
7	10	146
8	15	95

PPM = concentration in parts per million.
DP, degree of polymerization; FAL, furaldehyde.

24.7   SWEEP FREQUENCY ANALYSIS

The transformer is a complex RLC network, and any form of damage, inconsistency, or deformation anywhere in the active part results in change of this RLC circuit. When different frequencies are applied to the transformer, the RLC network offers different impedance path, and the transfer function at each frequency is a measure of the effective impedance of the network. Any geometrical change alters the RLC network, which in turn changes the transfer function at different frequencies, thereby highlighting the area of concern.

When a new transformer is manufactured and passes the tests, the SFRA of the transformer can be taken for using as a reference (signature). This can be compared with SFRA measurement in future, for example, after transportation to site, after installation and after an incident like a tripping and earthquake.

If SFRA of a transformer is not available, detection of abnormalities can be tired in any one of the following methods.

- Checking the SFRA of another transformer of the same design from the same batch of production, if available.
- Comparing SFRA of other phases of the transformer.

SFRA test is done typically between 20 Hz to 1 MHz.

The number of tests for various types of transformers is as per Table 24.10.

TABLE 24.10

Number of Tests for Various Types of Transformers

S. No.	Type of Transformer	Number of Tests
1	Two-winding transformer	15
2	Autotransformer without tertiary	18
3	Autotransformer with tertiary	33
4	Three-winding transformer	36

TABLE 24.11

Guideline for the Interpretation of Probable Reasons

S. No.	Frequency Range	Component	Probable Reasons for Deviation of SFRA
1	<2000 Hz	Main core Main winding Inductance	Core deformation, core joints becoming loose, open circuit of winding, shorted turns, residual magnetism
2	2–20 kHz	Shunt inductance	• Movement between winding and clamping structure • Loose clamping
3	20–400 kHz	Main windings	Deformation within the main or tap windings
4	400 kHz–1 MHz	Main windings Tap windings Internal leads	• Movement of the main and tap winding • Ground impedance variation

SFRA, sweep frequency analysis.

The response of SFRA for different ranges of frequencies can pinpoint the region/component of concern/problem. A guideline for the interpretation of probable reasons is given in Table 24.11.

24.8 DIELECTRIC FREQUENCY RESPONSE ANALYSIS

The traditional method of measuring the moisture content of the oil in the transformer does not give the oil absorbed in the cellulose insulation. This is because when the transformer is loaded, and the winding temperature rises, the moisture comes out from the cellulose and absorbs in oil. When the load comes down, the temperature comes down, and the moisture returns back to the cellulose. Though the moisture equilibrium curves can be used to determine the moisture in cellulose, the results are highly dependent on temperature of the oil, and it may give inaccurate results.

The dielectric frequency response is an advanced technique to ascertain the moisture in insulation. The frequency response of a dielectric is a unique characteristic of a particular insulation system. The increased moisture content changes the dielectric model and the frequency response.

By measuring the dielectric response over a wide frequency range, the moisture content of the insulation can be determined, and the condition assessed.

The test is normally done between the high-voltage and low-voltage winding of a transformer by applying frequencies from 0.001 to 1000 Hz to measure the dissipation factor. The voltage used for testing is up to 200 V. The moisture content and ageing are revealed at higher frequencies. The shape of the curve shows the relative moisture levels, and a typical response pattern is shown in Figure 24.1.

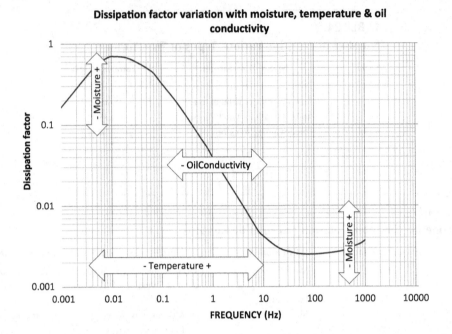

FIGURE 24.1   Dielectric frequency response pattern.

24.9   RECOVERY VOLTAGE MEASUREMENT

A step voltage is applied to the transformer, and a charging current flows. Then, it is short-circuited during the discharge time, and then it is open-circuited. A recovery voltage appears across the terminals. This is caused by the unfinished polarization.

The recovery voltage is measured by changing the charging time. The discharging time is kept as two times charging time for all measurements. These measurements at different charging times give the polarization spectra.

24.10   OTHER MONITORING AND DIAGNOSTIC METHODS

24.10.1   Optical Spectroscopy

The absorbance of ultraviolet (UV) light of fresh oil is low. When the oil ages, the absorption of UV light goes up. The absorption is the highest between the wavelengths of 200–380 nm. Figure 24.2 shows the pattern of absorbance of UV light in fresh oil and service aged oil.

24.10.2   Search Coil–Based Online Diagnostics of Transformer Internal Faults

Internal faults related to transformer windings such as interturn faults and winding displacement change the inductance of the winding when compared with the inductance at healthy conditions. If search coils are installed during manufacture, the induced voltages in the search coils can be analyzed to monitor the changes of the inductance. The faulty winding and location of the defect can be detected.

24.10.3   Polarization and Depolarization Current Test

The polarization current of a transformer winding is a function of the geometry, oil properties and ageing. Initial value of polarization current is determined by the oil conductivity. The depolarization

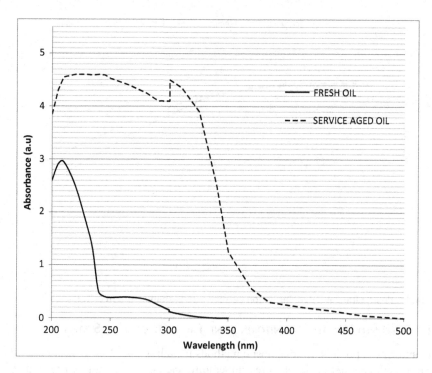

FIGURE 24.2 Absorbance of ultraviolet (UV) light in fresh oil and aged oil.

current comes down with the elapse of time when compared with polarization current depending on the moisture. A typical pattern of the polarization and depolarization current when the moisture content is low ($\approx$, %) is shown in Figure 24.3. The typical pattern when the moisture content is about 5% is shown in Figure 24.4.

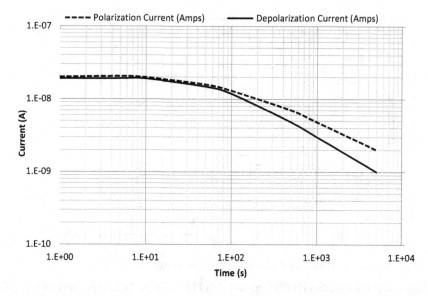

FIGURE 24.3 Polarization and depolarization current with low moisture in cellulose ($\approx$1%).

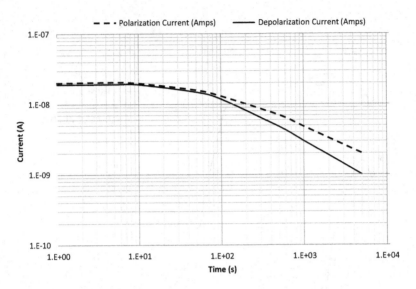

FIGURE 24.4   Polarization and depolarization current with high moisture in cellulose (≈5%).

24.10.4 EMBEDDED WIRELESS MONITORING AND FAULT DIAGNOSTIC SYSTEM

Online monitoring is possible for diagnostic methods including dissolved gases in oil, oil temperature, winding temperature, oil level, core temperature and cooler operation. Embedded system is
used, and the data are transmitted by RF transmitter to the control room. Schematic diagram of a
typical wireless monitoring system is shown in Figure 24.5.

24.10.5 FREQUENCY DOMAIN SPECTROSCOPY

When a dielectric material is subjected to an electric field, the electrical dipoles are oriented in the direction of the field. This orientation has a time delay, which depends on the material, the moisture content
and the geometry of the insulation. The frequency range used for analysis is normally 0.001–1000 Hz.

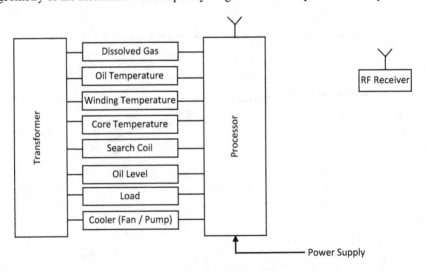

FIGURE 24.5   Schematic of embedded wireless monitoring system.

Source: Copyright permission from IEEE C 57–104–2019 IEEE Guide for Interpretation of Gases Generated
in Mineral Oil–Immersed Transformers.

24.10.6 MONITORING OF TEMPERATURE

Monitoring of the ambient, winding, oil and tank body temperatures can identify any excessive temperature above the permissible values. If the temperature exceeds the designed value applicable for the class of insulation, the ageing becomes rapid, and also this can be an indication of defects/faults, leading to failure.

The top oil temperature and winding temperature can be monitored online, and tools are available for remaining life estimation and load management.

Online thermography can monitor and detect temperature gradients on tank surface. The gradients can be compared with the tested values under defined operating conditions to take suitable action.

24.10.7 LOAD MONITORING

The loads of each phase of the primary, secondary and other windings (if applicable) are monitored for load management by including the data of ambient temperature, oil temperature and winding temperature. Where fans and pumps are used for transformer cooling, the input currents to these devices and the operation can be monitored.

24.10.8 VIBRATION MONITORING

Excessive vibration is often an indication of abnormal operation of fan, pump, loose coils, loose clamping, shorted turns, etc., which can lead to a failure.

For comparison, vibration measurement results of new transformer after installation are necessary.

24.10.9 MONITORING OF THE FUNCTIONING OF BUSHINGS AND ON-LOAD TAP CHANGER

The operation of condenser bushing is monitored by measuring the current through the test tap in operation. Changes in tan delta and the charging current are monitored to detect the abnormality/fault of the bushing.

- In oil-insulated on-load tap changer (OLTC), the contact temperature can be indirectly monitored by measuring the temperature of the oil in the OLTC compartment. This temperature is compared with the oil temperature of the main tank.

24.11 FUZZY INFORMATION APPROACH FOR INTERPRETATION OF RESULTS OF DIFFERENT DIAGNOSTIC METHODS

Good experience is essential to interpret the results of many of the diagnostic techniques with accuracy. Artificial intelligence (AI) tools can be applied to arrive at the most probable interpretation using the fuzzy information approach. Information about the relationship between different fault types and diagnostic results is made into a fault tree.

This approach consists of the following:

- Select the diagnostic method.
- Initiate the probability of all conclusions showing probabilities between 1 and 0.
- The measurement results are then applied to the fault tree.
- Perform conflict resolution.
- Calculate the basic assignment for all nodes of the fault tree.
- Repeat the procedure for all measurement methods and aggregate the results.

25 Carbon Footprint Calculation of Transformer

25.1 INTRODUCTION

Carbon footprint of a product is a "measure of exclusive total amount of carbon dioxide emissions that is directly and indirectly caused by an activity or is accounted over the life stage of a product" (Definitions by ISA Research Report 07-01-Weidman T and Minix J).

As per ISO 1406.7, the carbon footprint of a product is the sum of greenhouse gas emissions and removals in a product/system expressed as a CO_2 equivalent and based on a life cycle assessment.

The ever-increasing demand for electric power in the world is a major cause of the growth of CO_2 emissions of the planet. When more transformers are added to the power system, the contribution of CO_2 emission from transformer increases. It is estimated that nearly 4% of the present greenhouse gas emissions from the electric power system are contributed by the transformers, which amounts to 730 million tons/year.

The focus on increasing renewable power generation and the efforts to improve the efficiency of the transformer are expected to reduce the impact in future. An estimate of the carbon footprint of the transformer is a useful tool for the stakeholders such as manufacturers, users and policy makers for strategic planning and designing.

25.2 CARBON FOOTPRINT CALCULATION FLOWCHART

The total carbon emissions of a product from "cradle to gate" are considered for the calculation of the carbon footprint. The boundaries of the product life cycle considered for the analysis are schematically shown in Figure 25.1.

All the aforementioned stages have energy usage and CO_2 emissions. When recycled material is used in full or part for the production, the carbon footprint in the overall system comes down.

25.3 PERFORMANCE PARAMETERS OF THE TRANSFORMERS CONSIDERED FOR THE CALCULATION

The quantity of materials used for the production of a transformer depends on the parameters including the efficiency requirements specified by the user complying with the national or international standards. These factors affect the lifetime carbon footprint because the energy loss of the transformer in operation and the materials used in production are the major variables affecting the carbon footprint.

The efficiency level of the transformer was improving in the past as a result of the development of high-performance materials, advanced analytical techniques in design and mandatory regulations, guidelines, etc. Several power utilities tried to optimize the total owning cost of the transformer by specifying the loss evaluation formula. Governments also mandated regulations such as minimum-efficiency performance regulation/high-efficiency performance specification. The European Union carried out a study on the environmental and economic aspects of the performance of the transformer and issued EU Regulation 548/2014 in May 2014 where the maximum permissible value of the transformer losses was stipulated.

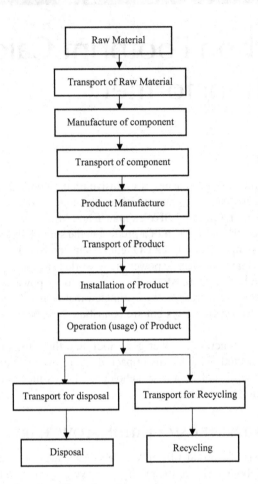

FIGURE 25.1 Boundaries of product life cycle.

The typical values of no-load losses and load losses prevalent prior to the issue of EU directive were higher and were broadly within the different levels of no-load losses and load loss levels specified by the standard EN 50464-1-2007.

The EU norms specified Tier 1 losses to be adopted from 1 July 2015 and Tier 2 losses, which are still lower, to be implemented from 1 July 2021.

The performance parameters of the transformers considered for the carbon footprint calculation are given in Table 25.1.

The carbon emissions at various stages of the life cycle of a 1000 kVA ONAN transformer are calculated in this section.

25.3.1 Raw Materials for Production

This stage involves the calculation of the carbon footprint of the raw materials and components for producing the transformer. The weights of components and minor materials are grouped together as miscellaneous materials.

The material category steel includes tank, cooling fins, bolts, nuts, clamping materials and so on. Electrical steel (Cold Rolled Grain Oriented) for transformer core is grouped separately. Table 25.2 gives the materials' quantities of three different designs complying with the losses specified in Table 25.1.

TABLE 25.1
Performance Parameters of the Transformers

Design Number	Design Norms for No-Load Loss and Load Loss	Performance Parameters (Maximum Losses)	
		No-Load Loss, W	Load Loss, W
1	EN 50464-1 $C_o C_K$ Level	1100	10500
2	EU Tier 1	770	10500
3	EU Tier 2	693	7600

Source: Copyright permission from Carbon footprint calculation of transformer and the potential for reducing CO_2 emissions. Paper number 1570575832 presented at 4th International Conference on Technology, Informatics, Management, Engineering and Environment (Time-E 2019), 13–15 November at Bali, Indonesia.

TABLE 25.2
Quantities of Materials of 1000-kVA ONAN Transformers

Material	Quantity per Transformer – kg		
	Design 1 EN-50464-1	Design 2 EU Tier 1	Design 3 EU Tier 2
Electrical steel (cold-rolled grain-oriented)	1020	1150	1320
Copper	510	720	1025
Steel	1110	1050	1150
Transformer oil	710	650	700
Densified wood	80	90	95
Insulation Materials	70	70	75
Miscellaneous Materials	60	60	60
Total kg	3560	3790	4425

Source: Copyright permission from Carbon footprint calculation of transformer and the potential for reducing CO_2 emissions. Paper number 1570575832 presented at 4th International Conference on Technology, Informatics, Management, Engineering and Environment (Time-E 2019), 13–15 November at Bali, Indonesia.

The carbon footprints of the raw materials are calculated by considering the CO_2 emissions for producing the materials as per Table 25.3.

The contribution of the carbon footprint of a transformer from the raw materials is calculated for the three different designs with parameters list of Table 25.1, and the results are given in Table 25.4.

25.3.2 TRANSPORT OF RAW MATERIALS

The raw materials for producing the transformer are transported by a combination of truck, rail, ship and so on. Considering the wide dispersion of the production centers of the materials and consumption points, the typical mode of carriage and the average CO_2 emission per ton of material transported are estimated as per Table 25.5.

TABLE 25.3
Carbon Footprint of Materials

Material	CO_2 Emission of Material (kg/kg)
Electrical steel	4.0
Amorphous steel	2.2
Copper	3.8
Steel	1.8
Transformer oil	3.0
Densified wood	1.5
Insulation Materials	1.0
Miscellaneous Materials	2.2

Source: Copyright permission from Carbon footprint calculation of transformer and the potential for reducing co₂ emissions. Paper number 1570575832 presented at 4th International Conference on Technology, Informatics, Management, Engineering and Environment (Time-E 2019), 13–15 November at Bali, Indonesia.

TABLE 25.4
Carbon Footprint from Raw Materials

Design No	Design Type	Carbon Footprint from the Raw Materials of a 1000-kVA Transformer (kg)
1	EN 50464-1	10343
2	EU Tier 1	11386
3	EU Tier 2	13485

Source: Copyright permission from Carbon footprint calculation of transformer and the potential for reducing CO_2 emissions. Paper number 1570575832 presented at 4th International Conference on Technology, Informatics, Management, Engineering and Environment (Time-E 2019), 13–15 November at Bali, Indonesia.

TABLE 25.5
Carbon Footprint of Transportation of Materials

Material	Mode of Transportation	Average CO_2 Emission/Ton (kg)
Electrical steel	Ship/truck	200
Amorphous steel	Ship/truck	560
Copper	Ship/truck	80
Steel	Ship/truck/train	60
Transformer oil	Ship/truck	80
Other raw materials	Truck	50

Source: Copyright permission from Carbon footprint calculation of transformer and the potential for reducing CO_2 emissions. Paper number 1570575832 presented at 4th International Conference on Technology, Informatics, Management, Engineering and Environment (Time-E 2019), 13–15 November at Bali, Indonesia.

The unit emissions per kilometer of transportation assumed for calculations are as follows;

Transport by truck	:	100 gm/km/ton
Transport by train	:	50 gm/km/ton
Transport by ship	:	20 m/km/ton

25.3.3 MANUFACTURE OF COMPONENTS AND SUBASSEMBLIES

This stage consists of the manufacturing of components of the transformer including bushings, tap changers, meters, indicators, valves and so on, which are purchased by the transformer manufacturers from specialized vendors. Subassemblies purchased in majority of cases include tanks, cooling radiators/fins, clamps, cable boxes and so on.

The energy intensity of the aforementioned activity is medium.

The average weight of subassemblies and components is estimated to be 1200 kg, and the power requirement for the production is calculated as 400 kWh. The CO_2 emission of this activity is estimated as 240 kg.

25.3.4 TRANSPORT OF COMPONENTS AND SUBASSEMBLIES

Most of the subassemblies are manufactured in facilities in the proximity of the transformer manufacturer. Transportation distance of 500 km on an average is considered.

The components are procured from manufacturers from specialized sources and involve transport by air, ship, and/or truck. The weight of these parts is estimated as 100 kg for a 1000 kVA transformer, and the average transportation distance is calculated as 2000 km.

25.3.5 MANUFACTURE OF THE PRODUCT

The production activity is mainly assembly process, and the energy intensity of the industry is low. The average energy consumption for the production has a wide variation depending on the place of manufacture, the automation level adopted, and energy conservation practices followed by the entity.

The average energy consumption of transformer manufacturers is analyzed, and 750 kWh of energy consumption for production of one 1000 kVA unit is estimated from this. This contributes to 450 kg of CO_2 emission/transformer.

25.3.6 TRANSPORTATION OF PRODUCT

The global demand for transformer is met by a combination of imports and exports, where some regions have exports and some regions have predominantly imports. When the product is used in the region where it is produced, the transportation distance is low, whereas when it is exported, it can be of the order of few thousands of kilometers. A global weighted average transportation distance of 1500 km is considered for the analysis, where 60:40 ratio is estimated for transport by ship and truck. The emissions for the three designs of transformer are given in Table 25.6.

25.3.7 INSTALLATION OF PRODUCT

The installation process has low energy intensity, and the emission estimated is 60 kg of CO_2 for a 1000 kVA transformer.

TABLE 25.6

Emission from Transportation of Product from Factory

Design Numbers	Weight of Product (kg)	CO_2 Emission from Transportation of Product (kg)
1	3560	278
2	3790	295
3	4425	345

Source: Copyright permission from Carbon footprint calculation of transformer and the potential for reducing CO_2 emissions. Paper number 1570575832 presented at 4th International Conference on Technology, Informatics, Management, Engineering and Environment (Time-E 2019), 13–15 November at Bali, Indonesia.

25.3.8   Operation (Usage) of the Product

This phase generates the maximum CO_2 emission because the no-load losses and load losses of the transformer are to be considered for the expected lifetime of 25 years.

The following calculations assume that the transformer operates at 50% load factor for the entire useful lifetime and the no-load losses are unchanged during this period.

The calculations shown in Table 25.7 assume that the CO_2 emissions rate/kWh remains unchanged for the lifetime. However, this will be reducing in future due to the fact that the additions of new generating capacity in future will be substantially from renewable sources.

25.3.9   Disposal and Recycling at the End of Life

This stage of the life cycle involves recycling of materials and disposal of wastage. The potential for recycling is calculated based on today's technology.

Table 25.8 shows the quantities of materials recycled and disposed. The recycling and usage at this stage have negative value of emissions, as the reclamation is an environmental profit as shown in Table 25.8. Disposal of other materials (nonrecycled) and leftover after recycling requires energy, and the emission is estimated as 200 kg/ton.

25.4   TOTAL LIFETIME CARBON FOOTPRINT OF 1000-KVA TRANSFORMERS

The net carbon footprints of the three designs of 1000 kVA transformers are summarized in Table 25.9.

TABLE 25.7

Carbon Emission from 1000 kVA Transformer from operation

Particulars	Design 1	Design 2	Design 3
No-load loss – watts	1100	770	693
Load loss – watts	10500	10500	7600
No load loss/year – kWh	9636	6745.2	6070.7
Load loss/year – kWh at 50% load factor	22995	22995	16644
Total loss/year – kWh	32631	29740.2	22714.7
Total CO_2 emission/year (tons)	19.578	17.844	13.629
Total CO_2 emission for 25 years (tons)	489.45	446.1	340.7

Source: Copyright permission from Carbon footprint calculation of transformer and the potential for reducing CO_2 emissions. Paper number 1570575832 presented at 4th International Conference on Technology, Informatics, Management, Engineering and Environment (Time-E 2019), 13–15 November at Bali, Indonesia.

TABLE 25.8
Recycling and Disposal of Material at the End of Life

Material	Design 1 Qty Recycled kg	Design 1 CO$_2$ emission kg	Design 2 Qty Recycled kg	Design 2 CO$_2$ Emission kg	Design 3 Qty Recycled kg	Design 3 CO$_2$ Emission kg	Remarks
Electrical steel	765	−3060	862	−3448	990	−3960	It is assumed that amorphous metal is not recyclable. Emission due to disposal is shown here.
Copper	380	−1444	540	−2052	770	−2926	
Steel	666	−1120	630	−1134	690	−1242	
Transformer Oil	639	−1917	585	−1755	630	−1890	
Other materials + wastage	1110	+2200	1173	+2346	1342	+2690	
Total Kg		−3981		−6043		−7328	

Source: Copyright permission from Carbon footprint calculation of transformer and the potential for reducing CO$_2$ emissions. Paper number 1570575832 presented at 4th International Conference on Technology, Informatics, Management, Engineering and Environment (Time-E 2019), 13–15 November at Bali, Indonesia.

TABLE 25.9
Lifetime Carbon Footprint of 1000 kVA Transformer

Life Cycle Boundary	Lifetime Carbon Footprint (Tons) Design 1	Design 2	Design 3
Raw material production	10.343	11.386	13.485
Transport of raw materials	0.38	0.44	0.483
Manufacturing of components and subassemblies	0.24	0.24	0.24
Transport of components and subassemblies	0.08	0.08	0.08
Manufacturing of the product	0.45	0.45	0.45
Installation of the product	0.06	0.06	0.06
Operation of the product for 25 years	489.45	446.10	340.70
Disposal of the product including recycling	−3.98	−6.043	−7.328
Total tons	497.022	452.713	348.89
%	100	91	70

Source: Copyright permission from Carbon footprint calculation of transformer and the potential for reducing CO$_2$ emissions. Paper number 1570575832 presented at 4th International Conference on Technology, Informatics, Management, Engineering and Environment (Time-E 2019), 13–15 November at Bali, Indonesia.

26 Seismic Response Calculation for Transformers

26.1 INTRODUCTION

Energy of the earthquake E (ergs) is related to the earthquake magnitude (M) by the following;

$$\log E = 9.9 + 1.9\,M + 0.024\,M^2 \tag{26.1}$$

Hypocenter = position of the earthquake in the earth's crest
   Epicenter = point at ground level, vertically above the hypocenter
   Figure 26.1 shows the relative position of epicenter and hypocenter.
   Majority of earthquakes occur on the fault lines. The sudden release of the distortion energy creates earthquake.
   The depth of hypocenter is typically few km to 100 km.

- Earthquake creates simultaneous horizontal and vertical ground accelerations.
- It is usually between 15 and 30 seconds.
- The frequency of the response is 1–35 Hz, and the most destructive effects are between 1 and 10 Hz.
- Horizontal ground acceleration is generally lower than 0.5 g.
- The amplitude of vertical acceleration is two-thirds of the amplitude of horizontal acceleration.

26.2 INTENSITY AND SEISMIC ZONES

The intensity of earthquakes is measured in different scales such as Richter scale and Mercalli scale. The approximate comparison of various scales, the horizontal ground acceleration and seismic zones and the effects felt are shown in Table 26.1.

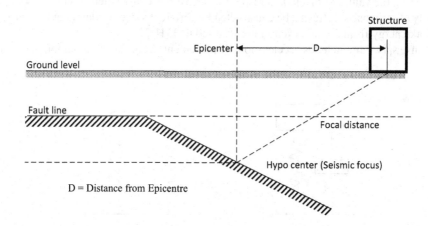

FIGURE 26.1   Epicenter and hypocenter.

TABLE 26.1

Intensity of Seismic Zones and Effects

Ritcher	Intensity (Mercalli)	Effect	Approximately Horizontal Acceleration m/s²	Seismic Zone
0–2	1	Tremor not felt		Zone 0
1–2	2	Tremor felt by those resting on top floors		
2–3	3	Hanging objects move. Slight vibrations		
3–4	4	Window panes and crockery rattle. Stationery vehicles sway	2	
4–5	5	Tremor felt outside building. Small objects fall down		Zone 1
5–6	6	Tremor felt by everyone. Broken glass. Slight cracks on plastering		
5–7	7	Tremor felt in moving vehicles. Those standing may fall off balance. Some brick houses collapse		Zone 2
6–8	8	Branches fall from trees, slight to substantial damages to brick buildings, prefabricated building	3	Zones 3 & 4
7–9	9	Fissure in ground. Destruction of nonreinforced brick buildings		
>8	10	Landslide and major ground destruction. Destruction to bridges and tunnels	5	
	11	Permanent ground deformation		
	12	Almost total destruction		

26.3 RESPONSE SPECTRUM

The response spectrum allows the earthquake to be characterized in terms of the effect it produces on equipment.

For this, an array of single degree of freedom oscillators is arranged.

$$\text{The resonant frequency, } F_r = \frac{1}{2\pi}\sqrt{\frac{k}{m}}\left(1-\lambda^2\right) \qquad (26.2)$$

K = spring constant, m = mass, λ = damping factor (Figures 26.2 and 26.3).

By changing the values of m and k, a response spectrum is constructed (Figure 26.4).

A family of response spectra can be constructed for different values of damping factors λ_1, λ_2, etc.

The peak energy from seismic excitation is limited to 35 Hz.

A response spectrum includes acceleration, displacement, frequency and damping.

FIGURE 26.2   Single degree of freedom system.

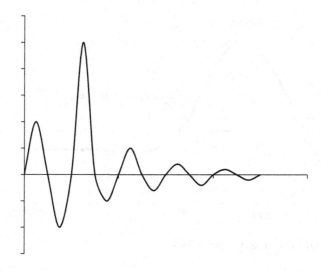

FIGURE 26.3 Response of single degree of freedom.

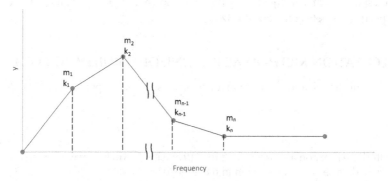

FIGURE 26.4 Response spectrum.

$$\text{Safe Shut Down Earthquake (SSE)} = \text{Maximum Historical Earthquake likelihood}$$

$$+ \text{one Degree extra on Mercalli Scale}$$

That is, if the maximum likelihood of earthquake intensity is 5, SSE = 6 (hence, this will fall in zone 1). Typical damping factors for 50% and 100% of maximum yield for various structures are given in Table 26.2.

TABLE 26.2
Damping Factors

Type of structure	Damping %	
	50% yield	100% yield
Welded steel structure	2	4
Bolted steel structure	4	7
Reinforced concrete structure	4	7
Switchgear cabinet, etc.	2	5
Racks	2	5

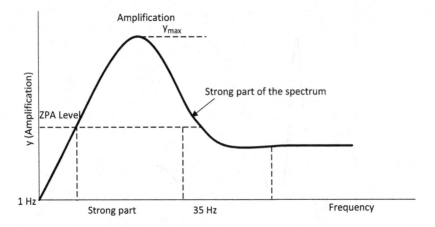

FIGURE 26.5   Response of earthquake oscillations.

A typical response of an earthquake oscillation is shown in Figure 26.5. It can be seen that strong part of the spectrum is between 5 and 35 Hz.

26.4   CALCULATION METHOD AS PER UNIFORM BUILDING CODE

The calculations of the seismic response involve various factors as per Tables 26.3–26.9.

Example

Calculate the base shear on a structure 31.2 m × 19 m with 32.8 m height with standard occupancy and constructed on sandy soil (Location in the zone 1).
   For zone 1, $Z = 0.075$
   For standard occupancy, $I = 1.0$ (Table 26.4)
   (structural system – lateral force resisting system)

Ω_o = seismic force amplification factor
R = response modification factor for lateral force resisting system
Soil profile type SD, Table 26.5
From Table 26.10, $R = 5.5$
From Table 26.6, $C_a = 0.12$
From Table 26.7, $C_v = 0.18$

$$\text{Time period, } t = 0.0488(H)^{\frac{3}{4}} \tag{26.3}$$

where H = height in m

$$t = 0.0488 \times 32.8^{0.75} = 0.67 \text{ s}$$

Total base shear V

$$V = \frac{C_v I \ W}{RT} = \frac{0.18 \ W}{5.5 \times 0.75} = 0.0436 \ W$$

TABLE 26.3

Seismic Zone Factors

Zone	1	2A	2B	3	4
Z	0.075	0.15	0.20	0.30	0.40

Source: Copyright permission from; UBC 1997 Design Guide to the 1997 Uniform Building Code.

TABLE 26.4

Occupancy Importance Factors I

Occupancy Category	I
Essential facilities	1.25
Hazardous facilities	1.25
Special occupancy structures	1.00
Standard occupancy structures	1.00
Miscellaneous structures	1.00

Source: Copyright permission from; UBC 1997 Design Guide to the 1997 Uniform Building Code.

TABLE 26.5

Soil Profile Types

Soil Type	Description	Average Soil Properties for Top 30 m		
		Shear Wave Velocity Vs, m/s	Standard Penetration N (blows/foot)	Undrained Shear Strength Su, K P_a
SA	Hard rock	> 1500	-	-
SB		> 60 to 1500	-	-
SC	Very dense soil and soft rock	360–760	> 50	> 100
SD	Stiff soil profile	180–360	15 to 50	50 to 100
SE	Soft soil profile	< 180	< 15	< 50
SF	Soil requiring site-specific evaluation			

Source: Copyright permission from; UBC 1997 Design Guide to the 1997 Uniform Building Code.

TABLE 26.6

Seismic Coefficient C_a

Soil profile	Seismic Zone Factor Z				
	Z = 0.075	Z = 0.15	Z = 0.2	Z = 0.3	Z = 0.4
SA	0.06	0.12	0.16	0.24	0.32 N_a
SB	0.08	0.15	0.20	0.30	0.40 N_a
SC	0.09	0.18	0.24	0.33	0.40 N_a
SD	0.12	0.22	0.28	0.36	0.44 N_a
SE	0.19	0.33	0.34	0.36	0.36 N_a
SF	Site-specific geotechnical investigation and dynamic response required				

Source: Copyright permission from; UBC 1997 Design Guide to the 1997 Uniform Building Code.

V shall not exceed $\dfrac{2.5\ C_a IW}{RT}$

$$V = \frac{2.5 \times 0.12 \times W}{5.5} = 0.0545W$$

The total design base shear shall not be less than

$$V = 0.11\ Ca\ I\ W = 0.11 \times 0.12\ W = 0.0132\ W$$

Design base shear

$$V = \frac{C_v IW}{RT} \tag{26.4}$$

Total design base shear shall not exceed

$$V = \frac{2.5\ C_v IW}{RT}$$

$$V = 0.11 C_a IW$$

The total design base shear shall not be less than

$$V = 0.11 C_a\ I\ W = 0.11 \times 0.12W = 0.0132W$$

For zone 4, the total base shear shall not be less than

$$V = \frac{0.8\ Z\ N_v I\ W}{R} \tag{26.5}$$

For RCC moment resisting frames,

$$T = 0.073(H)^{\frac{3}{4}} \tag{26.6}$$

26.5   DESIGN OF ANCHOR BOLT

A 100 kN equipment is installed at zone 4. It is attached by four anchor bolts on concrete slab (Figure 26.6).

Zone factor $= 0.24 = Z$
Height of point of attachment $= 16.2 = h$ m
Amplification factor $= 1$ (rigid component) $= a_p$
Response modification factor Rp $= 2.5$
Importance factor $= 1 = I_p$
$W_p =$ weight of equipment $= 100$ kN
$X =$ height of anchor bolt point

TABLE 26.7

Seismic Coefficient C_v

Soil profile	Seismic Zone Factor Z				
	$Z = 0.075$	$Z = 0.15$	$Z = 0.2$	$Z = 0.3$	$Z = 0.4$
SA	0.06	0.12	0.16	0.25	0.32 N_a
SB	0.08	0.15	0.20	0.30	0.40 N_a
SC	0.13	0.25	0.33	0.45	0.56 N_a
SD	0.18	0.32	0.40	0.54	0.64 N_a
SE	0.26	0.50	0.64	0.84	0.96 N_a
SF	Site-specific geotechnical investigation and dynamic response required				

Source: Copyright permission from; UBC 1997 Design Guide to the 1997 Uniform Building Code.

TABLE 26.8

Near Source Factor N_a

Seismic Source Type	Closest Distance to Known Seismic Source		
	≤ 2 km	5 km	≥ 10 km
A	1.5	1.2	1.0
B	1.3	1.0	1.0
C	1.0	1.0	1.0

Source: Copyright permission from; UBC 1997 Design Guide to the 1997 Uniform Building Code.

TABLE 26.9

Near Source Factor N_v

Seismic Source Type	Closest Distance to Known Seismic Source			
	≤ 2 km	5 km	10 km	≥ 15 km
A	2.0	1.6	1.2	1.0
B	1.6	1.2	1.0	1.0
C	1.0	1.0	1.0	1.0

Source: Copyright permission from UBC 1997 Design Guide to the 1997 Uniform Building Code.

TABLE 26.10

Seismic Force Amplification Factor and Response Modification Factor

Structural System	Lateral Force Resisting System	R	Ω 20
Bearing wall	Concrete shear walls	4.5	2.8
Building frame	Concrete shear walls	5.5	2.8
Steel moment resisting frame	SMRF	8.5	2.8
Intermediate moment resisting frame	IMRF	5.5	2.8
Ordinary moment resisting frame	OMRF	3.5	2.8
Cantilever column building	Cantilever column elements	2.2	2.0
Shear wall frame intersection	-	5.5	2.8

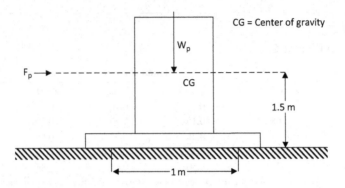

FIGURE 26.6   Details of 1000 kN equipment installed at zone 4.

Design seismic force

$$F_p = \frac{Z}{2}\left(1+\frac{X}{h}\right)\frac{a_p}{R_p}I_pW_p$$

$$F_p = \frac{0.24}{2}\left(1+\frac{16.2}{16.2}\right)\frac{1}{2.5}\times1\times100$$

$$= 9.6\ kN < 0.1\ Wp \tag{26.7}$$

∴ Design seismic force $Fp = 10$ kN
(0.1 Wp)
Anchorage of equipment shall be with two times the above force, i.e., 20 kN.

$$\text{Shear/anchor bolt} = V = \frac{2F_P}{4} = 5\,\text{kN}$$

overturning moment = 2 × FP × 1.5
  = 2 × 10 × 1.5 = 30 kN
(1.5 m is the height to CG)
Overturning moment is resisted by two anchor bolts on either side.
Tension per anchor bolt

$$F_t = \frac{30}{1\times2} = 15\ \text{kN}$$

26.6   CALCULATION OF NATURAL FREQUENCY OF VIBRATION OF TRANSFORMER

The natural frequency of vibrations is calculated by using the mass–spring oscillating system of the transformer schematically shown in Figure 26.7.
  Symbols used:

L = length of tank or width (m)
H = height of fluid in tank (m)
m_i = mass of core and coil assembly (kg)
m_c = mass of fluid (kg)

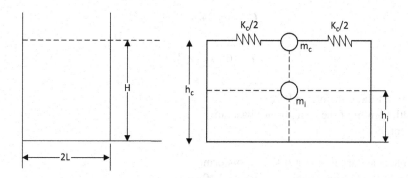

FIGURE 26.7 Equivalent mass–spring system of transformer.

h_i = height of center of gravity from bottom of tank (m)
h_c = height of center of fluid (m)
G = acceleration due to gravity
K_c = spring constant of the system

$$T_1 = 2\pi\sqrt{\frac{\pi L}{2G}\tanh\left(\frac{\pi H}{2L}\right)}$$ (26.8)

T_1 = first mode of vibration

$$F_1 = \frac{1}{T_1}$$

= frequency of first mode of vibration

The natural frequency of vibration of various components (e.g., radiator, conservator, bushings) and the stresses will have to be calculated separately.

The mounting of the radiator (radiator pipe), mounting of bushings and mounting of conservator will have to be analysed separately.

If the transformer is divided into subsystems and the natural frequencies of each item are calculated separately, the combined natural frequency of the transformer is calculated as

$$\frac{1}{n^2} - \frac{1}{n_1^2} + \frac{1}{n_2^2} - - -$$

n = natural frequency of the transformer

where n_1, n_2,... are the natural frequencies of subsystems (components).

If the natural frequency of the transformer is above 35 Hz, it is safe. However, it is not advisable to have the natural frequency of the transformer between 1 and 10 Hz where the seismic vibrations are maximum.

• Time period T as per Uniform Building Code (UBC) formula:

$$T = C_0(BD)^{0.3289\alpha}$$ (26.9)

$$C_0 = 0.0247 \ e^{0.1305H} \tag{26.10}$$

$$\alpha = 0.4773 \ e^{-0.441H} \tag{26.11}$$

B = length of the transformer base, m
D = width (depth) of the transformer base, m
H = height, m

- Sample calculation for 1000-kVA transformer
  B = length of the transformer base = 1.904 m
  D = width (depth) of the transformer base = 1.778 m
  H = height = 2.171 m
  $Co = 0.0247 \ e^{0.1305H}$
  = 0.03279
  $\alpha = 0.4773 \ e^{-0.441H}$
  = 0.1832
  $T = Co \ (BD)^{0.3289\alpha}$
  $= 0.03279 \times (1.904 \ \times 1.778)^{0.3289 \times 0.1832}$
  = 0.03528 S

$$f = \frac{1}{T} = 28.35 \ Hz$$

26.7   SEISMIC QUALIFICATIONS METHOD AS PER CLAUSE D-3 IEEE 693

Seismic withstand capability shall be demonstrated as follows:

- Transformer and liquid-filled reactors (kV referenced are the high side of the transformer and oil-filled reactors), except bushings and surge arrestors (Table 26.11).

TABLE 26.11
Method of Validation of Seismic Withstand Capability

	Voltage Class	Method of Validation	Clause No. of IEEE 93
(a)	115 kV and above	By static analysis	D 4.1
(b)	35 kV to less than 115 kV	By load path	D 4.2
(c)	Less than 35 kV	By inherently acceptable	D 4.3
		Bushings	
(a)	161 kV and above	By time history shake table test	D 4.4
(b)	35 kV less than 161 kV	By static pull test	D 4.5
(c)	Less than 35 kV	By inherently acceptable	D 4.3
		Surge Arresters	
(a)	90 kV duty cycle voltage and above	By time history	K 4.1
(b)	54 kV DCV to < 90 kV DCV	By dynamic analysis	K 4.2
(c)	35 kV DCV to < 54 kV DCV	By static coefficient analysis	K 4.3
(d)	Less than 35 kV DCV	By inherently acceptable	K 4.4

26.8 GENERAL SEISMIC CAPABILITY CALCULATION

The calculation of seismic response or the test is based on the applied excitation spectrum as explained here.

26.8.1 STANDARD AMPLITUDE METHOD

This procedure is applied, when information regarding the installation location and the details of the supporting method to the ground/foundation are not available.

- The earthquake zone is first identified.
- A floor acceleration level is obtained based on the relevant seismic zone.
- The relevant performance levels are identified, and the corresponding horizontal and vertical floor acceleration levels are determined from Table 26.12 for analysis or test.

The aforementioned acceleration levels do not consider the effect of super elevation, direction, damping, etc., as these are not available.

26.8.2 CALCULATED AMPLITUDE METHOD

This procedure is applicable when the information of the location of installation and the parameters of the supporting structure/foundation are available. The floor accelerations of the three axes (x, y and z) are calculated using the following formula.

$$A_{fx} = A_g. k. D_x$$

$$A_{fy} = A_g. k. D_y$$

$$A_{fz} = A_g.k.D_z$$

where
A_{fx}, A_{fy}, A_{fz} = floor acceleration in x, y, z axes, m/s^2
A_g = ground acceleration level, m/s^2
(The values of A_g are as per Table 26.13)
D_x, D_y, D_z = direction factors
(Factors for the three axes of vibration are as per Table 26.15)

TABLE 26.12
Acceleration Levels and Performance Levels

Acceleration Level m/s²	Seismic Performance Level	Floor Acceleration – AF	
		Horizontal m/s²	Vertical m/s²
2	S I	6	3
3	S II	9	4.5
5	S III	15	7.5

Source: Copyright permission from IEEE 693–2019 IEEE Recommended Practice of Seismic Design of Substations.

TABLE 26.13
Ground Acceleration Levels

A_g m/s²	Earthquake Level	Approximate Earthquake Magnitude		
		Ritcher Scale	UBC	MSK Intensity
2	Light to medium	<5.5	Zones 1, 2	<VIII
3	Medium to strong	5.5–7	Zone 3	VIII–IX
5	Strong to very strong	>7.0	Zone 4	>IX

K = Super elevation factor (This depends on the type of mounting/installation. The values of k are given in Table 26.14).

Example of Calculation of A_{fx}, A_{fy}, A_{fz}

ASSUMPTIONS

Seismic zone 2 (light to medium earthquake)
  Transformer is mounted on rigid foundation
  The transformer is mounted on ground level
  From Table 26.13, $A_g = 2$
  From Table 26.14, $k = 1$
  From Table 26.15, $D_x = 1$, $D_y = 1$, $D_z = 0.5$

$$\therefore A_{fx} = A_g. k. D_x = 2 \times 1 \times .5 = 1 \text{ m/s}^2$$

$$A_{fy} = A_g. k. D_y = 2 \times 1 \times .5 = 1 \text{ m/s}^2$$

$$A_{fz} = A_g.k.D_z = 2 \times 1 \times 1 = 2 \text{ m/s}^2$$

The analysis/test will have to consider the aforementioned ground acceleration levels in the x, y and z axes.

26.9   ANCHOR BOLT DESIGN CONSIDERING STATIC, WIND AND SEISMIC LOADS

The anchor bolts of the transformers shall be designed by considering the static, wind and seismic loads.

26.9.1   STATIC LOAD – D

The anchor bolts of the transformers shall be designed by considering the static, wind and seismic loads.

TABLE 26.14
Super Elevation Factors – K

K	Mounting/Installation Details
1.0	Mounting of equipment on rigid foundation or on structures of high rigidity
1.5	Installations rigidly connected to buildings
2.0	Installations on stiff structures connected rigidly to buildings
3.0	Installations on low rigidity
	Structures connected to building

TABLE 26.15
Direction Factors *D*

Axis of Vibration	*D* Factor	Remarks
x	$D_x = 1$	-
y	$D_y = 1$	-
z	$D_z = 0.5$	When the altitude of mounting is specified
z	$D_z = 1$	This is applicable when the mounting altitude is not specified

26.9.2 WIND LOAD – W

Wind load acting on the transformer depends on wind velocity, surface roughness coefficient, gust coefficient and pressure coefficient. The force resulting from the wind shall be calculated as per Section 30.14.

26.9.3 SEISMIC LOAD – E

The horizontal and vertical components of the seismic load are considered to be acting at the centre of gravity of the transformer. Table 26.16 gives the vertical and horizontal seismic loads applicable for the calculation of the strength of anchor bolts of transformer.

The seismic loads depend on the natural frequency of vibration of the transformer, which affects the load magnification.

D is the static load (total weight) in kg. Figure 26.8 shows the point of application of the vertical and horizontal load.

TABLE 26.16
Vertical and Horizontal Seismic Loads

Natural Frequency of Vibration of Transformer	Horizontal Load (kg)	Vertical Load (kg)
15 Hz	0.46 *D*	0.33 *D*
15–35 Hz	0.75 *D*	0.33 *D*

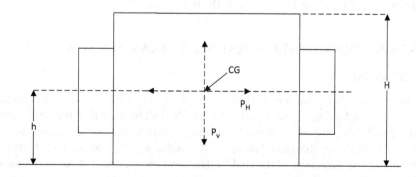

FIGURE 26.8   Point of application of forces.

P_H = horizontal seismic coefficient × D
Pv = vertical seismic coefficient × D
D = total weight of equipment, kg
h = height from base to centre of gravity
The overturning moment is created by the sum of $P_H + W$.
W is calculated by
$W = 1.95 \times L \times H$ (kg)
L = length of the transformer exposed to wind (m)
(If width (breadth) is higher than length, the value of width is used.)
H = height exposed to wind (m)

Note: The exposed surface area to wind is the value to be used for calculation.
Anchor bolt strength is calculated as follows:

$$\text{Total horizontal load on the transformer} = (P_H + W) \tag{26.12}$$

$$\text{Horizontal load/Anchor bolt} = \left(\frac{P_H + W}{N}\right) \tag{26.13}$$

N = Number of anchor bolts
Stress on anchor bolt

$$\sigma = \left(\frac{P_H + W}{Na}\right) \frac{\text{kg}}{\text{m}^2} \tag{26.14}$$

a = area of cross section of one anchor bolt (m²)
n = factor of safety

$$n = \frac{S}{\sigma} \tag{26.15}$$

S = maximum permissible stress on the anchor bolt (select the value based on the material of the bolt used)

Bending strength of base frame structure of transformer shall be checked. The maximum stress will be at the point where the anchor bolt is fixed to the base frame.

26.10 SEISMIC QUALIFICATION TESTING OF TRANSFORMER

26.10.1 SHAKE TABLE TESTING

When a transformer is to be qualified by testing, the seismic qualification testing is carried out by using shake table. The loading capacity of the shake table shall be more than the weight of the transformer to be tested. All components of the transformers shall be fitted on the transformer.

Biaxial or triaxial shake table can be used, and the frequency range of the shake table is 1–33 Hz. The lower limit of the frequency of the shake table, however, can be 0.75 times the resonant frequency of the transformer subject to a minimum of 1 Hz.

26.10.2 Response Spectra

a. This is the response spectra issued by the customer as specification for qualification, which depends on the level of qualification required. Based on the ground acceleration levels expected at the installation site, three levels of qualification are specified in the IEEE 693 "IEEE Recommended Practice for Seismic Design of Substations". Table 26.17 gives the ground acceleration levels and the corresponding seismic qualification levels.

For low qualification level, analysis or testing is not required.

The required response spectra (RRS) for moderate and high qualification levels depend on the damping factor. Typical damping factors are 2%, 5% and 10%. However, the RRS based on 2% damping is used for seismic qualification testing of transformer.

Table 26.18 gives the RRS values for 0.5 and 0.25 g.

b. Test response spectrum (TRS)

The TRS of the motion of the test table at the test facility is calculated, and the calculated TRS envelops the RRS with amplitudes, frequency and energy levels.

26.10.3 Test Procedure

The transformer to be tested shall be fully assembled with all the components fitted as in service. Control panels or components that are not mounted on the transformer to form an integral structure are excluded from the testing. However, these components or panels need to be tested for seismic qualification if it is specified.

TABLE 26.17

Ground Acceleration Levels and Seismic Qualification Levels

Ground Acceleration	Level of Seismic Qualification
Less than 0.1 g	Low qualification level
>0.1 g and less than 0.5 g	Moderate qualification level
>0.5 g	High qualification level

TABLE 26.18

RRS Values for 0.5 and 0.25 G

	Required Response Spectra (RRS)					
	0.5 G			0.25 G		
	Damping %			Damping %		
Frequency f	2	5	10	2	5	10
0–1.1	1.48	1.144	0.889	0.740	0.572	0.444
>1.1–8	1.629	1.250	0.971	0.809	0.625	0.486
10	1.323	1.052	0.846	0.661	0.526	0.423
15	0.929	0.788	0.681	0.465	0.394	0.340
20	0.733	0.656	0.598	0.366	0.328	0.299
25	0.614	0.577	0.548	0.307	0.288	0.274
30	0.536	0.524	0.515	0.268	0.262	0.256
>33	0.5	0.5	0.5	0.25	0.25	0.25

Horizontal and vertical seismic brazing of radiators is to be welded directly to the transformer tank. Bushings and surge arrestors can be qualified separately.

Bushings of 161 kV and above shall be qualified by shake table testing.

Qualification by static pull test is acceptable for bushings of rating from 35 to 161 kV.

When a transformer is tested on a biaxial shake table, it is tested by mounting in four positions.

Position 1 – The depth of the equipment is mounted parallel to the horizontal actuator of the shake table.

Position 2 – The width of the equipment is mounted parallel to the horizontal actuator of the shake table. (This position is 90° to position 1.)

Position 3 – The width of the equipment is mounted at an axis parallel to 45° with the horizontal actuator of the shake table.

Position 4 – This position is 90° from position 3.

The response of the transformer is measured during the test by placing accelerometers at different locations.

A typical test sequence applicable for a distribution transformer is given in Table 26.19.

OBE is operating base earthquake which is an earthquake that could reasonably be expected to occur at the site during the operating life of the equipment. This earthquake produces the ground motion for which the features of the equipment at the site are designed for continued safe operation.

SSE is safe shutdown earthquake, which is based on an evaluation of the maximum earthquake potential considering the geology, seismology and surface conditions of the site.

26.10.4 Acceptance Criteria and Test Report

On completion of the sequence of tests, there shall be no evidence of damage such as broken or loose components, hardware, insulators or leakage of oil.

The transformer shall pass the routine tests, and the meters and indicators shall function normally.

The test report shall contain the following information:

a. Description of the product
b. Details of anchoring of the equipment to the shake table

TABLE 26.19
Test Sequence for Seismic Qualification

Mounting Position Number	Test Sequence	Remarks
1	One test of OBE at less than 100% RRS	
	One test of OBE at 100% RRS	
	One test of SSE at 100% RRS	
2	Repeat above tests	
-	Inspect transformer	If there is no visible damage continue with test
1	One test OBE at less than 100% RRS	Record the acceleration of all the accelerometers
	Five test of OBE at 100% RRS	
	One test of SSE at 100% RRS	
2	Repeat above tests	
3	Repeat above tests	
4	Repeat above tests	

OBE, operating base earthquake; SSE, safe shutdown earthquake.

   c. Configuration of the equipment such as switch positions (weather closed or open), with or without control cabinet, cable box covers, wiring, etc.

   d. Level up to which the equipment is qualified

   e. Observations of abnormalities, if any, observed during test

   f. Description of the shake table used including size, loading capacity and frequency range

   g. Comparison of TRS to RRS

   h. Test sequence

   i. Summary showing the maximum stress, loads, displacements and acceleration

   j. A video of the test

   k. Drawings and data sheet of the transformer

27 Solid-State Transformer

27.1 INTRODUCTION

The power grid is becoming increasingly complex due to the usage of renewable energy distributed generating system, electric vehicle (EV) charging, rapid growth of nonlinear loads, demand for direct current (DC) power, etc. In order to meet the complexity, the grid will have to become "smart and Intelligent", and the transformers also will have to be suitable for the emerging "smart grid". The major demands of the smart grid and the customer from the transformers are as follows:

- Voltage sag compensation
     The conventional distribution transformer cannot correct the voltage levels and ensure constant voltage at the customer terminal.
- Harmonic isolation and filtering
     Nonlinear loads produce harmonics, and the transformers shall be capable of maintaining clean output, without harmonics.
- DC output
     In addition to the power frequency output, the transformer will have to give DC output of EV charging and other DC loads.
- Fault isolation
     The transformer shall be capable of isolating the grid from a fault on the load side and also isolate itself when a fault on the incoming side occurs.
- Capable of outage compensation by drawing power from the energy system
- Other features required including reactive power compensation, voltage balancing, protection from single phasing, reduced weight and footprint and elimination of oil/fluids

27.2 DESIGN FEATURES OF THE TRANSFORMER FOR SMART GRID

The transformer to meet the requirements of the smart grid is essentially solid state using power electronics. Pioneering work on solid-state transformer (SST) for utility application was initiated by Electric Power Research Institute (EPRI), United States, named as "Universal Intelligent Transformer".

Basic structure of the SST is given in Figure 27.1.

The incoming medium voltage from the grid at power frequency is first converted into medium-voltage DC by power electronic converters.

The power factor corrections, voltage sag compensation and harmonic isolation from the incoming side are done at this stage.

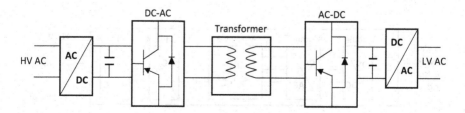

FIGURE 27.1 Basic structure of solid-state transformer.

The medium-voltage DC is inverted into alternating current (AC) at a comparatively high frequency (typically 5–20 kHz) and is fed to the SST where it is stepped down to low-voltage (LV) AC at the same frequency. The frequency of the transformer is selected by optimizing the total owning cost (TOC) of the transformer, based on the type of core material used, the cost of raw materials of the transformer and the loss evaluation (capitalization) rates for losses. The core materials used include amorphous metal, nanocrystalline metal and ferrite to give low loss at high frequency.

The output of the transformer is LV AC at medium/high frequency which is then converted to LV DC. A DC port is taken for DC loads from the stage. The DC voltage is further inverted to LV AC for the end use. The entire system in bidirectional.

27.3 DESIGN OF A 11KV/415V THREE-PHASE SOLID-STATE TRANSFORMER

27.3.1 BASIC STRUCTURE

The design consists of five modules as shown in Figure 27.2.

Module 1 – multilevel AC to DC converter
Module 2 – multilevel DC to AC inverter
Module 3 – medium-frequency transformer
Module 4 – AC to DC converter (LV)
Module 5 – DC to AC inverter (LV)

27.3.2 DESIGN OF MODULE 1: (MV AC TO DC)

Module 1 rectifies the 11 kV AC to DC by power electronic converters.

Based on the voltage rating of the power conversion device available, multilevel cascade-connected converter as shown in Figure 27.3 is used in module 1.

The rectifier output voltage is calculated here.

Input AC voltage is 11 kV line to line.

$$V_{dc} = \text{output of full-wave rectifier} = \frac{3\sqrt{2}}{\pi} \ V$$

where V = line voltage

$$V_{dc} = \frac{3}{\pi}\sqrt{2} \ \times 11000 = 14855 \text{ volts}$$

The rectifiers are cascade connected, depending on the voltage class of the commercially available insulated gate bipolar transistors (IGBTs) or silicon carbide (SiC) rectifiers. The number of rectifiers to be cascaded depends on the voltage class of the MV power system, the impulse test voltage and AC test voltages of the transformer.

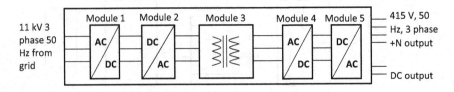

FIGURE 27.2 Five modules in solid-state transformer.

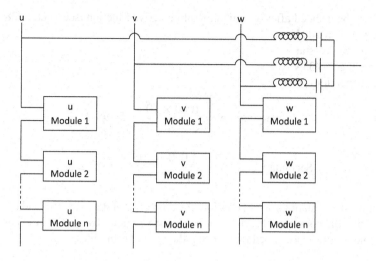

FIGURE 27.3 Three-phase full-bridge cascade converter.

27.3.3 Design of Module 2

Module 2 consists of a pulse width–modulated inverter where the input DC voltage is received from module 1.

∴ Input to inverter = 14,855 volts DC.

The rms value of the AC voltage from the inverter $= \dfrac{V_{dc}}{\sqrt{2}} = \dfrac{14855}{\sqrt{2}} = 10504.$

Say 10,500 V.

27.3.4 Design of Module 3

Medium-frequency transformer (MFT):

The input voltage to the transformer is 10500 volts AC.

The output voltage of the transformer is decided on the basis of the AC power frequency voltage required on the LV side of the SST (i.e., on the output side of module 5). The power frequency output line voltage is 415.

∴ 415 is the inverter output voltage.

Input to the inverter = $415 \times \sqrt{2}$ = 586.9 volts DC, which is the output of module 4.

$$\text{Input to module 4 (output from MFT)} = \frac{586.9 \times \pi}{3\sqrt{2}} = 434.6 \text{ volts AC}$$

Say 435 volts.

The connections of the three-phase MFT is considered as star/star.

The control of voltage sag and incoming voltage variations will be compensated in modules 1 and 2, and therefore, the transformer is without taps.

The frequency of the transformer is decided based on the optimization considering the initial cost and the lifetime owning cost.

The selection of the core material and winding material also depends on the aforementioned cost criterion.

The transformer parameters are summarized here.

kVA – 1000

Frequency – to be decided after optimization of objective function (see Section 28.4)
Primary voltage – 10500 V
Primary connection – star
Secondary voltage – 435 V
Secondary connection – star

$$\text{Primary current} = \frac{1000 \times 10^3}{\sqrt{3} \times 10500} = 54.98 \text{ A}$$

$$\text{Secondary current} = \frac{10,000 \times 10^3}{\sqrt{3} \times 435} = 1327.2 \text{ A}$$

The aforementioned calculations do not consider the efficiency of the converter and inverter systems. Due to the fact that the converter/inverter efficiencies are comparatively high, the basic parameters of the transformers can be calculated by ignoring the losses from this system.

27.3.5 Design of Module 4 (Low-Voltage Rectifier)

Input to LV rectifier = 435 V AC
  Frequency = Frequency of the medium-voltage transformer

$$\text{Output of Low voltage rectifier} = \frac{435 \times 3\sqrt{2}}{\pi}$$

$$= 587 \text{ V DC}$$

27.3.6 Design of Module 5 – (Low-Voltage Inverter)

Input to inverter = 587 V
Output from inverter = $\frac{587}{\sqrt{2}}$ = 415 volts AC
The output is three-phase AC at power frequency, 50 Hz in this case.

27.4 DESIGN OF MEDIUM-FREQUENCY TRANSFORMER

The specifications of the transformer are calculated in Section 27.3.4.
    The transformer is dry type with class F or class H insulation if the winding is cast in epoxy. Higher-temperature classes of dry-type design using 220° class materials like Nomex can be used. When the higher thermal class is used (e.g., 220°C), the thermal withstand capacity of the power electronics components of modules 1, 2, 4 and 5 will have to be checked.
    If the effect of the temperature rise of the transformer affects the components of any of the other modules, suitable isolation or reduction of the maximum allowable temperature rise of the transformer module will have to be made.

- Objective function for design optimization
      The design objective is to minimize the TOC of the transformer.

The basic formula is given here.

$$Q = A \times \text{No Load Loss} + B \times \text{Load Loss} + C \times \text{Core weight} + D \times \text{Conductor weight}$$

$$+ E \times \text{Insulation weight} + F \qquad (27.1)$$

where

Q = TOC of the transformer

A = capitalization rate of no-load loss, $/W

B = capitalization rate of load loss, $/W

C = cost of core material, $/kg

D = cost of conductor material, $/kg

E = cost of insulation material, $/kg

F = cost of other materials, $

Minimizing Q, subject to the specified and design constraints, is required.

The constraints are as follows:

- Maximum and minimum permissible levels of reactance
- Minimum efficiency level, if specified
- Temperature rise limits specified
- Limitation of total weight or footprint, if any

The cost optimization is done by considering the various design options available, as given in the following:

- The cost changes if the operating frequency is changed. When frequency is increased, the core weight comes down, and the overall weight and material cost come down. But the increase in frequency increases the core loss and also the eddy current and stray losses of the winding material.

    The material cost comes down, and the losses go up by increasing frequency for a given core material.
- The core loss characteristics change with the types of magnetic materials. Consequently, designs using different types of core materials need to be tried.
- Core materials and loss characteristics.

The commercially available magnetic materials suitable for the SST are as follows:

- Cold Rolled Grain Oriented (CRGO) silicon steel
- Amorphous alloy steel
- Nanocrystalline material
- Ferrite

The materials have different loss characteristics and saturation flux densities. The main characteristics are given in Table 27.1.

Ferrite materials find suitability at higher frequencies and lower power ratings.

TABLE 27.1
Core Material Characteristics

No	Material Category	Typical Type (Trade Name)	Saturation Flux Density Tesla	Density g/cm³	Thickness mm	Stacking Factor
1	Ultrathin silicon steel	10J NEX 900 (JFE STEEL)	1.8	7.5	0.1	0.9
2	Amorphous alloy steel	2605 SAI (METGLAS)	1.56	7.2	0.025	0.86
3	Nanocrystalline alloy	Vitroperm 500F (Vacuumschmelze)	1.2	7.3	0.023	0.7

The optimum working flux density and frequency for the minimization of the objective functions are not the same for the different materials.

Design iterations are required for deciding the best operating flux density and operating frequency for a given magnetic material.

The losses of the magnetic material consist of hysteresis loss, eddy current loss and anomalous loss.

The losses at medium frequencies can be calculated using the formula given in Table 27.2.

P_{si} = losses in W/cm³ for ultragrain silicon steel
P_{am} = losses in W/cm³ for amorphous steel
P_{NC} = losses in W/cm³ for nanocrystalline alloy

The losses of magnetic materials can be calculated by other methods where the loss is considered as the sum of hysteresis, eddy current and anomalous losses. The general equations and the parameters used are given here.

$$P_c = a_1 B^d f + a_2 B^2 f^2 + a_3 B^{1.5} f^{1.5} \tag{27.2}$$

where
P_c = specific core loss, w/kg
B = Flux density, Tesla
f = Frequency
a_1, a_2, a_3, d are parameters as per Table 27.3

The selection of working flux density depends on the type of material used and the frequency of operation. Another consideration is the specific heat generation of the core material, which is directly related to the temperature rise of the core. When the w/kg of the core goes up, the temperature rise also increases, which shall be less than the operating temperature limit of the insulation material used.

TABLE 27.2
Losses of Magnetic Material

Material	Formula for losses of magnetic material
Ultrathin silicon steel	$P_{si} = 0.159\, B^{1.82}\, f^{1.496}$
Amorphous alloy	$P_{am} = 0.0536\, B^{1.89}\, f^{1.117}$
Nanocrystalline alloy	$P_{NC} = 0.0111\, B^{2.16}\, f^{1.4286}$

TABLE 27.3
Loss Parameters of Magnetic Materials

Material	$a_1 \times 10^{-3}$	d	$a_2 \times 10^{-6}$	$a_3 \times 10^{-3}$
Ultrathin silicon steel, 0.1 mm	10.47	1.88	2.678	0.133
Nonoriented steel, 0.1 mm	20.91	1.78	4.135	0.179
HiB core, 0.23 mm	1.40	1.51	4.875	0.703
Nanocrystalline alloy	0.575	2.09	0.072	0.013
Vitroperm 500 F				

If the w/kg of core is limited to approximately 12, the temperature rise permissible for class F insulation is not exceeded. However, this will have to be calculated accurately as the temperature rise depends on several parameters.

27.5 WINDING DESIGN

27.5.1 WINDING ARRANGEMENT

Foil winding is preferred for reduction of eddy current losses, especially because the SST is operating at medium frequency. Litz wire is used where foil winding is not feasible.

In order to get the desired reactance value, interleaving of the LV winding as shown in Figure 27.4 may be necessary. 50% of the LV winding is wound after the core, and the remaining 50% is wound after the high-voltage (HV) winding.

Interleaving of the HV is generally not adopted as this requires more insulation gap.

27.5.2 CALCULATION OF LOAD LOSSES

- For foil winding:

    The ratio of AC resistance to DC resistance is calculated by the following formula:

$$\frac{R_{AC}}{R_{DC}} = Y \left[M(Y) + \frac{2}{3}\left(m^2 - 1\right) D(Y) \right] \qquad (27.3)$$

where
R_{DC} = DC resistance of the foil
R_{AC} = AC resistance of the foil

$$Y = \frac{d}{\delta}$$

d = thickness of the foil, m
δ = skin depth at operating frequency

$$\delta = \sqrt{\frac{2}{2\pi\, f\, \mu o\; \sigma}}$$

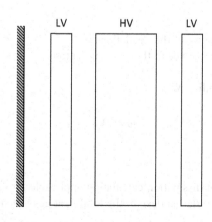

FIGURE 27.4 Interleaved winding.

f = frequency
σ = specific conductance of foil material S/m

$$\mu o = 4\pi \times 10^{-7}$$

m = number of layers of the winding

$$M(Y) = \frac{\text{Sinh}(2Y) + \text{Sin}(2Y)}{\text{Cosh}(2Y) - \text{Cos}(2Y)} \tag{27.4}$$

$$D(Y) = \frac{\text{Sinh}(Y) - \text{Sin}(Y)}{\text{Cosh}(Y) - \text{Cos } Y} \tag{27.5}$$

Y is in radians

For a three-phase transformer, the winding loss is $3 \times I^2$ RAC, where I is the phase current.

For round wire:

$$\frac{R_{AC}}{R_{DC}} = \frac{P}{2}\left\{ \left(\frac{\text{Sinh }P + \text{Sin }P}{\text{Cosh }P - \text{Cos }P} \right) + (2m-1)^2 \left(\frac{\text{Sinh }P - \text{Sin }P}{\text{Cosh }P + \text{Cos }P} \right) \right\} \tag{27.6}$$

where

$$P = \frac{\sqrt{\pi}}{2} \frac{d}{\delta}$$

d = bare diameter of the wire, m

27.5.3 INSULATION DESIGN

Class F or class H epoxy insulation suitable for vacuum casting of the coil is generally used. The clearances applicable for dry-type cast resin transformer can be used for SST winding insulation.

The dielectric losses in the insulation may become significant at higher frequencies and is calculated by the following formula:

$$P_i = K\omega CV^2 \tag{27.7}$$

where
K = dielectric dissipation factor of the insulation
C = winding to ground capacitance of the winding, Farads
V = phase voltage
P_i = Dielectric loss per winding W

$$\omega = 2\pi f$$

27.5.4 COOLING DESIGN

The temperature rise and heat dissipation calculation applicable for cast resin transformer is used for the cooling design. For ventilation design, the total heat generated by all the five modules will have to be considered.

27.6 MATERIALS REQUIRED FOR SOLID-STATE TRANSFORMERS

The major materials required to produce SST include active power conversion materials, conducting material to carry the current, magnetic materials to develop the flux and insulation materials to isolate the voltages. Characteristics of the major materials are given in the following sections.

27.6.1 ACTIVE POWER CONVERSION MATERIALS (SEMICONDUCTORS)

The conversion and inversion of power is achieved by using solid-state power electronics components.

A. IGBTs

6500-V class IGBTs are commercially available now and have 150°C junction temperature. A wide safe operating area (SOA) and comparatively low losses are achieved now.
B. SiC power devices

SiC power devices have significantly higher blocking voltages, higher switching frequencies and operating temperature up to 200°C. Recently, M/s. Wolfspeed, United States, have developed 24-kV SiC IGBT module for US Army. Trials have shown that at 24-kV/22 A, switching speed of 46 kV/µs is achieved and overall the 24-kV SiC IGBT has exhibited extremely fast switching speeds.
C. Wide-bandgap (WBG) materials

Continuous research work is being done to develop HV, high power conversion/inversion devices using WBG materials including SiC, gallium nitride (GaN) and diamond. The highest bandgap and thermal conductivity are offered by diamond. The comparison of the properties of these materials is shown in Table 27.4.

27.6.2 MAGNETIC MATERIALS

A vast majority of transformers are produced using CRGO steel which works at commercial frequency (50/60 Hz).

The SST for smart grid works at higher frequencies in order to reduce the weight, material quantity and footprint. The core losses of CRGO steel at high frequencies are considerably high, and the transformer efficiency comes down, which is undesirable.

Amorphous alloy steel offers low core losses at commercial frequencies and finds application for today's distribution transformers. The core losses of the material are comparatively less than CRGO steel.

TABLE 27.4
Comparison of Properties of wide-bandgap Materials

Properties	Silicon (Si)	Silicon Carbide (SiC)	Gallium Nitride (GaN)	Diamond
Bandgap E_g (eV)	1.1	3.23	3.45	5.45
Dielectric constant E_r	11.8	9.8	9	5.5
Thermal conductivity – W/cm k	1.5	5	1.5	22
Drift velocity e – Vs (10^{-7} cm/s)	1	2	2.2	2.7

Source: Copyright permission from Innovative materials for the transformers of the future. Paper presented at the International Electrical Engineering Congress (IEECON 2020), Chiang Mai, Thailand, March 4–6, 2020-Paper P1019.

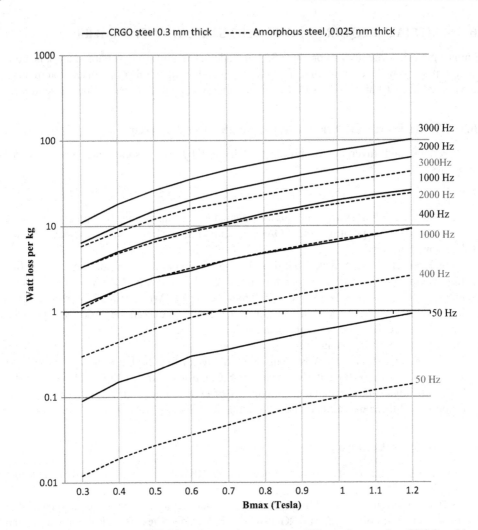

FIGURE 27.5 Specific losses of CRGO and amorphous steel at different frequencies. CRGO, cold-rolled grain-oriented.

Source: Copyright permission from; Innovative materials for the transformers of the future. Paper presented at the International Electrical Engineering Congress (IEECON 2020), Chiang Mai, Thailand, March 4–6, 2020-Paper P1019.

The comparative losses of CRGO steel and amorphous steel at different operating frequencies are given in Figure 27.5.

Amorphous nanocrystalline materials and soft ferrites are suitable for high-frequency applications because of the low losses. Soft magnetic composites are being developed, which are suitable for additive manufacturing methods (3D printing). Table 27.5 shows the comparison of magnetic properties of conventional CRGO steel, amorphous steel, nanocrystalline material and ferrite.

27.6.3 CONDUCTING MATERIALS

Copper and aluminium are used as conducting materials for commercial frequency transformers now. The increase in losses at high frequency due to skin effect and proximity effect is controlled by using Litz wire or continuously transposed conductors (CTCs) for high-power applications. The emerging technology developments of conducting materials are as follows:

TABLE 27.5

Comparison of Magnetic Properties

Material	Typical Composition	Permeability	Flux Density, Tesla		Typical Operating Frequency- Hz
			Typical	Max	
CRGO steel	Fe96.5Si3.5	1500–2000	1.5	1.8	50–2000
Permalloy	Nickel iron	25,000	0.65	0.8	1000–25,000
Amorphous steel	Fe78B13Si9	1500	1.4	1.5	250k
Nanocrystalline	Fe Cu NbSiB	30,000	1	1.2	250k
Ferrite	Mn Zn	0.75–15k	0.3	0.5	10–2000k
Ferrite	Ni Zn	0.2–1.5k	0.3	0.4	0.2–100 M

CRGO, cold-rolled grain-oriented.

Source: Copyright permission from Innovative materials for the transformers of the future. Paper presented at the International Electrical Engineering Congress (IEECON 2020), Chiang Mai, Thailand, March 4–6, 2020-Paper P1019.

k = kilo, M = mega.

- Use of high-temperature superconducting materials
- Carbon nanotubes (CNTs)
- Covetic materials
- Sandwiched layers of graphene

A. High-temperature Superconductors

Superconductivity at low temperature (4 K) was first observed by Heike Onnes in 1911. In the 1980s, superconducting metal alloys were developed, which became superconductors at 77 K. These were bismuth-based materials and were called the first-generation superconductors.

The second-generation superconductors were developed by 2005, which had critical temperature of more than 90 K. Yttrium–barium–copper oxide $YBa_2Cu_3O_{7-y}$ $(0.4 < Y < 0.9)$ conducting wires are manufactured for commercial use, and several manufacturers have produced transformers using YBCO wires cooled by liquid nitrogen.

Development work to achieve higher levels of critical temperature close to 300 K is ongoing. Some of the superconductors with promising potential under development are shown in Table 27.6.

These materials have several advantages as shown here when compared with copper/aluminium winding wires.

- Very low load losses due to the high current carrying capacity of 10–50 times of copper
- High overload capacity
- Low CO_2 emission during the product life cycle, i.e., low global warming potential
- Capability to withstand overloads without loss of life

B. Carbon nanotubes

Carbon nanotubes (CNTs) are isotopes of carbon, which have a cylindrical nanostructure. CNTs show very high conductivity, high mechanical strength and very low thermal coefficient of resistance. Rice University, United States, has developed a process for producing high-conductivity CNT winding wires.

A Japanese–Dutch Company, Teijin-Aramid in collaboration with Rice University, has developed a spinning technology to produce multifiber yarn suitable for production of transformers, cables and motors. A comparison of CNT yarn with copper and aluminium is given in Table 27.7.

TABLE 27.6

Superconductors Development in Future

Description and Composition	Critical Temperature, Kelvin	Critical Current Density, A/mm²
Yttrium–barium–copper oxide $YBa_2Cu_3O_{7-y}$ $(0.4 < Y < 0.9)$	90	200–2000
Bismuth–strontium–calcium–copper oxide $Bi_2Sr_2Ca_{an-1}Cu_nO_{4+2n+x}$	110	400–40,000
Thallium–barium–calcium–copper oxide $Tl_2Ba_2Ca_{n-1}Cu_nO_{2n+1}$	124	$10^4 - 3 \times 10^5$

Source: Copyright permission from Innovative materials for the transformers of the future. Paper presented at the International Electrical Engineering Congress (IEECON 2020), Chiang Mai, Thailand, March 4–6, 2020-Paper P1019

TABLE 27.7

Comparison of CNT Yarn with Copper and Aluminium

Material	Density kg/m³	Electrical Conductivity at 75°C μs/m	Electrical Resistivity at 20°C Ω/m	Temperature Coefficient of Resistivity per°C	Thermal Conductivity w/m/c
CNT	1500	100	$I \times 10^{-8}$	−0.001 to 0.002	635
Copper	8960	50	1.68×10^{-8}	−0.00386	400
Aluminium	2760	30	2.65×10^{-8}	0.00429	235

CNT, carbon nanotube.

Source: Copyright permission from; Innovative materials for the transformers of the future. Paper presented at the International Electrical Engineering Congress (IEECON 2020), Chiang Mai, Thailand, March 4–6, 2020-Paper P1019.

The conductivity of CNT given in Table 27.7 is based on the present development. However, the theoretical level of conductivity is several 100 times of copper, and it is expected that high-conductivity CNT material for production of transformers will be available in the next 3–5 years. The density of the material is nearly one-sixth of copper, and therefore, the overall weight of the transformer is reduced.

C. Covetic materials

Covetic (covalent metal) materials are produced by electrical infusing of nanocarbon particles to copper, aluminium, etc. to improve the electrical conductivity, strength and thermal conductivity.

Experiments made by GDC Industries, Ohio, United States, have shown that the addition of 3% by weight of carbon to copper has improved the electrical conductivity by 7% and thermal conductivity by 10%. It is reported at the "Ultrawire workshop 2017" (12–7–2017) that nearly 30% increase in electrical conductivity and approximately 15% increase in thermal conductivity were measured on covetic copper.

D. Graphene

Graphene made in a single layer of carbon atom is a better conductor than copper. A team of Massachusetts Institute of Technology (MIT) found that two layers of graphene can become superconductor when the layers are twisted by the "magic angle". This discovery may open up the development of room-temperature superconductors based on graphene.

27.6.4   INSULATION MATERIALS AND OTHER PARTS

The transformers of the future need high-performance solid insulation materials for the power distribution system. The ideal material shall have very high electrical strength, high thermal conductivity, very low dielectric losses, high mechanical strength, extremely low flammability and high biodegradability. Other desirable characteristics are self-healing property, hydrophobic property and suitability for additive manufacture. None of the presently available materials have all these properties. There is also need for developing materials for efficient heat transfer, e.g., nonconducting heat pipes. Some of the emerging materials under development are discussed here.

A. Self-healing electrical insulation

Damage of the electrical insulation leads to the failure of the equipment, and therefore, it is essential to use insulation materials of high strength and integrity. If the insulation can heal on its own similar to a biological system, it would be possible for it to repair the incipient faults. The self-healing property is introduced in some of the epoxy polymers by the addition of a healing agent in microcapsules during the formation of the system. Monomers that are compatible with the epoxy polymers are used for this purpose.

The failure of the insulation starts with treeing, and the microcapsules melt and fill the cracks. A premixed catalyst in the polymer reacts with the monomer, and the resulting polymerization stops the propagation of the treeing and heals the system. This technology needs further development to bring out stable self-healing insulation systems.

B. Hydrophobic insulation

Hydrophobicity of a material is a measure of its resistance to water flowing along its surface. It is a highly desirable property for an insulation material, especially when it is necessary to operate in contaminated or moist environment. At present, hydrophobic cycloaliphatic epoxy systems are available for outdoor application.

The emerging scenario is the development of hydrophobic, nonflammable, nontoxic insulation with high temperature index (e.g., 220°C).

C. Nonconducting heat pipes

Heat pipe uses a "phase change fluid" to efficiently conduct large amount of heat between two solid surfaces. It consists of an enclosed tube (the evaporator) containing a fluid with high specific heat in vacuum. The fluid absorbs thermal energy from the heat source and boils. The vapour flows to the external condensing area. Natural airflow cools the condenser section, and the vapour condenses and flows back to the evaporating area. The process is in a sealed system and is self-sustaining. Conventional heat pipes are made of electrically conducting materials. The future material for the pipe will be electrically nonconducting having high heat transfer capability for embedding it in the heat producing area of the power electronics, magnetic and conducting systems of the SST.

28 Transformer Design Optimization

28.1 INTRODUCTION

Design optimization is the selection of the best option within the available resources ensuring compliance with the specified and mandated constraints and limitations. The selection necessarily is based on certain objectives, and it is called the objective function. For a transformer design, the objective function can be any one of the following;

a. Lowest initial material cost of the product
b. Lowest initial cost of the product including materials and manufacturing expenses
c. Lowest lifetime owning cost of the product, which includes the initial product cost plus the lifetime operating cost considering the cost of losses
d. Lowest lifetime owning cost including the initial product cost plus lifetime cost of losses and cost of CO_2 emissions for the lifetime

The optimization of the transformer design for minimizing a, b or c is common. However, design optimization with due consideration of the cost of "carbon footprint" has not become popular yet.

Several optimization techniques are available to the designer, and the following sections cover some of the optimization methods employed for liquid-filled transformers.

28.2 OBJECTIVE FUNCTIONS FOR DESIGN OPTIMIZATION

28.2.1 OPTIMIZATION OF LOWEST INITIAL MATERIAL COST

The costs of the main materials considered for optimization algorithms include the following:

- Magnetic core
- Winding conductor (all windings)
- Transformer oil
- Insulation materials
- Tank and parts
- Cooling system

The components such as bushings, temperature indicators and tap changer are fixed for a particular specification, and therefore, the costs of these items are considered as constant in the algorithm.

28.2.2 LOWEST INITIAL PRODUCT COST

The manufacturing cost of the product is added to the cost of materials and components as per Section 28.2.1. If the manufacturing cost is considered as a fixed amount for a particular rating of transformer, this cost becomes an element similar to the cost of a component. If the manufacturer considers it as a function of the cost of materials (including components), the material and component costs need to be considered for optimization.

28.2.3 Lowest Total Owning Cost

In this option, the objective function consists of the total cost of materials and components plus the lifetime cost of the losses. The product cost is considered as in Section 28.2.2. The lifetime cost of no-load losses and load losses is calculated from the loss capitalization rates. Annexure A5 gives the procedure for calculating total owning cost.

28.2.4 Lowest Total Owning Cost Including the Cost of Lifetime CO_2 Emission

The objective function consists of the cost as in Section 28.2.3 plus the cost of the lifetime carbon emissions from the transformer. When more and more renewable energy generation is added to the power system, the cost of carbon emission attributed to the transformer losses comes down, and therefore, it is a dynamic variable. The cost of CO_2 emission is evaluated as \$/ton of CO_2.

28.3 CONSTRAINTS OF DESIGN OPTIMIZATION

The constraints to be considered are the specified guaranteed and design limits including the following:

- Upper limit of no-load loss is not exceeded (for objectives function as per Section 28.2.1 & Section 28.2.2).
- Upper limit of load loss is not exceeded (for objectives function as per Section 28.2.1 & Section 28.2.2).
- Impedance is within tolerance limits.
- Ratio error is within tolerance limits.
- Top oil temperature rise is within limit.
- Average winding temperature rise is within limit.
- The short-circuit temperature rise of the windings is not exceeded.
- The dynamic short-circuit stresses of conductors and windings are within safe limits.
- The electrical stresses are not exceeded in any part of the windings, between windings, between turns, between layers/disc, to earth, etc. (This is included in the optimization algorithm if a subroutine for electrical stress optimization using finite element method is also part of the program. For distribution transformers and medium power transformers, the clearances are kept as standard, and therefore, the electrical stress constraint is not included in the algorithm.)
- Dimensional or weight limitations, if any.

28.4 DESIGN OPTIMIZATION METHODS

28.4.1 Brute Force Method

This is similar to the brute force program used for password cracking. This procedure iterates the entire combinations of variables sequentially. It calculates the objective function (i.e., cost) of each and every combination and selects the optimum one. This becomes tedious and time-consuming, when the number of variables becomes high, especially for high-voltage (HV) power transformers, where the number of iterations can be several billions.

The iterations required for a 1000 kVA, two-winding ONAN, 11/0.4 kV distribution transformer are given as an example. Table 28.1 shows the variables, the upper and lower limits of iterations and the incremental variation of the parameters for each step. The total number of iterations required, when five different variables are considered, are

$60 \times 60 \times 150 \times 110 \times 56 = 3,326,400,000$ (3.3 billions)

TABLE 28.1

Variables and Numbers of Iterations

Variable Parameter	Lower Limit	Upper Limit	Step Size for Iteration	Number of Iteration
Core diameter	180	240	1	60
Copper foil width	300	600	5	60
Copper foil thickness	0.5	2.0	0.01	150
Rectangular conductor width	5	16	0.1	110
Rectangular conductor thickness	1.2	4.0	0.05	56

It can be noted that if the limits are widened and the step sizes are reduced, the iterations become enormous. Many of the design outputs may not comply with the constraints, and therefore, the number of feasible designs can be few thousands.

When the number of variables increases and the ranges are not clear, the iterations required are enormous, and there is an obvious need for practical solutions. From experience of previous designs, it is possible to make better approximation of the upper and lower limits and the step sizes for iteration. This reduces the requirement of the iterations to reasonable levels, and the procedure is called heuristic method, which solves the problem in a faster and efficient manner. Heuristic algorithms generally give near optimum solution.

28.4.2 OPTIMUM DESIGN OF DISTRIBUTION TRANSFORMERS

The design of a distribution transformer is comparatively less complex, though the optimization involves the solution of multivariable, nonlinear functions. A simple method is worked out in this section which renders a "near-optimum" solution directly for a two-winding ONAN distribution transformer. The solutions consider options where the upper limits of no-load loss and load losses are specified and also where the loss capitalization rates are available.

28.4.2.1 When Upper Limits of No-Load Loss and Load Loss Are Specified

In this case, the design objective is to minimize the total cost of transformer subject to the following conditions;

a. The no-load loss is less than the specified value; W_{Fe} Watt
b. The load loss at reference temperature and at rated current is less than the specified value; W_c Watt
c. Other parameters are meeting the specified requirements, which are as follows:
   - Top oil temperature rise is less than θ_{oil} (°C).
   - The average temperature rise of any winding is less than θ_{wdg} (°C).
   - The percentage impedance is Z (subject to applicable tolerance).
   - The excitation current is less than I_o (if specified).
   - The flux density in core is less than B_m (Tesla).
        (If this is not specified, B_m shall be selected such that overfluxing does not occur under any of the specified operating conditions and that the no-load loss and no-load current do not exceed the specified values.)
   - The winding conductors withstand the dynamic and thermal short-circuit stresses.

The objective function to be minimized can be written as follows:

$$C = A_1 \ W_{core} + B_1 \ W_{wdg} + C_{coil} \ Q_{oil} + D_1 \ W_{steel} + Ew_{rad}/Ew_{fin} \qquad (28.1)$$

where
  C = cost of raw material, which is variable with respect to different designs
  A_1 = unit cost of core material ($/kg)
  W_{core} = weight of core (kg)
  B_1 = unit cost of conductor material ($/kg)
  W_{wdg} = weight of conductor material (kg)
  C_{oil} = unit cost of transformer oil ($/litre)
  Q_{oil} = quantity of transformer oil (litre)
  D_1 = unit cost of steel for tank ($/kg)
  W_{steel} = weight of steel for tank (kg)
  E = unit cost of cooling fin/radiator ($/kg)
  W_{rad} = weight of radiator (kg)
  W_{fin} = weight of fin (kg)

The optimum design is the one with the minimum value of C, satisfying the specified conditions as per Section 28.2.1 a–c.

28.4.2.2   When Capitalizations of No-Load Loss and Load Loss Are Specified

The objective function to be minimized is

$$P = V_F\left(A_1 W_{core} + B_1\ W_{wdg} + C_{coil}\cdot Q_{oil} + D_1 W_{steel} + E W_{rad}/E W_{fin}\right) + F\ X\ W_{Fe} + G\ X\ W_{Cu} \qquad (28.2)$$

where
  P = capitalized cost (total owning cost)
  V_F = value addition factor
  (This factor changes from manufacturer to manufacturer depending on the overheads and other
      costs + profit.)
  F = capitalization rate for no-load loss ($/W)
  W_{Fe} = no-load loss (W)
  G = capitalization rate for load loss ($/W)
  W_{cu} = load loss (W)

Other symbols have the same meanings as explained in Section 28.4.2.1.
  If the excitation current is capitalized, the term $H.I_0$ also will have to be added to the right-hand side of equation (28.2), where

  H = capitalization rate for no-load current ($/A)
  I_0 = no-load current (A)

28.4.2.3   Derivation of Parameters of Objective Function

Symbols used:
  The symbols used are explained with the aid of Figure 28.1. The most commonly used units for practical designs are employed, for the calculations.

  D = core circle diameter (mm)
  a = core to low-voltage (LV) gap (mm)
  W_1 = radial depth of LV winding (mm)
  g = HV − LV gap (mm)
  W_2 = radial depth of HV winding (mm)
  h = axial height of winding (mm)
  (It is assumed that the heights of HV and LV are equal.)
  u = clearance of winding to yoke (mm)

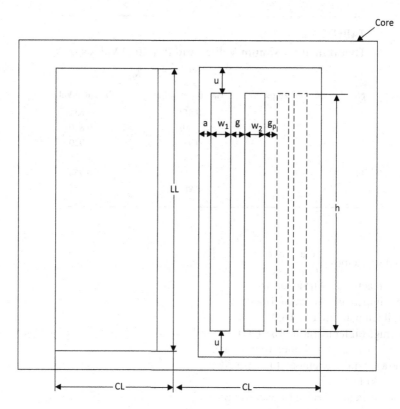

FIGURE 28.1 Core and winding dimension (windings are shown in one window only).

L_L = core window height (mm)
$(L_L = h + 2u)$
g_p = phase to phase clearance (mm)
CL = distance of core leg centers (mm)
α = LV winding space factor
β = HV winding space factor
A = area of cross section of conductor (mm²)
N = number of turns
Q = kVA = $3 \, V_{ph} \, I_{ph} \times 10^{-3}$
V_{ph} = phase voltage (V)
I_{ph} = phase current (A)
δ = current density (A/mm²)
V_t = voltage per turn
V_L = line voltage
f = frequency (Hz)
B = flux density in core (Tesla)
K_f = core fill factor

The theoretical maximum values and practical values of K_f for various number of core steps are given in Table 28.2. For calculations, the practical values can be used.

K_s = core stacking factor = 0.97
w/kg = specific loss of core material
W_{BF} = building factor (destruction factor) for core loss
W = width of maximum step in core (mm)

TABLE 28.2

Theoretical Maximum Values and Practical Values of K_f

| | K_f | |
Number of Core Steps	Maximum Value	Practical Value
3	0.851	0.835
5	0.908	0.890
7	0.934	0.920
9	0.948	0.935
11	0.958	0.950
13	0.962	0.955

K_{st} = stray loss factor = $\dfrac{\text{Load Loss}}{I^2 R \text{ Loss}}$

X = percent reactance of transformer

Z = percent impedance of transformer

θ_{oil} = top oil temperature rise (°C)

$\theta_{oil\,av}$ = average oil temperature rise (°C)

$\theta_{oil\,av}$ = 0.8 × θ_{oil} (for distribution transformer)

θ_{wdg} = average winding temperature rise (°C)

K_R = Rogowski factor

D_m = mean diameter of HV–LV gap (mm)

$D + 2a + 2w_1 + g$

H, M,b, c, d, p, q, R – constants

A_{Tank} = winding to tank clearance on HV side (mm)

B_{Tank} = winding to tank clearance on LV side (mm)

C_{Tank} = winding to tank clearance on tank width side (mm)

V_A / kg = specific excitation V_A for core material

E_{BF} = building factor for excitation V_A of core

H_{TC} = height of off-circuit tap changer from top surface core (mm)

(when off-circuit tap changer is not provided H_{TC} = 0)

(when cable operated off-circuit tap changer is proved H_{TC} = 0)

K_1 = oil height above tap changer (mm)

K_2 = tank bottom to core clearance (mm)

K_3 = free space required over oil level at the time of oil filling

(applicable for sealed type transformer with gash cushion)

T_L = tank length (mm)

T_W = tank width (mm)

T_H = tank height (mm)

A_T = cooling surface area of tank (m²)

t_{rad} = thickness of radiator (mm)

t_{fin} = thickness of corrugated fin (mm)

L = length of radiator/fin = $[T_H - K_3 - K_2 - W - 150]$

(round off to lower multiple of 50)

D_{fin} = depth of corrugated fin (mm)

(round off to nearest 10)

28.4.2.4 Derivation of Equations

For a three-phase two-winding transformer:

$$Q = 3 \ V_{ph} \ I_{ph} \times 10^{-3} \tag{28.3}$$

$$V_{ph} = V_t N \tag{28.4}$$

$$I_{ph} = \delta \ A \tag{28.5}$$

$$Q = 3 V_t N \delta A \times 10^{-3} \tag{28.6}$$

$$NA = \alpha \ w_1 \ h \tag{28.7}$$

Substituting in Q,

$$Q = 3 V_t \delta \ \alpha \ W_1 \ h \times 10^{-3} \tag{28.8}$$

Also

$$X = \frac{1.24 \times Q \ D_m \left[(W_1 + W_2 \div 3) + g \right] K_R}{30 \ V^2 \ h} \tag{28.9}$$

Substituting for Q,

$$X = \frac{1.24 \times 3 \ V_t \ \delta \ \alpha \ W_1 \ h \times 10^{-3} \ D_m \left[(W_1 + W_2 \div 3) + g \right] K_R}{30 \ V_t^2 \ h} \tag{28.10}$$

Assuming that equal current density is used for both HV and LV,

$$\alpha \ W_1 \ h = \beta \ W_2 \ h \tag{28.11}$$

$$\therefore W_2 = \frac{\alpha W_1}{\beta} \tag{28.12}$$

(If δ_1 is current density of LV and δ_2 is the current density of HV, $W_2 = \dfrac{\alpha W_1}{\beta} \times \dfrac{\delta_1}{\delta_2}$)

Also,
$$V_t = 4.44 \ fB \left(\frac{\pi}{4} D^2 K_f K_s \right) \times 10^{-6} \tag{28.13}$$

Substituting for W_2 and V_t in equation and simplifying, a cubic equation for W_1 is obtained.

$$X = \frac{W_1^3 + W_1^2 \left[D(\alpha + \beta) + 2a(\alpha + \beta) + g(7\beta + \alpha) \right]}{2(\alpha + \beta)} + \frac{W_1 \left(6ag + 3g^2 + 3D \ g \right)}{2(\alpha + \beta)}$$
$$- \frac{42.184 \times 10^{-3} \ f \ BD^2 \ K_f \ K_s \times \beta}{\alpha \ \delta \ K_R \ (\alpha + \beta)} \tag{28.14}$$

Solution of the cubic equation for real value of W_1 gives the radial depth of LV winding.
The real value of W_1 is solved as follows:
Let

$$b = \frac{\left[D(\alpha+\beta)+2a\,(\alpha+\beta)+g(7\beta+\alpha)\right]}{2(\alpha+\beta)} \tag{28.15}$$

$$c = \frac{\beta\left(6ag+3g^2+3Dg\right)}{2(\alpha+\beta)} \tag{28.16}$$

$$d = \frac{42.184\times10^{-3}\ fB\ D^2\ K_f\ K_s\times\beta}{\alpha\ \delta\ K_R\ (\alpha+\beta)} \tag{28.17}$$

where

$$H = \sqrt[3]{\frac{-q}{2}+\sqrt{R}} \tag{28.18}$$

$$m = \sqrt[3]{\frac{-q}{2}-R} \tag{28.19}$$

$$R = \left(\frac{p}{3}\right)^3+\left(\frac{q}{2}\right)^2 \tag{28.20}$$

$$p = \frac{1}{3}\left(3c-b^2\right) \tag{28.21}$$

$$q = \frac{1}{27}\left(27d-9bc+2b^3\right) \tag{28.22}$$

Radial depth of LV winding $= W_1$

a. Winding axial height "h"

$$h = \frac{9.559\times10^7 Q}{\delta\alpha w_1 f\ B\ D^2\ K_f\ K_s} \tag{28.23}$$

b. LV inside diameter, outside diameter and mean diameter
    LV inside diameter $= D+2a$
    LV mean diameter $= D+2a+w_1$
    LV outside diameter $= D+2a+2\,w_1$
c. HV inside diameter, outside diameter and mean diameter

HV inside diameter $= D+2a+2w_1+2g$

HV mean diameter $= D+2a+2w_1+2g+w_1\left(\dfrac{\alpha}{\beta}\right)$

$$\text{HV outside diameter} = D + 2a + 2w_1 + 2g + 2w_1\left(\frac{\alpha}{\beta}\right)$$

d. Weight of conductor

Assuming equal current density in HV and LV windings:

For copper winding:

Weight of copper in kg;

$$\left(W_{\text{cu wdg}}\right) = \frac{8\times10^3 Q\left(2D + 4a + 3w_1 + 2g + \dfrac{\alpha}{\beta}w_1\right)}{\delta\, f\, B\, D^2\, K_f\, K_s} \tag{28.24}$$

For aluminium winding

Weight of aluminium in kg;

$$\left(W_{\text{Al wdg}}\right) = \frac{2.425\times10^3 Q\left(2D + 4a + 3w_1 + 2g + \dfrac{\alpha}{\beta}w_1\right)}{\delta f\, B\, D^2\, K_f\, K_s} \tag{28.25}$$

e. I²R loss and load loss

$$I^2R \text{ loss for copper winding at 75 °C} = 18.88\times10^3 Q \times \frac{\left(2D + 4a + 3w_1 + 2g + \dfrac{\alpha}{\beta}w_1\right)}{\delta\, f\, B\, D^2\, K_f\, K_s} \tag{28.26}$$

$$I^2R \text{ loss for aluminum winding at 75°C} = 31.51\times10^3 Q \times \frac{\left(2D + 4a + 3w_1 + 2g + \dfrac{\alpha}{\beta}w_1\right)}{\delta\, f\, B\, D^2\, K_f\, K_s} \tag{28.27}$$

$$\text{Load loss}\left(W_c\right) = K_{\text{st}}I^2R \tag{28.28}$$

f. Core weight

$$W_{\text{core}} = 6.008\times10^{-6} + K_f K_s D^2$$

$$\times\left[\frac{2.8677\times10^8 Q}{\delta\alpha\, fBK_f K_s D^2 w_1} + \left(6u + 8a + 8g + 4g_1 - 10\right) + 6D + 8w_1\left(1 + \frac{\alpha}{\beta}\right)\right] \tag{28.29}$$

g. No-load loss (core loss)

$$W_{\text{Fe}} = W_{\text{core}} \times \text{specific loss of core material} \times \text{building factor}$$

$$W_{\text{Fe}} = W_{\text{core}} \times W/\text{kg} \times W_{\text{BF}} \tag{28.30}$$

h. No-load current

$$I_0 = \frac{W_{\text{core}} \times V_A/\text{kg} \times E_{\text{BF}}}{\sqrt{3}\, V_L} \tag{28.31}$$

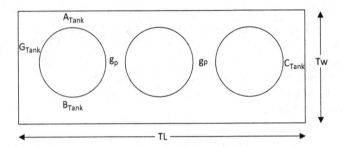

FIGURE 28.2   Size of tank.

i. Size of tank (Figure 28.2)

T_L = Tank length = 3 (HV OD) + 2gp + 2 C_{Tank}

T_w = Tank width = HV OD + A_{Tank} + B_{Tank}

T_H = Tank height = $L_L + 2w + H_{TC} + K_1 + K_2$ (for transformer with conservator)

T_H = Tank height = $L_L + 2w + H_{TC} + K_1 + K_2 + K_3$ (for sealed transformer)

Calculation of weight of radiator

W_{Total} = Total losses (Watt)

A_T = Cooling surface area of tank

$$A_T = \begin{array}{l} \text{Area of 4 sides + Area of top cover (for transformer with conservator)} \\ \text{Wetted area of 4 sides (area of 4 sides in contact with oil)} \end{array} \qquad (28.32)$$

For transformer with conservator

$$A_T = 2T_H\left(T_L + T_w\right) \qquad (28.33)$$

$$A_T = 2T_H\left(3\text{HVOD} + 2gp + 2C_{Tank} + \text{HVOD} + A_{Tank} + B_{Tank}\right)$$

$$A_T = \left(LL + 2w + H_{TC} + K_1 + K_2\right)\left(3\text{HVOD} + 2gp + 2C_{Tank} + \text{HVOD} + A_{Tank} + B_{Tank}\right)$$

$$A_T = \left(LL + 2w + H_{TC} + K_1 + K_2\right)\left(4\text{HVOD} + 2gp + 2C_{Tank} + A_{Tank} + B_{Tank}\right)$$

For sealed transformer with gas cushion

$$AT = 2T_H\left(T_L + T_w\right)T_L + T_w \qquad (28.34)$$

$$= \left(LL + 2w + H_{TC} + K_1 + K_2 + K_3\right)\left(4\text{HVOD} + 2gp + 2C_{Tank} + A_{Tank} + B_{Tank}\right) + T_L + T_w$$

$$= \left(LL + 2w + H_{TC} + K_1 + K_2 + K_3\right)\left(4\text{HVOD} + 2gp + 2C_{Tank} + A_{Tank} + B_{Tank}\right)$$

$$+ \left(3\text{HVOD} + 2gp + 2C_{Tank}\right)\left(\text{HVOD} + A_{Tank} + B_{Tank}\right) \qquad (28.35)$$

Mean oil temperature rise = θ_m

= $0.8 \times \theta_{oil}$ (OR)

$(W_{wdg} - \theta w)$ whichever is lower:

θw = winding gradient

W_{wdg} = Average winding temperature rise

Weight of radiator

$$W_{rad} = \frac{0.3568 \ L^{0.2} \ t_{rad} \ (W_{Fe} + W_{Cu})}{\theta_{oil}^{1.1}} \qquad (28.36)$$

Weight of corrugated fin

$$W_{fin} = \frac{0.987 \ D_{fin}^{0.2} \ t_{fin} \ (W_{Fe} + W_{Cu})}{\theta_{oil}^{1.2}} \qquad (28.37)$$

Oil quantity

Q_{oil} = Oil in main tank + oil in radiator or corrugated fin + oil in conservator

(whichever applicable)

Where,
Q_{oil} = Oil quantity

(For sealed transformers, oil in conservator is not applicable.)
Oil in main tank is calculated by subtracting the displacement volumes of core and conductor.
(Displacement volumes of clamping structures, paper, pressboard and other insulating materials are ignored in the above calculation. Due to this assumption, the calculated oil volume will be slightly higher than the actual quantity required for filling.)

$$\text{Oil quantity in main tank} = T_L . T_w . T_H \times 10^{-6} - \frac{W_{wdg}}{\Delta} - \frac{W_{core}}{7.65} \qquad (28.38)$$

Δ = density of conductor (g/cc)
= 2.7 for aluminium
= 8.9 for copper
Oil in radiator element = $L_{Rad} \times 0.9$ (l)
Oil in conservator = 0.03 (oil in main tank + oil in round cooling tube or radiator element)

This procedure requires iterations of a few number of variables only as follows:

δ – current density
α – LV winding space factor
β – HV winding space factor
B – flux density
D – core diameter

The first iteration is done by using arbitrary values of α and β, and these factors are updated by calculating the LV and HV fill factor of first design.

28.4.3 GENETIC ALGORITHMS

Genetic algorithms (GAs) are computerized search and optimization techniques. These are based on the natural phenomena of genetics and natural selection GAs can navigate through very large search space and identify optimal selection of variables and solution. The major elements involved in GA are as follows:

TABLE 28.3

Comparison of Genetic Algorithm and Sequential Calculation Methods

Normal Sequential Method	Genetic Algorithm
Calculates a single point at each iteration	Generates a population of points at each iteration
The sequential calculation reaches a solution for the selected point	The best point in the population approaches an optimal solution
Selects the subsequent point in the sequence by deterministic computation	The next population by computation is obtained by random number generator

Note: The flowchart of a genetic algorithm is given in Figure 28.3 for clarity of the algorithm.

- Defining the objective functions
- Definition and implementation of genetic representation
- Definition and implementation of genetic operators

Objective function

The function requiring optimization is the objective function. The parameters of the objective function can have upper and lower limits. The objective functions applicable for transformers are described in Section 28.1.

Search space

This is the space for each feasible solution, which is marked by its value of fitness. The solution searches for the minimum value if the objective function is to be minimized and vice versa.

Population

Population is the set of solutions in the search space. The set of solutions is represented as chromosomes. The initial solution is a random number in the search space.

GA is different from the normal derivative-based optimization where the solution process is carried out in step-by-step sequence. The comparison of GA with the normal sequential method is shown in Table 28.3. The flow chart of GA is given in Figure 28.3 for better understanding of the algorithm.

28.4.4 HARMONY SEARCH ALGORITHM

Harmony search (HS) algorithm is based on the principles of the improvisation of the harmony by the musicians, where they try various possible combinations of the pitch stored in their memory. This is analogous to finding the optimum solution of a transformer design. The steps involved in HS are as follows:

a. Initiate the HS memory by selecting randomly generated solutions to the optimization problem.
b. Improvise a new solution from the harmony memory HM. The pitching adjust rate (PAR) determines the probability of a candidate from the HM to be muted.
c. Update the HM. The new solution from step b is evaluated. If it yields a better fitness than the worst one from the HM, it is selected. If not, it is rejected.
d. Repeat steps b and c until a preset termination criterion is reached. Similar to GA, HS is a random search method. It does not require gradient information of the objective function. HS method has simplicity of computations.

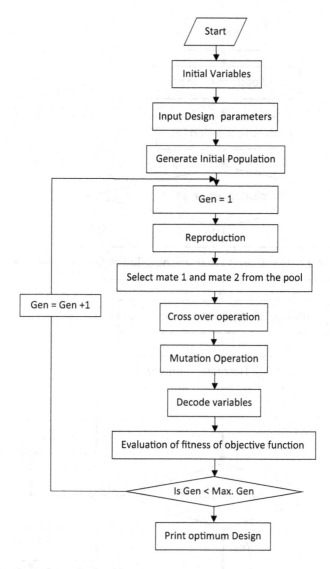

FIGURE 28.3 Flowchart of genetic algorithm.

Figure 28.4 shows the flowchart of an HS algorithm for clear understanding.
   Symbols used in the flowchart;

Rand = random
HMCR = harmony memory considering rate
PAR = pitching adjust rate
LL = load loss
NLL = no-load loss
Z = impedance
θ_{oil} = top oil temperature rise
θ_{wdg} = average winding temperature rise
Sc Thermal = temperature rise of conductor under short circuit
Sc hoop = hoop stress on conductor under dynamic short circuit

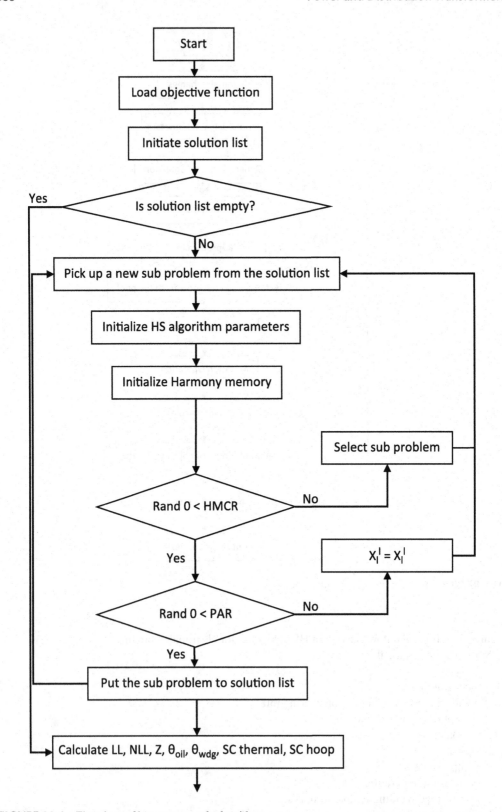

FIGURE 28.4 Flowchart of harmony search algorithm.

(*Continued*)

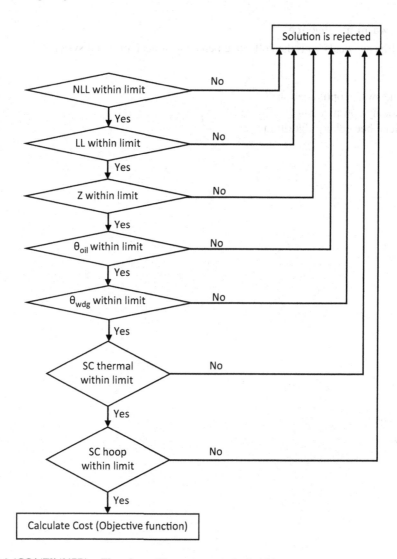

FIGURE 28.4 (*CONTINUED*)   Flowchart of harmony search algorithm.

28.4.5   OTHER OPTIMIZATION TECHNIQUES

28.4.5.1   Simulated Annealing

Simulating annealing algorithm is based on the annealing process in metallurgy where the solution converges in the same way as metals are brought to minimum energy configuration by increasing the grain size. It is useful for global optimization where functions with a large search space need solution. At each iteration, a new point is randomly generated. The distance from the new point from the current point or the extent of the search is dependent on the probability distribution of the temperature.

28.4.5.2   Tabu Search Algorithm

Tabu search algorithm is a heuristic method using dynamically generated tabus to guide the search of the solution. It searches potential neighbours for an improved solution. It also sets up dynamic rules so that the system will not search around the same area redundantly. This method is useful when there are more than one equally fit solutions. It is good when the search is stuck in suboptimal areas.

28.4.5.3 Swarm Intelligence

Swarm intelligence deals with the collective behavior of decentralized systems. The examples are as follows:

- Particle swarm optimization
- Ant colony optimization
- Artificial bee colony algorithm

29 Corrosion Protection

29.1 TYPICAL PAINTING SYSTEMS FOR TRANSFORMER TANKS AND RADIATORS

The transformer is an outdoor equipment, and it is exposed to the atmosphere for the lifetime. Depending on the corrosion level of the atmosphere, suitable painting system will have to be selected.

The standards applicable for this is ISO 12944 – paints and varnishes – corrosion protection of steel structures by protective paint systems (parts 1–8).

ISO 12944 has classified the environment based on the level of corrosion, which are applicable for steel exposed to temperatures less than 120°C. Table 29.1 gives the classifications.

29.2 DESIGN LIFE (DURABILITY) OF THE COATING SYSTEM

ISO 12944 defines the durability as the time to first major maintenance. The durability and design life are as per Table 29.2.

The paint systems for external surfaces for transformer tank and parts for high durability are given in Tables 29.3–29.5 corresponding to C4, C5-I and C5-M corrosion categories.

Prior to applying the paint, the steel structures will have to be cleaned, and the rust will have to be removed. The blast cleaning of the surface to standard Sa 2½ or Sa 3 as per BS EN ISO 8501–4:2007 is good for the commercial-grade steels.

TABLE 29.1
Classification of Environment

ISO 12944 Classification	Typical Environment
C1	Rural areas with low pollution
C2	Neutral atmosphere
C3	Urban and Industrial atmosphere
C4	Industrial and coastal chemical processing plant
C5-I	Industrial areas with high humidity and aggressive atmosphere
C5-M	Marine, offshore, coastal areas with high humidity

TABLE 29.2
Durability Classification

Durability Classification	Design Life – Years
High durability	>15
Medium durability	5–15
Low durability	<5

TABLE 29.3

Paint System for Corrosion Level C4 – High Durability

System Number	Binder	Type of Coats	Number of Coats	NDFT Micron	Binder Types	Number of Coats	NDFT Micron
		Primer Coats			Final Coats	Paint System	
A_4. 09	EP	Mis	1	80	EP, PUR	2–3	280
A_4. 15	EP, PUR	Zn (R)	1	60	EP, PUR	3–4	240

Note: The surface preparation for rust grade A, B or C as per ISO 8501, for the above paint system is Sa 2½

TABLE 29.4

Paint System for Corrosion Level C5-I – High Durability

System Number	Binder	Type of Coats	Number of Coats	NDFT Micron	Binder Types	Number of Coats	NDFT Micron
		Primer Coats			Final Coats	Paint System	
A_5 I.02	EP, PUR	Misc	1	80	EP, PUR	3–4	320
A_5 I.05	EP, PUR	Zn (R)	1	60	EP, PUR	3–5	320

Note: The surface preparation for rust grade A, B or C as per ISO 8501 for above paint system is Sa 2½ or Sa 3.

TABLE 29.5

Paint System for Corrosion Level C5-M – High Durability

System Number	Binder	Type of Coats	Number of Coats	NDFT Micron	Binder Types	Number of Coats	NDFT Micron
		Primer Coats			Final Coats	Paint System	
A_5 M.02	EP, PUR	Misc	1	80	EP, PUR	3–4	320
A_5 M.04	EP, PUR	Misc	1	250	EP, PUR	2	500
A_5 M.06	EP, PUR, ESI	Zn (R)	1	60	PP, PUR	4–5	320
A_5 M.07	EP, PUR, ESI	Zn (R)	1	60	EPC	3–4	400

Note: The surface preparation for rust grade A, B or C as per ISO 8501 for the above system is Sa 2½ or Sa 3.

Meaning of the symbols used:

EP – epoxy – 2 packs
PUR – polyurethane, aliphatic – 1 or 2 packs
Misc – primers with miscellaneous types of anticorrosive pigments
Zn (R) – zinc-rich primer
AY – acrylic – 1 pack
ESI – ethyl silicate – 1 or 2 packs
NDFT – nominal dry film thickness
EPC – epoxy combination – 2 packs

TABLE 29.6

Painting System of Internal Air-Filled Areas

Coating	Material	NDFT Microns
Primer	Two-component inorganic zinc ethyl silicate coating	75
Intermediate coating	Micaceous iron oxide–pigmented two-pack epoxy	75
Final coating	Two-component alipathic acrylic polyurethane coating	75
Total NDFT		225 μm

29.3 PAINTING CRITERION

29.3.1 PAINTING OF INTERNAL AREAS (AIR FILLED)

Internal air-filled surfaces require surface cleaning and corrosion-preventing high durability paint system as follows (Table 29.6).

Surface preparation is Sa 2½ as per ISO 8501.

29.3.2 PAINTING OF INTERNAL OIL FILLED AREAS

Surface preparation for rust grade A, B or C as per ISO 8501 for these surfaces is Sa 2½

The paint system used for internal oil-filled areas will have to be compatible with the transformer oil or other insulation fluid used for the transformer. A typical system compatible with mineral oil is two-component phenolic/novolac epoxy coating of NDFT 50 μm applied after surface preparation.

29.4 TIN COATING (ELECTRODEPOSITED COATING OF TIN)

Tin coating is used to provide a low contact resistance surface and to protect against corrosion.

In normal indoor applications, tin is protective on iron, steel, nickel, aluminium, copper and alloys of these metals.

If the tin coating has discontinuities such as pores and cracks, corrosion can be expected in outdoor application. The galvanic couples formed between the tin and the base metal creates the corrosion when there are discontinuities.

The minimum thickness values of tin coating recommended by ASTM-B545-97 (2004) "Standard Specification for Electrodeposited Coating of Tin" for various applications are given in Table 29.7

TABLE 29.7

Minimum Thickness of Tin Coating

Service Class	Minimum Coating Thickness (μm)	Typical Application
A	2.5	For indoor application where the surface is not used for soldering
B	5.0	Mild service conditions
C	8.0 10 for steel	Moderate exposure conditions such as electrical hardware; class C can be used for transformer components
D	15 20 for steel	Severe service including exposure to dampness and industrial environment
E	30	Very severe service conditions including elevated temperature

29.5   ZINC COATING ON IRON OR STEEL

Electrodeposited coating of zinc on iron or steel surface is used to protect the surface from corrosion. ASTM standards applicable for zinc coating on iron or steel are as follows:

ASTM B183 – Practice for Preparation of Low-Carbon Steel for Electroplating
ASTM B242 – Practice for Preparation of High-Carbon Steel for Electroplating
ASTM B254 – Practice for Preparation and Electroplating of Stainless Steel
ASTM B 322 – Practice for Cleaning Metals Prior to Electroplating
ASTM B 633–98 – Standard Specification for Electrodeposited Coatings of Zinc on Iron and Steel

The coating thickness is given in four classes as per Table 29.8.
   Four different finish types for the coating are defined by ASTM B-633–98 as given in Table 29.9.
   The classification and finish required are selected based on environment and aesthetic finish required.

TABLE 29.8
Thickness Classes of Zinc Coating on Steel

Classification Number	Service Condition	Coating Thickness Micron
Fe/Zn 25	Sc 4 – very severe	25
Fe/Zn 12	Sc 3 – severe	12
Fe/Zn 8	Sc 2 – moderate	8
Fe/Zn 5	Sc 1 – mild	5

Source: Copyright permission from ASTM 633–98 Standard Specification for Electrodeposited Coatings of Zinc on Iron and Steel.

TABLE 29.9
Finish Types of Zinc Coating

Finish Type	Description
I	As plated – without further treatment
II	With coloured chromate conversion coating
III	With colourless chromate conversion coating
IV	With phosphate conversion coating

30 Calculation of Miscellaneous Technical Parameters

30.1 OVERLOADING OF OIL-IMMERSED TRANSFORMERS

30.1.1 INTRODUCTION

Transformers are designed as per the specifications and the applicable national and international standards where the maximum ambient temperature is specified. The temperature rise limits specified are also based on continuous loading at full rated capacity when operating at the standard ambient temperature. In reality, the ambient temperature changes in the course of the day, and it has a cyclic variation over the year. The load on the transformer has variation depending on the application. Thus, the temperature rise of oil and winding has a dynamic variation determined by the ambient and load variations.

It is evident that the average winding temperature rise, top oil temperature rise and consequently the hot spot temperature change continuously in operation. The ageing of the insulation and the loss of life of the transformer also are dynamic variables, which can be used for the optimum loading strategy of the transformer. The following paragraphs give the concepts of insulation ageing, the classifications of overloading and the calculation methods to ascertain that the loss of life of insulation is within safe limits.

The standards applicable for loading guides are as follows:

- IEC 60076-7-2017 "Loading guide for oil-immersed power transformers"
- IEEE C.57.91–2011 "Guide for loading mineral oil–filled transformers and step regulators"

30.1.2 LOSS OF LIFE OF INSULATION

The loss of life of an insulation system is dependent on the hot spot temperature. Two different types of insulation systems are generally used for mineral oil–filled transformers, which are conventional kraft paper insulation and thermally upgraded (TU) insulation.

It is considered that the oil-immersed kraft paper insulation has a normal life when operated at 98°C hot spot temperature. For TU paper, this is 110°C. In both cases, a 6°C increase or decrease of the hot spot temperature will halve or double the normal lifetime. Table 30.1 shows the hot spot temperature and the corresponding loss of life per unit values.

If a TU paper insulation works at a hot spot temperature of 122°C for 1 hour, the paper consumes 4 hours of life. However, if it works for 1 hour at 92°C, the loss of life is only 0.125 hours. This is the basis of normal cyclic loading where the overall loss of life is the integral sum of it over the time period.

30.1.3 EFFECTS OF LOADING ABOVE THE RATED LOAD

When a transformer is operating above its rated loading capacity, the effect can vary depending on the design of the transformer, the ambient conditions, ratings of the components used, etc. The moisture equilibrium of the oil paper insulation may get affected, and there can be possibilities of bubbling when the hot spot temperature goes above 140°C, mainly for smaller transformer.

TABLE 30.1

Hot Spot Temperature and Loss of Life

Hot Spot Temperature °C	Normal Kraft Paper Loss of Life, PU	Thermally Upgraded Paper Loss of Life, PU
128	32.00	8.00
122	16.00	4.00
116	8.00	2.00
110	4.00	1.00
104	2.00	0.50
98	1.00	0.25
92	0.50	0.125
86	0.25	0.0625
80	0.125	0.03125
74	0.0625	0.015625

The increase of the leakage flux density of some transformers can cause hot spots on tank and clamping structures. The current ratings of the components and connections may also be exceeded.

30.1.4 CATEGORIES OF OVERLOADING

30.1.4.1 Normal Cyclic Loading

In this type of loading, the loss of life from a higher ambient temperature or a higher than rated current operation for a particular time duration is balanced by the gain in life from the operation at a low ambient temperature or low load during the loading cycle. For small and medium transformers, the maximum loading allowed is 1.5 PU, and for large transformers above 100 MVA, the loading limit is 1.3 PU. The hot spot temperature of the winding and the top oil temperatures are limited to 120°C and 105°C, respectively. The loading duration or the magnitude of a given load shall be limited by the above.

30.1.4.2 Long-Time Emergency Loading

Long-time emergency loading is imposed on a transformer when there is outage of some element in the system and the transformer is subjected to overloading for a long time.

The hot spot temperature and top oil temperature in this type of overloading are limited to 140°C and 115°C, respectively. The maximum permitted loading shall be 1.8 PU for small transformers, 1.5 PU for medium transformers, and 1.3 PU for large transformers above 100 MVA

30.1.4.3 Short-Time Emergency Loading

Short-time emergency loading arises out of unusual events in the system, and the loading duration is less than 30 minutes. The hot spot temperature and top oil temperature are 140°C and 115°C, respectively. The maximum loading for small transformers is 2.0 PU which is 1.8 PU for medium transformers and 1.5 PU for large transformers.

30.1.5 CALCULATION OF HOT SPOT TEMPERATURE AND TOP OIL TEMPERATURE

The basic method of calculation of hot spot temperature and top oil temperature under overloading conditions can be done by the solution of the exponential equation or the solution of differential equation.

The solutions are based on the following equations.

a. Top oil temperature

$$\theta_o(t) = \theta_a + \Delta\theta_{or} \left[\frac{W_2}{W_1} \right]^x + \left[\Delta\theta_{oi} - \Delta\theta_{or} \left(\frac{W_2}{W_1} \right)^x \right] e^{-\frac{t}{K_{11}T_o}} \qquad (30.1)$$

where

$\theta_o(t)$ = top oil temperature at time "t" min
θ_a = ambient temperature °C
$\Delta\theta_{or}$ = top oil temperature rise at rated load °C
W_1 = no-load loss + load loss at rated load (watts)
W_2 = no-load loss + load loss at new load (watts)
(The load loss is considered as the $(PU)^2 \times Rated\ Load\ Loss$)
x = oil exponent = 0.8
$\Delta\theta_{oi}$ = Initial oil temperature rise °C
t = time duration in minutes
K_{11} = 1.0 for small transformers
0.5 for medium and large transformers
T_0 = time constant of the transformer (minutes)
= 180 for small transformers
= 210 for medium transformers
= 150 for large transformers

b. The hot spot temperature of the winding is calculated by the following formula:

$$\theta_h(t) = \theta_a + \theta_o(f) + H \times g(t) \qquad (30.2)$$

where

$\theta_h(t)$ = hot spot temperature at time (t)
θ_a = ambient temperature
$\theta_o(f)$ = top oil temperature rise at time (t)
$g(t)$ = gradient at time (t)
= difference between average winding temperature rise and average oil temperature rise
H = hot spot factor
H = QS

Factor Q depends on the loss concentration on any part of the winding.

Factor Q depends on the leakage flux distribution. For accurate determination, finite element analysis is required for each winding.

The approximate values of Q are given in Table 30.2.

Factor s depends on the efficiency of the oil flow through the winding.

For normal windings without oil directing washers, 's' can be taken as 1.0.

For more accuracy, the hydraulic circuit of oil flow needs to be solved.

30.2 ALTITUDE CORRECTION FACTORS

As per IEC 60076, the altitude of installation for normal service conditions is 1000 m above sea level. When the altitude goes up, the air density comes down, and this affects the voltage withstand capability and heat dissipation. Considering this, correction factors are specified by the standards. For internal insulation, it is assumed that the atmospheric air conditions do not influence the insulation or heat transfer properties. The correction factors applicable for oil-filled transformers are as follows.

TABLE 30.2

Conductor Width and Loss Concentration Factor

	Value of Q		
Conductor Width, mm	Transformer Up to Small 10 MVA Transformers	Transformer >10 MVA Up to 100 MVA	100 MVA
8	1.09	1.15	1.2
10	1.15	1.25	1.3
12	1.20	1.30	1.5
14	1.25	1.45	1.6
16	1.30	1.55	1.75

30.2.1 CORRECTION FACTOR FOR TEMPERATURE RISE

If the installation site is more than 1000 m above sea level, but the factory laboratory is not 1000 m above sea level, the allowable temperature rise shall be reduced as follows.

a. For natural cooling, the limit of temperature shall be reduced by 1 K for every 400 m by which the installation exceeds 1000 m.
b. For forced cooled transformers, the reduction shall be 1 K for every 250 m by which the installation exceeds 100 m.
c. A corresponding reverse correction can be applied if the altitude of the factory is above 1000 m and the altitude of installation is below 1000 m sea level.
d. The altitude correction factor is that the approximate values all be rounded to nearest whole number degrees.
e. For oil-immersed water-cooled transformers, the altitude correction factor is not required, if the cooling water temperature is not above 25°C.

30.2.2 CORRECTION FACTORS FOR EXTERNAL CLEARANCES

If the transformer is specified for operation at an altitude higher than 1000 m, the external clearances shall be increased by 1% for every 100 m by which the altitude exceeds 1000 m.

- This requires increased creepage distance of bushing and higher voltage class of bushing.
  - Altitude correction factors for impulse and power frequency test voltages are given in Table 30.3:
    where
    k_a = altitude correction factor
    H = altitude above sea level, m

TABLE 30.3

Altitude correction factors for test voltage as per IEC and ANSI/IEEE

IEC	ANSI/IEEE
$k_a = e^{\frac{H-1000}{8150}}$	$k_a = \dfrac{1}{-0.0001 \times H + 1.1}$
$u_p = k_a u_{pt}$	$u_p = k_a u_{pt}$
$u_d = k_a u_{dt}$	$u_d = k_a u_{dt}$

u_p = impulse test voltage to be applied, kV
u_{pt} = rated impulse test voltage at < 1000 m altitude, kV
u_{dt} = rated short duration power frequency test voltage at <1000 m altitude, kV

30.2.3 ALTITUDE CORRECTION FACTOR AS PER CIGRE REPORT 659—JUNE 2016 (TRANSFORMER THERMAL MODELLING)

Air density as a function of altitude:

$$\rho = \frac{352.984}{T}\left[1-2.2558H\right]^{5.25588} \tag{30.3}$$

where
ρ = air density at a height H meter above mean sea level, kg/m^3
T = temperature, K
H = altitude above mean sea level, m

a. For natural convection with laminar flow:

$$\frac{\Delta\theta_2}{\Delta\theta_1}=\left[\frac{\rho_1}{\rho_2}\right]^k=\left[\frac{1-2.2558\times10^{-5}H_1}{1-2.2558\times10^{-5}H_2}\right]^{5.25588} \tag{30.4}$$

$k = 0.5$–0.66 for AN cooling
$= 0.8$ for AF cooling
b. For forced convection:

$$\frac{\Delta\theta_2}{\Delta\theta_1}=\left[\frac{\rho_1}{\rho_2}\right]^{0.8}=\left[\frac{1-2.2558\times10^{-5}H_1}{1-2.2558\times10^{-5}H_2}\right]^{4.204704} \tag{30.5}$$

Example

Cooling	Air Forced
Average oil temperature rise	45°C
Altitude of installation	2500 m
Correction as per IEC	$\frac{2500-1000}{250}=6°C$

As per heat transfer formula,

$$\frac{\Delta\theta_2}{\Delta\theta_1}=\left[\frac{1-2.2558\times10^{-5}H_1}{1-2.2558\times10^{-5}H_2}\right]^{4.204704}$$

$H_1 = 0$, $H_2 = 2500$, $\Delta\theta_1 = 30$

$$\frac{\Delta\theta_2}{\Delta\theta_1}=\left[\frac{1}{1-2.2558\times10^{-5}\times2500}\right]^{4.204704}=1.276$$

$\Delta\theta_2 = 30 \times 1.276 = 38.28$

Altitude correction	$= 38.28-30$
	$= 8.28°C \approx 8°C$

30.3 EFFECT OF SOLAR RADIATION ON THE TEMPERATURE RISE OF OIL-FILLED TRANSFORMERS

30.3.1 SOLAR RADIATION

The top oil temperature rise and the winding temperature rise are affected by the solar radiation on the transformer surface, if the transformer is exposed directly to the sunlight. The incident power from the sun changes depending on the geographical location, the day of the year, the time of the day, and the atmospheric conditions prevailing at any given time.

The amount of solar radiation striking a perpendicular surface located at the mean sun earth distance above the earth's atmosphere is called the solar constant. This is approximately 1366 w/m². However, this value is reduced while passing through the atmosphere.

The maximum values at noon at extreme climate areas can reach values like 1200 w/m² in a clear day.

30.3.2 SOLAR RADIATION ON THE SURFACE OF THE TRANSFORMER

The solar radiation on the transformer surface at a given time is given by the following formula:

$$P_{sun}(t) = \varepsilon AR(t) \qquad (30.6)$$

where
   ε = solar radiation constant of the transformer surface. This depends on the surface condition, and typical values are given in Table 30.4.
   A = surface of the transformer exposed to sunlight, m².
   (This is taken as the area of the top cover.)
   This is assumed to be a constant irrespective of the movement of the sun.
   $R(t)$ = The radiation from the sun on the surface at time t, w/m².
   $R(t)$ is calculated by the following formula:
   $R(t) = K_s(\text{Sin } \alpha)^{1.15}$
   $K_s = 951.39$
   (Constant based on Adnot model for calculating global radiation in clear sky conditions)
   α = the elevation angle of the sun is calculated by the following equation:

$$\alpha = \left\{ \text{Sin} \left[\text{Sin}\delta.\text{Sin}(\text{Lat}) + \text{Cos } \delta.\text{Cos}(\text{Lat})\text{Cos}(\text{HRA}) \right] \right\}^{-1}$$

TABLE 30.4
Surface Condition and Solar Radiation Constant

Surface Condition	Solar radiation constant
White paint	0.25
Light cream paint	0.35
Aluminium paint	0.55
Light grey paint	0.75
Dark grey paint	0.95
Black paint	0.97

TABLE 30.5

Parameters Used for Calculating $\propto$

Parameter	Value of Parameter
LST (local solar time)	Local time + correction factor
HRA – hour angle°	15° (LST – 12)
δ = declination angle°	$23.45 \sin\left[\dfrac{360}{365}(\text{day} - 81)\right]$

- LST is calculated by adding a correction factor to adjust for the local time, which can differ from LST because of adjustments a day light saving.
- The parameter "day" for calculation of δ is the number of days in the year. 1st January is taken as 1, and 31st December is taken as 365.

---

where
δ = declination angle of the sun in degrees
(The angle between the equator and a line drawn from the earth to the sun)
Lat = latitude of the location in degrees
HRA = hour angle in degrees at the location
(This converts the local solar time [LST] into the number of degrees with the sun)

The parameters used for calculating $\propto$ are given in Table 30.5.

The solar power incident on the transformer is calculated using equation (30.6) on an hourly basis.

30.3.3 CALCULATION OF THE EFFECT OF SOLAR RADIATION ON OIL TEMPERATURE RISE AND WINDING TEMPERATURE RISE

If the solar radiation is not considered, the steady state hot spot temperature of the transformer is calculated as follows:

$$\theta_{hs}(t) = \theta_a(t) + \Delta\theta_{or}\left[\frac{1 + R\left(K(t)\right)^2}{1 + R}\right]^x + \Delta\theta_{hr}\left(K(t)\right)^y \qquad (30.7)$$

where
$\theta_{hs}(t)$ = steady-state hot spot temperature at time (t)
$\theta_a(t)$ = ambient temperature at time (t)
R = ratio of full-load loss to no-load loss
$K(t)$ = ratio of load current at time (t) to the rated current
$\Delta\theta_{hr}$ = hot spot gradient
= H.g_r
H = hot spot factor (as per IEC 60076-7)
g_r = average winding to average oil temperature gradient
x = oil exponent (as per IEC 60076-7)
y = winding exponent (As per IEC 60076-7)
θ_{or} = top oil rise at rated load

The effect of solar radiation is introduced to equation (30.7) by adding the solar radiation on the transformer surface normalized to the rated total loss of the transformer as per equation (30.8)

$$\theta_{hs_1}(t) = \theta_a(t) + \Delta\theta_{or}\left[\frac{1+R\left(K(t)\right)^2}{(1+R)} + \frac{P_{sun}(t)}{w}\right]^x + \Delta\theta_{hr}\left(K(t)\right)^y \qquad (30.8)$$

where

$\theta_{hs_1}(t)$ = steady-state hot spot temperature rise including effect of solar radiation

w = rated total losses

= no-load loss + rated full-load loss

From equation (30.8), the top oil rise and hot spot temperature rise of the transformer at any given time are estimated.

No-load loss of transformer	=	1.7 kw
Load loss of transformer	=	7 kw
Area of top cover	=	2 m^2
Top oil temperature rise at full rated load	=	45°C
Average winding rise at rated full load	=	50°C
Oil exponent (x)	=	0.8
Winding exponent y	=	1.6
For dark grey paint shade, ε	=	0.95
Latitude		24.45° (latitude corresponds to the location where transformer is installed)

Example

$\theta_a(t) = 50°C$ (ambient temperature)

Calculate the effect of solar radiation on 30th June at 11 am:

$$P_{sun}(t) = 0.95 \times 2\left(R(t)\right)$$

$$R(t) = K_s\left(Sin\alpha\right)^{1.15}$$

$K_s = 951.39$

$\alpha = \left\{Sin\left[Sin\delta.Sin(Lat) + Cos\delta.Cos(Lat).Cos(HRA)\right]\right\}^{-1}$

$\delta = 23.45\ Sin\left[\dfrac{360}{365}(day - 81)\right]$

Day = 30th June = 31 + 28 + 31 + 30 + 31 + 30 = 181

$$\therefore \delta = 23.45\ Sin\left[\frac{360}{365}(181 - 81)\right]$$

= 23.18

Lat = 24.45°

HRA = 15 [LST − 12] = 15[11−12] = −15°

Calculation for 11 am:

$\alpha = \left\{Sin\left[Sin\ 23.18.Sin\ 24.45 + Cos\ 23.18.Cos\ 24.45\right]Cos\ (-15)\right\}^{-1}$

$= \left\{Sin\left[0.3936 \times 0.4138 + 0.919 \times 0.91\right]Cos\ (-15)\right\}^{-1}$

$= Sin^{-1}\ 0.9649 = 74.8°$

$R(t) = 951.39 \left(\text{Sin } 74.8\right)^{1.15} = 0.96 \times 951.39 = 913.2$

$P_{sun}(t) = 0.95 \times 2 \times 913.2 = 1735 \text{ W}$

Hot spot temperature at full load and without considering the solar radiation at 11 am:

$$\theta_{hs}(t) = \theta_a(t) + \Delta\theta_{or} \left[\frac{1 + R(K(t))^2}{1 + R}\right]^x + \Delta\theta_{hr}(k(t))^y$$

$\theta_a(t) = 50°C$

$\Delta\theta_{or} = 45°C$ (top oil temperature rise)

$R = \dfrac{7}{1.7} = 4.11$

$K = 1$

$g_r = \text{gradient} = \text{average winding rise} - \text{average oil rise}$

$\Delta\theta_{hr} = H.gr$

$H.gr = 1.1 \times [50 - 0.8 \times 45] = 1.1 \times 14 = 15.4$

$$\therefore \theta_{hs}(t) = 50 + 45 \left[\frac{1 + 4.11}{1 + 4.11}\right]^{0.8} + 15.4$$

$= 50 + 45 + 15.4 = 110.4°C$

When the solar radiation of 1735 W is added to the losses, the hot spot temperature is calculated using the parameters as follows;

$$\theta_{hs_1}(t) = \theta_a(t) + \Delta\theta_{or} \left[\frac{1 + R\,(K(t))^2}{1 + R} + \frac{P_{sun}(t)}{W}\right]^x + \Delta\theta_{hr}(K(t))^y$$

$\theta_a(t) = 50°C$

$W = 7000 + 1700 = 8700$

$P_{sun} = 1735 \text{ W}$

$$\therefore \theta_{hs_1}(t) = 50 + 45 \left[\frac{1 + 4.11}{1 + 4.11} + \frac{1735}{8700}\right]^{0.8} + 15.4$$

$= 50 + 45 \left[1 + 0.199\right]^{0.8} + 15.4$

$= 50 + 52.05 + 15.4 = 117.45°C$

The comparison of temperatures is given in Table 30.6.

TABLE 30.6

Comparison of Temperatures

Parameter	Without Considering Solar Radiation	Considering Solar Radiation
Top oil rise at full load, °C	45	52.05
Average winding rise at full load, °C	50	50
Hot spot temperature rise at full load, °C	60.4	66.05
Top oil temperature, °C	95	102.05
Hot spot temperature, °C	110.4	117.45

30.4 CIRCULATING CURRENT IN TRANSFORMER

30.4.1 INTRODUCTION

Circulating current can be generated in transformers from different sources. The most common causes of circulating current and the possible solutions are given in the following paragraphs.

30.4.2 CIRCULATING CURRENT IN TANK

The leakage flux from the winding can set up different voltages at different parts of the tank, thereby creating localized circulating currents.

The mirror currents due to the high current loads or bushing can create circulating current in tank. These circulating currents can be brought down to nearly zero by providing magnetic or electromagnetic shields on the tank.

30.4.3 CIRCULATING CURRENT ACROSS TANK BAND

A heavy nonuniform leakage flux distribution can create a circulating current across the tank band, which can overheat the bolts and even cause melting of the bolts in extreme cases. Isolating the tank bolts or bypassing the bolts using high conductivity paths, is the solution

30.4.4 CIRCULATING CURRENT IN WINDING WITH PARALLEL CONDUCTORS

When parallel conductors are used for the winding, unequal impedance between the parallel strands is likely to arise, and this can cause circulating currents in the winding due to the unequal voltages induced in parallel strands from the radial and axial leakage flux. If the leakage flux distribution is not symmetrical along the axial direction, the magnitude of circulating current is increased.

30.4.5 CIRCULATING CURRENT IN CORE CLAMPING FRAME

The leakage flux distribution on metallic core clamping frame can create circulating current, if it is connected to earth at more than one point. If the core clamping frame is earthed or touching to the tank at more than one point, it causes circulating current. Single-point earthing of the metallic core clamping frame is essential to avoid it.

30.4.6 CIRCULATING CURRENT FROM THE TANK TO EARTH

The circulating current from the tank to earth can reach high values especially if the transformer supplies nonlinear loads, unbalanced loads, etc. Isolating the tank from the ground and single-point earthing are required to avoid this circulating current.

30.4.7 CIRCULATING CURRENT FROM CORE LAMINATIONS

The mirror current from the winding nearest to core can create circulating current in the core lamination. This can be solved by designing the winding arrangement with cancellation effect.

30.4.8 CIRCULATING CURRENT FROM GEOMAGNETICALLY INDUCED CURRENTS

Once in 10–12 years, turbulence in the sun causes change in the atmosphere near the polar region, and this generates direct current (DC) flow to earth, which enters through transformer neutral. This current travels through the transmission line and returns to earth through another transformer on the same line. This current flow creates core saturation, noise, high current, etc.

The problem of geomagnetically induced current is solved by installing DC filters at transmission line neutrals.

30.5 ELECTRICAL AND MAGNETIC FIELDS OUTSIDE THE TRANSFORMERS

30.5.1 ELECTRIC FIELD

The electric field coming outside the transformer is very low as the metallic tank/enclosure acts as a shield (Faraday cage).

Dry-type transformers of IP00 execution will be necessarily installed in secured closed rooms, and the interference caused by electric field is low.

30.5.2 MAGNETIC FIELD

Interferences can be caused by magnetic fields. Typically the magnetic field beyond 3 m distance of a distribution transformer is considerably low.

The Federal Environmental Protection Law allows a maximum magnetic flux density of 100 micro Tesla at the exposure point for a 50 Hz field. However, computer monitors can get interference at 1 micro Tesla also.

The approximate value of magnetic field of a distribution transformer can be calculated using the following empirical equation.

a. For oil-filled distribution transformers:

$$B = 14 \ Z \ \sqrt{\frac{kVA}{1000}} \ x^{-2.8} \tag{30.9}$$

where
B = magnetic field at a distance of X m from the transformer tank surface, micro Tesla
Z = percent impedance of the transformer
x = distance from transformer, m

b. For dry-type distribution transformer:

$$B = 22 \ Z \ \sqrt{\frac{kVA}{1000}} x^{-2.8} \tag{30.10}$$

x is the distance from enclosure in meters.

30.6 FAULT CURRENT OF TRANSFORMERS

The following table shows the fault currents (per unit values) on high-voltage and low-voltage (LV) sides of three-phase distribution transformers. The system impedance is neglected for this calculation. The fault current for the connections Dyn11 and Yzn5 used in some distribution systems are shown in Tables 30.8–30.10.

The assumptions made are as follows:

a. The source voltage is 1.0 PU.
b. The calculations are done for the rated tap position.

c. Neutral conductor impedance from transformer terminal to fault location is neglected.
d. Three-legged core form construction is assumed.

Zero sequence reactance and resistance values are assumed as per Table 30.7.
The aforementioned results are obtained by solving the equations as follows:

a. Bolted three-phase fault on LV:

$$I_{a1} = \frac{V}{Z_{s1} + Z_t}$$

(30.11)

$$I_{a2} = 0$$

$$I_{a0} = 0$$

$$I_{a1} = I_s = \frac{1 \angle 90°}{1 \angle 60°} = 1.0 < 30° (\text{PU})$$

TABLE 30.7
Zero Sequence Reactance and Resistance Ratio

Vector Group	X_0/X_t	R_0/R_t
Dyn11	0.95	1
Yzn5	0.1	0.4

TABLE 30.8
Three-Phase Secondary Fault

Vector Group	Primary Line fault Current (PU)	Secondary Line fault Current (PU)
Dyn11	$I_A = 1.0 \angle 180°$	$I_a = 1.0 \angle 30°$
	$I_B = 1.0 \angle 60°$	$I_b = 1.0 \angle 270°$
	$I_C = 1.0 \angle 300°$	$I_c = 1.0 \angle 150°$
Yzn5	$I_A = 1.0 \angle 180°$	$I_a = 1.0 \angle 30°$
	$I_B = 1.0 \angle 60°$	$I_b = 1.0 \angle 270°$
	$I_C = 1.0 \angle 300°$	$I_c = 1.0 \angle 150°$

TABLE 30.9
Line to Line Fault on Secondary

Vector Group	Primary Line fault Current (PU)	Secondary Line fault Current (PU)
Dyn11	$I_A = 0.5 \angle 120°$	$I_a = 0.0 \angle 0°$
	$I_B = 0.5 \angle 120°$	$I_b = \sqrt{3}/2 \angle -60°$
	$I_C = 1.0 \angle -120°$	$I_c = \sqrt{3}/2 \angle 120°$
Yzn5	$I_A = 0.5 \angle 120°$	$I_a = 0.0 \angle 0°$
	$I_B = 0.5 \angle 120°$	$I_b = \sqrt{3}/2 \angle -60°$
	$I_C = 1.0 \angle -120°$	$I_c = \sqrt{3}/2 \angle 120°$

TABLE 30.10

Line to Neutral Fault on Secondary

Vector Group	Primary Line fault Current (PU)	Secondary Line fault Current (PU)
Dyn11	$I_A = 0.0 \angle 0°$	$I_a = 0.0 \angle 0°$
	$I_B = 0.585 \angle 90°$	$I_b = 1.013 \angle 270°$
	$I_C = 0.585 \angle 270°$	$I_c = 0.0 \angle 0°$
Yzn5	$I_A = 0.0 \angle 0°$	$I_a = 0.0 \angle 0°$
	$I_B = 0.795 \angle 91.6°$	$I_b = 1.377 \angle 273°$
	$I_C = 0.795 \angle 271.6°$	$I_c = 0.0 \angle 0°$

b. Bolted phase to phase fault on LV:

$$I_a = I_{a1} + I_{a2} = 0$$

$$I_{a1} = -I_{a2} = \frac{V}{2/Z_s + 2Z_t}, I_{a0} = 0$$

$$I_{a1} = -I_{a2} = \frac{1 \angle 90°}{2 \angle 60°} = 0.5 < 30°(PU)$$

c. Bolted phase to neutral fault on LV:

$$I_a = I_c = 0.0$$

$$a^2 I_{a1} = a\, I_{a2} = I_{a0}$$

$$I_b = 3\, a^2 I_{a1} = \frac{3\, a^2\, V}{2\, Z_s + 2\, Z_t + Z_{t0}}(Z_{s1} = 0)$$

$$I_b = \frac{a^2\, V}{Z_t} \times \frac{3}{2 + \dfrac{Z_{t0}}{Z_t}}$$

For Dyn11 transformer:

$$\frac{Z_{T0}}{Z_T} = 1 - \frac{0.5}{\left(\dfrac{R_t}{X_t}\right)^2 + 1} \times \left[1 + J\, \frac{R_t}{X_t}\right] \qquad (30.12)$$

For Yzn5 transformer:

$$\frac{Z_{T0}}{Z_T} = 0.4 - \frac{0.3}{\left(\dfrac{R_t}{X_t}\right)^2 + 1} \times \left[1 + J\, \frac{R_t}{X_t}\right] \qquad (30.13)$$

30.7 CONVERSION OF LOSSES, IMPEDANCE AND NOISE LEVEL MEASURED AT 50 HZ FOR TRANSFORMERS DESIGNED FOR 60 HZ AND VICE VERSA

30.7.1 INTRODUCTION

The transformer designed for operation at 60 Hz needs to be tested at the rated frequency of 60 Hz. However, if the test facility in the factory has only 50 Hz power source, the tests can be carried out at 50 Hz, and the test results can be converted by calculation to 60 Hz subject to mutual agreement between the manufacturer and the purchaser. Conversely, if the transformer is designed for 60 Hz operation and tested at 50 Hz power supply also, the test results can be converted by applying conversion factors.

30.7.2 CONVERSION FACTORS FOR NO-LOAD LOSSES AND NO-LOAD CURRENT

Factors for converting no-load loss measured at 50 Hz to 60 Hz are as per Table 30.11.

Factors for converting no-load loss measured at 60 Hz to 50 Hz are as per Table 30.12

For core losses of CRGO grades without domain refinement such as M5, M4 and MOH, the factors for conversion is 1.32–1.33 for the entire flux density range for 50 Hz to 60 Hz conversion.

For HiB-type core material and domain-refined core material, the factors for conversion from 50 to 60 Hz are about 1.31 for flux density up to 1.4 Tesla. However, the factor is reduced to 1.29 at 1.7 Tesla.

The factors as per the aforementioned tables are based on the average values. The excitation current conversion factor is considered as unity.

Where test voltage is applied at 50 Hz for a transformer designed for 60 Hz operation, the test voltage applied at 50 Hz shall be five-sixths of the rated voltage.

TABLE 30.11

Conversion Factors of No-Load Loss from 50 Hz to 60 Hz

	Conversion Factors for 50 Hz to 60 Hz	
	B = Flux Density	Factor
Single-phase transformer	B < 1.4 T	1.32
	B > 1.4 T	1.32–0.05 (B-1.4)
Three-phase transformer	B < 1.4 T	1.33
	B > 1.4 T	1.33–0.05 (B-1.4)

TABLE 30.12

Factors for Converting No-Load Loss Measured at 60 Hz to 50 Hz

	Conversion Factors for 50 Hz to 60 Hz	
	B = Flux Density	Factor
Single-phase transformer	B < 1.4 T	$\dfrac{1}{1.32}$
	B > 1.4 T	$\dfrac{1}{1.32 - 0.05\,(B-1.4)}$
Three-phase transformer	B < 1.4T	$\dfrac{1}{1.33}$
	B > 1.4 T	$\dfrac{1}{1.33 - 0.05\,(B-1.4)}$

Example

A three-phase transformer is designed for 60 Hz operation, and the tests are carried out by applying 50 Hz supply.

Rated voltage at 60 Hz = 34,500

Design flux density at 60 Hz = 1.6 Tesla = B

Test voltage to be applied at 50 Hz = $\dfrac{34,500 \times 5}{6}$ = 28,750

Measured no-load loss at 50 Hz = 15 kW

Correction factor for converting to 60 Hz = 1.33 − 0.05 (1.6 − 1.4) = 1.32

Corrected no-load loss at 60 Hz = 1.32 × 15 = 19.8 kW

No-load current measured at 50 Hz = 4 A

No-load current at 60 Hz = 4 A

30.7.3 IMPEDANCE OF TRANSFORMER

Impedance of the transformer consists of the resistive and reactive components. The resistive component is unchanged with frequency. The reactive component is proportional to the frequency.

Using this relation, the conversion of impedance measured at 50 Hz can be converted to 60 Hz.

For measuring the impedance voltage, applying five-sixths of the rated voltage gives very small error for transformers with high X/R ratios. The approximate error, for low X/R ratios, is given in Table 30.13.

TABLE 30.13

$\dfrac{X}{R}$ **Ratio and Error of Impedance Voltage**

$\dfrac{X}{R}$ Ratio	% Error of Impedance Voltage
5	3
10	1
15	0.5
20	0.25

30.7.4 CONVERSION FACTORS FOR LOAD LOSS

Factors for converting load loss measured at 50 Hz to 60 Hz are given in Table 30.14.

Factors for converting load loss measured at 60 Hz to 50 Hz are given in Table 30.15.

The aforementioned factors are based on average of different types of design. The factors have minor variation depending on whether the tank is unshielded or shielded. The approximate stray loss factors for different shielding arrangements are given in Table 30.16.

TABLE 30.14

Factor for Converting Load Loss from 50 Hz to 60 Hz Conversion Factors

	Conversion Factor
Winding eddy current loss	1.44
Stray loss	1.23
Winding eddy current + stray loss	1.34

TABLE 30.15

Factors for Converting Load Loss from 60 Hz to 50 Hz

	Conversion Factor
Winding eddy current loss	$\dfrac{1}{1.44}$
Stray loss	$\dfrac{1}{1.23}$
Winding eddy current + stray loss	$\dfrac{1}{1.34}$

TABLE 30.16

Type of Shield and Stray Loss

Type of Shield	Factors for Conversion from 50 Hz to 60 Hz
No tank shield	1.28
Magnetic shielding of tank	1.19
Conductive shielding of tank	1.23

If winding eddy current loss is calculated using established numerical methods, the factors for eddy current loss and stray loss can be applied by separating these losses. In the absence of accurate estimate of eddy current loss, the factor for eddy current + stray loss can be used.

Example

A transformer designed for 50 Hz operation is tested at 60 Hz, and the measured load loss and calculated I^2R loss at reference temperatures are as follows:

   Load loss at reference temperature = 80 kW
   (measured at 60 Hz)
   Calculated I^2R loss at reference temperature = 60 kW
   $\therefore$ Winding eddy current and stray loss at 60 Hz = 20 kW
   Corrected losses for 50 Hz:

Winding eddy current + stray loss $= 20 \times \dfrac{1}{1.34} = 14.92$ kW

I^2R loss = 60 kW
Total load loss = 60 + 14.9 = 74.9 kW

30.7.5   CONVERSION FACTOR FOR SOUND LEVEL

Sound level conversion for ONAN Transformer:

   When testing a transformer designed for 60 Hz operation at 50 Hz, the test shall be done by applying the voltage to get the design flux density, which is $\dfrac{5}{6}$ times the voltage corresponding to 60 Hz operation.

   For ONAN transformer, the sound level for 60 Hz operation is calculated by adding 3.6 dB to the 50 Hz measured value.

Similarly if the test is done at 60 Hz for a transformer designed for operation at 50 Hz, 3.6 dB is subtracted from the measured value at 60 Hz.

Sound level conversion for ONAF transformer:

If a ONAF transformer designed for 60 Hz is tested at 50 Hz, the sound level measured can be converted to 60 Hz by the following method.

It is assumed that the fans are operated at 60 Hz when the transformer is excited with 50 Hz supply for the sound level measurement.

Measured values;

L 50 ONAN = sound level at ONAN measured at 50 Hz.

L 50 ONAF = sound level measured at ONAF cooling by exciting transformer at 50 Hz and supplying fans at 60 Hz.

Calculations:

L60 ONAN = L50 ONAN + 3.6

$$L60 \text{ ONAF} = 10 \log_{10} \left[10^{0.1 \text{ L50 ONAF}} - 10^{0.1 \text{ L50 ONAN}} + 10^{0.1 \text{ L60 ONAN}} \right]$$

Example:

Measured sound level at 50 Hz:

L50 ONAN = 55 dB (A)

L50 ONAF = 61 dB (A)

Calculated sound levels at 60 Hz:

L60 ONAN = 55 + 3.6 = 58.6 dB

$$L60 \text{ ONAF} = 10 \log_{10} \left[10^{0.1 \times 61} - 10^{0.1 \times 55} + 10^{0.1 \times 58.6} \right]$$

$$= 62.2 \text{ dB}$$

30.8   IEC RATING OF TRANSFORMER DESIGNED FOR OPERATION AT HIGHER THAN IEC AMBIENT TEMPERATURE

When a transformer is designed for operation at an ambient temperature exceeding the limit specified by IEC, the top oil temperature rise and the average winding temperature rise are correspondingly reduced from the maximum specified by the standard. In case the customer specifies a reduced temperature rise limit for ambient temperature conforming to IEC, the transformer has inherent extra capacity.

These transformers are capable of delivering higher loads when working at ambient temperature as per the IEC specification. The ratings for ONAN and ONAF transformers can be calculated using the following formula.

Symbols used:

W_i = no-load loss, watts

W_c = full-load loss at 75°C, watts

θ_{o-iec} = top oil temperature rise as per IEC = 60°C
(for IEC ambient conditions)

θ_{w-iec} = average winding temperature rise as per IEC = 65°C
(for IEC ambient conditions)

θ_{oil} = top oil temperature rise of transformer (Design value) (°C)

θ_{wdg} = average winding temperature rise of transformer (Design value) (°C)

x = kVA (per unit value at IEC rating)
(per unit value)

$$x = \sqrt{\frac{(W_i + W_c)\left(\dfrac{\theta_{o-iec}}{\theta_{oil}}\right)^{1.25} - W_i}{W_c}}$$ (30.14)

$$x = \sqrt{\left(\dfrac{\theta_{w-iec}}{\theta_w}\right)^{0.625}}$$ (30.15)

The IEC rating is the lower value of x, from equation (30.14) or (30.15)
The aforementioned calculation is applicable to oil-immersed distribution transformers.

Example

A 1000 kVA three-phase ONAN transformer is designed to operate at a maximum ambient temperature of 55°C. The top oil temperature rise and average winding temperature rise specified are 45°C and 50°C, respectively. Calculate the rating of this transformer when it is operating at IEC ambient temperature. The transformer no-load loss is 770 watts, and load loss is 10,500 watts.

CALCULATION

$W_i = 770\ W_c = 10{,}500$
   $\theta_{o-iec} = 60°C\ \theta_{w-iec} = 65°C$
   $\theta_{oil} = 45°C\ \theta_w = 50°C$
   Applying equation (30.1),

$$x = \sqrt{\frac{(770 + 10500)\left(\dfrac{60}{45}\right)^{1.25} - 770}{10500}} = 1.21$$

Applying equation (30.2),

$$x = \left(\frac{65}{50}\right)^{0.625} = 1.178$$

The IEC rating is 1.178 PU (lower value of x)
   = 1178 kVA

30.9 HEAT DISSIPATION FROM STEEL PREFABRICATED SUBSTATION

This procedure gives the calculation of heat dissipation from a prefabricated ventilated steel-fabricated substation. These substations are called unit substation, package substation, pocket substations, etc. where the configurations can have differences.

   This section shows a sample calculation of the heat dissipation from a ventilated prefabricated steel substation (PSS)

Description and Dimensions of the PSS
The substation includes a transformer, a LV panel and ring main unit interconnected.

Transformer = 1600 kVA
Phases = 3

Voltage ratio = 11000/400
Cooling = ONAN
Frequency = 50 Hz
No-load loss of transformer = 1800 W
Load loss of transformer = 15300 W
Loss at full load from LV panels = 2436 W
Length × width × height of the enclosure = 3000 × 3000 × 2800
Inlet louver area = 4.6 m²
Outlet louver area = 5.3 m²
Mid height of the transformer tank from enclosure bottom = 0.81 m
Mid height of the outlet louver from the tank bottom = 2.06 m
∴ Difference between midheight of the air outlet and midheight of the transformer = 1.25 m

$$\Delta\theta = \left[\frac{w^2}{(0.1 \times A_1)^2 \, H}\right]^{\frac{1}{3}}$$

(30.16)

where
$\Delta\theta$ = the temperature difference between the air outlet and air inlet in the enclosure
(the temperature class of enclosure)
w = total loss to be dissipated, kw
$H = 1.25$ m
(difference between the midheight of the air outlet to the midheight of the transformer)
A_1 = louver area of the inlet m²
= 4.6 m²
w = total loss to be dissipated
= 1800 + 15,300 + 2436 = 19536 W = 19.536 kW

$$\Delta\theta = \left[\frac{19.536^2}{(0.1 \times 4.6)^2 \, 1.25}\right]^{\frac{1}{3}} = 11.3°C$$

This temperature rise calculation is valid only if the area of the outlet louvers is at least 1.1 times the inlet louver.
Here, $A_2 = 5.3$ m² and $A_1 = 4.6$ m².

$$\therefore \frac{A_2}{A_1} = \frac{5.3}{4.6} = 1.152 > 1.1, \text{ hence OK}$$

30.10   TRANSFORMER RATING FOR SUPPLYING NONSINUSOIDAL LOAD

The rating of a transformer is calculated based on the sinusoidal values of current and voltage. All the parameters including losses and temperature rise are based on the sinusoidal values.

When a transformer supplies a nonsinusoidal load, the I^2R loss remains unchanged, but the eddy current losses and stray losses increase, depending on the magnitudes and frequencies of the different harmonics.

The harmonics are created by nonlinear loads.

Two approaches are followed by the standards to ensure that the transformer is suitable for supplying power to a nonsinusoidal load.

ANSI/IEEE C57.110 standard and UL 1562 approach this by defining the K factor, which gives the loss enhancement factor from the harmonics. A K factor of 1 means pure linear load. Standard K factor ratings of transformers are K4, K9, K13, K20, K30, K40, and K50.

For example, K13 rated transformer is capable of working with enhanced eddy current losses of 13 times the normal eddy current losses.

An example of the K factor calculation is shown in Table 30.17.

The symbols used are as follows:

I_1 = magnitude of fundamental current

I_n = RMS current including harmonics

$$I_n = \sqrt{\sum_1^n \left(\frac{I}{I_1}\right)^2} \tag{30.17}$$

The K factor is given by the summation of the last column of Table 30.17, which is 12.507.

The next higher standard K factor of 13 is selected. The transformer will have to be designed to supply nonsinusoidal loads consisting of the harmonic magnitudes shown in Table 30.17, without exceeding the temperature rise limits. The temperature rise test shall be carried out by applying the losses including the harmonic losses.

The European practice is to specify a K factor, which is for derating the transformer designed for supplying sinusoidal load. Alternatively the enhanced rating of the required transformer can be calculated from the K factor.

Table 30.18 shows the calculation of factor K, where the same magnitudes of harmonics used for the calculation of K factor in Table 30.17 are used.

q is taken as 1.7 (for foil winding, q = 1.5).

q is the exponent for the harmonic frequency for eddy current loss calculation

$$I_n = \sqrt{\sum_1^n \left(\frac{I}{I_1}\right)^2} = \sqrt{2.21} = 1.486$$

K factor is now calculated as follows:

TABLE 30.17

Calculation of K Factor

Harmonic Number n	Magnitude of Current I	$\frac{I}{I_1}$	$\left(\frac{I}{I_1}\right)^2$	$\frac{I}{In}$	$\left(\frac{I}{In}\right)^2$	$n^2\left(\frac{I}{In}\right)^2$
1	1	1	1	0.673	0.453	0.453
3	0.79	0.79	0.624	0.532	0.283	2.547
5	0.62	0.62	0.384	0.417	0.174	4.347
7	0.40	0.40	0.160	0.269	0.072	3.545
9	0.20	0.20	0.040	0.135	0.018	1.476
11	0.05	0.05	0.0025	0.034	0.0016	0.139
Total	-	-	2.21	-	-	12.507
			$I_n = \sqrt{2.21} = 1.486$			K factor

TABLE 30.18

Calculation of Factor K

Harmonic Number n	Magnitude of Current I	$\dfrac{I}{I_1}$	$\left(\dfrac{I}{I_1}\right)^2$	n^q	$n^q\left(\dfrac{I}{I}\right)^2$
1	1	1	1	1	1.00
3	0.79	0.79	0.624	6.473	4.039
7	0.40	0.40	0.160	27.332	4.373
9	0.20	0.20	0.040	41.900	1.676
11	0.05	0.05	0.0025	58.934	0.147
Total	-	-	2.21	-	17.158

$$\text{Factor K} = \left[1+\left(\frac{e}{1+e}\right)\left(\frac{I}{I_n}\right)^2 \sum_{n+1}^{n} n^q\left(\frac{I}{I_n}\right)^2\right]^{0.5} \tag{30.18}$$

e = eddy current loss from fundamental frequency as a ratio of I^2R loss from fundamental. e is assumed as 0.1 for the calculation.

$$\therefore \text{Factor K} = \left[1+\left(\frac{0.1}{1.1}\right)\left(\frac{1}{1.486}\right)^2 17.158\right]^{0.5}$$

$= 1.306$

$$\text{Derating factor} = \frac{1}{1.306} = 0.766$$

That is, a transformer designed for sinusoidal load application is to be derated by 76.6% in order to supply a nonsinusoidal load as per Table 30.18.

Alternatively if the customer specifies the total harmonic distortion (THD) in PU of the fundamental current, the PU values of harmonics from 3rd to 25th can be calculated by the following formula:

$$I_h = \frac{\alpha\, I_1}{h} \tag{30.19}$$

$$I_1 = \left(\frac{1}{1+(\text{THD})^2}\right)^{0.5} \tag{30.20}$$

$$\alpha = \frac{\text{THD}}{0.463} \tag{30.21}$$

I_1 = fundamental current.

   h = harmonic number (3, 5,7, 9, 11, 13, 15, 17, 19, 21, 23, and 25).

   The eddy loss enhancement factor is calculated from the following formula;

$$K = \left[1 + \sum_{h=2}^{25} \left(I_{\text{hpu}} \right)^2 h^{1.5} \right] \tag{30.22}$$

The exponent of h can be 2 instead of 1.5 as per UL, which gives a higher value of enhancement factor and higher factor of safety.

30.11 FUSES FOR TRANSFORMER PROTECTION

Fuses offer simple and economical protection of distribution transformers up to 36-kV class. The two classes of fuses are "backup current-limiting fuse" and "general-purpose fuse." Backup fuses are capable of interrupting from maximum interrupting rating to the minimum interrupting rating. General-purpose fuses can interrupt all currents, maximum interrupting current down to the current that causes the fuse element to melt in less than 1 hour.

   The fuse ratings are defined for 40°C ambient temperature. Under normal operating conditions and rated current, the body temperature rise shall be limited to 115°C, and the temperature rise of contacts shall be limited to 65°C.

   Outdoor-type fuses are used for protecting overhead transformers, whereas indoor-type fuses are installed on the primary of the transformer.

   The fuse current ratings shall be derated when the ambient temperature is above 40°C using the following formula.

$$k_\theta = \frac{1}{\sqrt{\dfrac{120 - \theta_a}{80}}} \tag{30.23}$$

where

   K_θ = derating factor

   θ_a = ambient temperature, °C

When the fuses are for application at an altitude above 1000 m above mean sea level, the correction factors for rated current as per Table 30.19 shall be used.

   The fuses shall be capable of withstanding 8–15 times the transformer rated current.

   The rated current of a fuse for protecting a distribution transformer can be calculated by the following formula:

$$I_n = 1.7 \; K_1 \; K_\theta$$

TABLE 30.19
Correction Factor for Rated Current

Altitude m	Correction Factor for Rated Current
1000	1
1500	0.99
2000	0.98
2500	0.97
3000	0.96

where

I_n = rated current of fuse

K_1 = transformer overload factor, which is for allowing short-time or long-time overload capability of the transformer

K_θ = temperature correction factor

The I_n is selected from the nearest standard fuse rating.

Table 30.20 gives the rated current ratings of fuse for various voltages of the transformer installed at altitude of maximum 1000 m and operating at a maximum ambient temperature of 40°C.

TABLE 30.20

Fuse Ratings for Three-Phase Outdoor Transformer

	Fuse Rating – A						
Three-Phase	Operating Voltage – kV						
Transformer (kVA)	3.3	6/6.6	10/11	13.8	15	20/22	30/33
25	16	10	6.3	6.3	6.3	6.3	6.3
50	25	16	10	10	10	6.3	6.3
63	25	20	16	10	10	6.3	6.3
80	31.5	25	16	16	10	10	6.3
100	40	25	20	16	16	10	6.3
125	50	31.5	25	16	16	16	10
160	50	31.5	25	20	16	16	10
200	63	40	31.5	20	20	16	16
250	80	50	40	25	25	20	16
315	100	63	50	31.5	25	25	16
400		80	63	40	31.5	25	20
500		100	63	50	40	31.5	25
630			80	50	50	40	31.5
800			100	63	63	50	1.5
1000				80	80	50	40
1250				100	100	63	50
1600						80	63
2000						100	80

30.12 TRANSFORMER SOUND LEVEL

30.12.1 INTRODUCTION

Sound is produced by vibrating objects, which reaches the ears of the listener through air or other media. Noise and sound are used interchangeably. However, the difference is that noise generally refers to unwanted sound. An energized transformer produces unwanted noise, which can create disturbance or nuisance to the people in the vicinity. National and international standards on the transformer have specified the maximum permissible sound levels of the transformer. The trend is toward reducing the sound levels of transformers, especially for applications in populated areas.

30.12.2 DETERMINATION OF SOUND LEVEL

The frequency spectrum of the sound is very wide. However, the audible frequency range to the human ear is between 20 Hz and 20 kHz. The measuring instrument is equipped with a filter to cover this range, and the measured sound level is called "A" weighted sound level. The frequency range is divided into octave bands.

The sound is determined by three parameters:

a. Sound pressure level L_p

$$L_P = 20 \log \frac{P}{P_O} \; dB \qquad (30.24)$$

where
L_P = sound pressure level in decibels
P_o = base sound pressure level = 20×10^{-6} pa
P = sound pressure level, pascals

b. Sound power level L_w

$$L_w = 10 \log \frac{W}{W_O} \; dB \qquad (30.25)$$

L_w = sound power level, dB
W_o = base sound power level, 10^{-12} watts
W = sound power level, watts

c. Sound intersity L_i

$$L_i = 10 \log \frac{I}{I_O} \; dB \qquad (30.26)$$

L_i = Sound intersity in dB
I_o = Base sound intersity 10^{-12} w/m²
I = Sound intensity w/m²

When the sound pressure level is measured by "A" weighted meter, it is designated as $L_p(A)dB$.

30.12.3 SOURCES OF SOUND FROM A TRANSFORMER

Sound is generated from various sources in a transformer. The magnetic core material has "magnetostriction," which changes cyclically with the frequency. The approximate relationship of the flux variation and magnetostriction over a cycle is shown in Figure 30.1. The magnetostriction of CRGO steel is 0.1 μm/m depending on the flux density.

The frequency spectrum of the sound consists of 2f, 4f, 6f, etc., where f is the supply frequency. Thus, the sound consists of frequencies, which are even multiples of the supply frequency. If the natural frequency of the core is near to any of these frequencies, there can be resonance effect and magnification of the sound.

Vibration of conductor produces sound, and this is another source of the transformer noise. The sound produced by the operation of the fans and pumps adds to the sound level of the transformer.

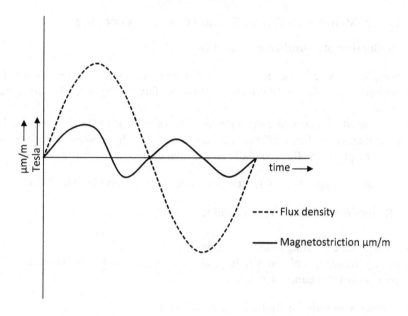

FIGURE 30.1 Magnetostriction and flux.

30.12.4 CALCULATION OF SOUND LEVEL OF TRANSFORMER

The following empirical formulas can be used for estimating the noise level of a transformer.

$$LWA_{core} = K_1 \log_{10} W_c + K_2 B + K_3 \tag{30.27}$$

where
 LWA_{core} = A weighted sound power level from core
 W_c = weight of core
 B = flux density of core
 K_1, K_2, K_3 = Constants

$$LWA_{winding} = P_1 + P_2 \log_{10}(KVA) + 40 \log_{10} \frac{I}{I_r} \tag{30.28}$$

where
 $LWA_{winding}$ = A weighted sound power level from winding
 P_1, P_2 = constants
 KVA = rating of transformer, kVA
 I = load current, A
 Ir = rated full-load current, A

$$LWA_{Fan} = LWA_o + 10 \log_{10} n \tag{30.29}$$

where
 LWA_{Fan} = A weighted sound power level from fan
 $LWAo$ = normal sound power level of one fan
 n = number of fans

30.12.5 Design Methods to Reduce Sound Level of Transformer

30.12.5.1 Reduction of Sound Level from Core

- The sound level reduces with reduction of flux density. However, it increases the cost of the transformer. A reduction of 0.1 Tesla of the core flux density gives 2–3 dB reduction of sound level.
- When compared with normal core material, HiB core offers a reduction of 2–3 dB.
- Step-lap construction of core offers a reduction of 5–6 dB when compared with mitred core.
- The clamping pressure of the core affects the sound level.

Applying epoxy at core edges (except for top yoke) reduces the sound level by 2–3 decibels.

30.12.5.2 Reduction of Sound from Winding

- Prestressing of the winding reduces the sound level.
- The natural frequency of the winding structure shall not be the multiples of power frequency so that resonance is avoided.

30.12.5.3 Other Methods for Reducing Sound Level

- Isolating core and coil assembly from tank by using antivibration pads give a reduction of 2–4 dB.
- Shielded tanks increase the sound level, if the shields are loose and vibrating.
- Increasing the tank wall mass gives reduction of sound level. A tank with stiffners has more rigidity, and it reduces the sound level.
- When substantial reduction of sound level is required, double-wall tank with sound absorbant material in between the double walls can be used, which gives a reduction of 12–15 dB.
- Use of cooling fans with low sound level reduces the transformer sound level.

The fans operating at low speed have low sound level. However, the airflow of a low speed fan is low when compared with a high-speed fan.

The impeller diameter, the speed, the airflow, and the sound level are interrelated.

30.12.6 Other Factors Affecting Sound Level of Transformer

30.12.6.1 DC Bias in Magnetization

If the core has DC bias current, the sound level of the transformer increases. This has no effect on the transformer in service as the initial DC bias at the time of energization disappears in few cycles of operation of the transformer. However, the DC bias current, if present, affects the sound level measurement. Table 30.21 gives the approximate increase of sound level with different magnitudes of DC bias.

TABLE 30.21
DC Bias Current and Increase of Sound Level

DC Bias (Per Unit of No-Load Current)	Increase in Sound Level dB(A)	
	1.6 Tesla	1.7 Tesla
0.2	2	4
0.4	3	6
0.6	5	8
0.8	6	10

30.12.6.2 Harmonics in Load Current

The sound power from the transformer depends on the magnitude of the currents of all frequencies. The fundamental and the most significant harmonics contribute to the total sound level.

30.13 DESIGN OF CONNECTION BUS BARS INSIDE TRANSFORMERS

This section gives the design procedure of connection bus bars of oil-filled and dry-type transformers of rated frequency of 60 Hz or less.

30.13.1 PERMISSIBLE CURRENT IN A BUS BAR

$$I = \frac{0.025 \times k_1 \times k_2 \times \theta^{0.6} \times S^{0.5} \times P^{0.4}}{\sqrt{\rho\left(1 + 0.004(\theta a + \theta - 75)\right)}} \tag{30.30}$$

where
I = maximum permissible current through the bus bar, A
ρ = resistivity of bus bar material at 75°C, Ohm cm
$= 2.10 \times 10^{-6}$ for copper
$= 3.12 \times 10^{-6}$ for aluminium
P = perimeter of bus bar $= 2(T + B)$, cm
T = thickness of bus bar, cm
B = width of bus bar, cm
θ_a = ambient temperature
θ = permissible temperature rise of bus bar
= top oil temperature rise for oil-filled transformers or
= average winding temperature rise $-$ 10 (for dry-type cast resin transformers)
S = cross-sectional area of bus bar $= B \times T$ cm^2
k_1 = factor for number of parallel bus bars per phase
= 1 for single bus bar per phase

$= 1.65 + \dfrac{1.24xT}{B}$ for two parallel bus bars per phase

$= 2.40 \dfrac{1.9xT}{B}$ for three parallel bus bars per phase

$k_2 = 0.75$ for nonventilated bus bar (example is bus bar inside cable box)
= 1.00 for oil-immersed transformer with natural oil cooling
= 1.55 for oil-immersed transformer with forced oil cooling

30.13.2 SHORT-TIME THERMAL CAPABILITY OF BUS BAR

The bus bar shall be capable of withstanding the thermal effect of the short-time fault current. The typical duration of this is 1 second, 3 seconds, 5 seconds, etc. The temperature of the bus bar after the specified duration shall be less than 250°C for copper and 200°C for aluminium.

$$\theta_{sc} = \theta a + \theta + \frac{0.24 \times Kc \times t(I_{sc})^2}{C \times D(n \times s)^2} \tag{30.31}$$

where

θ_{SC} = temperature of bus bar after the fault duration, °C
θ_a = ambient temperature, °C
θ = temperature rise of the bus bar under normal rated current, °C
$K_c = 2.10 \times 10^{-6}$ for copper, 3.12×10^{-6} for aluminium
$D = 8.9$ for copper, 2.7 for aluminium
C = specific heat of bus bar material
= 0.091 kcal/kg °C for copper
= 0.230 kcal/kg °C for aluminium
I_{sc} = short-time current (fault current through bus bar), A rms value
t = fault duration, seconds
n = number of parallel bus bars/phase.

30.13.3 NATURAL FREQUENCY OF BUS BAR

The intrinsic natural frequency of the bus bar shall not be within ±15% of the rated frequency and twice the rated frequency of the transformer.

$$f_n = 112\sqrt{\frac{EI}{ml^4}} \tag{30.32}$$

where
f_n = natural frequency of the bus bar, Hz
E = modulus of elasticity of the bus bar material, kg/cm^2
= 1.3×10^6 for copper, 0.67×10^6 for aluminium
m = mass/cm length of bus bar, kg
l = length between two supports/clamps of bus bar, cm.

30.13.4 SHORT-CIRCUIT FORCE BETWEEN BUS BARS

The dynamic peak short-circuit force between bus bars is calculated by the following formula:

$$F = 2\frac{l}{d}(I_{scd})^2 \times 10^{-8} \tag{30.33}$$

where
F = dynamic short-circuit force, kg
L = distance between two adjacent supports of the bus bar, cm
d = phase to phase center to center distance of bus bar, cm
I_{scd} = dynamic peak short-circuit current, A

For the primary side bus bars, the dynamic peak current corresponding to the primary system fault level shall be used for the calculation.

30.13.5 MECHANICAL STRENGTH OF BUS BAR UNDER SHORT CIRCUIT

The maximum bending stress of the bus bar against the dynamic peak short-circuit current is calculated by the following formula:

$$f_s = \frac{F \times l \times T}{24I} \tag{30.34}$$

where

f_s = maximum bending stress on the bus bar, kg/cm^2

F = force on the bus bar, kg

l = distance between adjacent supports of bus bar, cm

T = thickness of bus bar, cm

I = moment of inertia of the bus bar, cm^4, $= \dfrac{B}{12}T^3$ for rectangular bus bar

B = width of bus bar, cm

The maximum allowable stress on the bus bar shall be less than 1200 kg/cm^2 for copper and 500 kg/cm^2 for aluminium.

Sample calculation

Specifications;

Rated current through bus bar = 1333 A

Frequency = 50 Hz

Bus bar material copper

Bus bar length = 120 cm

Maximum permissible temperature rise of bus bar at rated current = 90°C

Ambient temperature = 40°C

Thermal fault withstand capacity = 25 kA for 3 seconds

Dynamic peak short-circuit current = 50 kA

The bus bar is for application on the secondary side of ventilated dry-type transformer.

   a. Design calculations

      Bus bar size selected = 8 × 0.8 cm copper

      Number of bus bars per phase = 1

      Center to center distance between bus bars = 4 cm

      Number of bus bar supports including supports at both ends = 5

      Distance between two adjacent supports $= \dfrac{120}{(5-1)} = 30$ cm

   b. Calculation of maximum permissible current through the bus bar

      $k_1 = 1$ (single bus bar)

      $k_2 = 1$ (ventilated bus bar)

      $\theta = 90°$

      $S = 8 \times 0.8 = 6.4\,cm^2$

      $P = 2(8 + 0.8) = 17.6$ cm

      $\rho = 2.1 \times 10^{-6}$ copper bus bar

      $\theta_a = 40°C$

      Applying the equation for the permissible current I,

$$I = \frac{1 \times 1 \times 0.025 \times 90^{0.6} \times 6.4^{0.5} \times 17.6^{0.4}}{\sqrt{2.1 \times 10^{-6}\left(1 + 0.004(40 + 90 - 75)\right)}} = 1851 \text{ A.}$$

      This is safe as the rated current is 1333 A.

   c. Calculation of short-time temperature rise

      $t = 3$ seconds, $I_{sc} = 25000$ A, $D = 8.9$, $C = 0.091$ kcal/kg°C, $S = 6.4\,cm^2$, $n = 1$,

      $k_c = 2.1 \times 10^{-6}$

Substituting in the equation for θ_{SC},

$$\theta_{sc} = 40 + 90 + \frac{0.24 \times 2.1 \times 10^{-6} \times 25000^2 \times 3}{(1 \times 6.4)^2 \times 0.091 \times 8.9} = 158.5°C$$

This is within the permissible short-time temperature of copper.

d. Calculation of natural frequency of the bus bar

The parameters for calculation of the natural frequency are

$l = 30$ cm, $E = 1.3 \times 10^6$ kg/cm^2, $m = $ mass/cm $= 8 \times 0.8 \times 8.9 \times 10^{-3} = 0.569$ kg/cm

$$I = \frac{8 x 0.8^3}{12} = 0.341 \ cm^4$$

Substituting in the equation for natural frequency,

$$f_n = 112\sqrt{\frac{1.3 x 10^6 x 0.341}{0.0569 x 30^4}} = 347.4 \, Hz$$

This is a safe frequency.

e. Short-circuit force between bus bars

The parameters for calculation are

$d = 4$ cm, $l = 30$ cm

$$F = \frac{2 \times 30 \times 50,000^2 \times 10^{-8}}{4} = 375 \, kg$$

f. Short-circuit stress of bus bar

Substituting the parameters for the stress,

$$f_s = \frac{375 \times 30 \times 0.8}{24 \times 0.341} = 1099.7 \text{ kg/cm}^2$$

This is within the safe limit of 1200 kg/cm^2 for copper bus bar.

30.14 CALCULATION OF WIND LOAD ON TRANSFORMER

30.14.1 INTRODUCTION

Wind flow has beneficial aspect of generating power. However, it creates a load on all structures, buildings, equipment, etc. exposed to it. This section gives the procedure for calculating the forces/load produced by wind on a transformer and the consequent design features required for the design of the transformer.

30.14.2 FACTORS AFFECTING THE WIND LOAD

In order to calculate the effect of the wind on a transformer, the following information are required.

- Wind velocity at the place of installation

    The data given by the customer or in the absence of this the wind velocity of the wind zone published by the national/international authorities will have to be used.
- The terrain and topography factor relevant to the place of installation, if this factor is available.

Structure importance factor

    This factor is related to the type of service/load the transformer is delivering.

TABLE 30.22

Terrain Types and Surface Friction Length

Terrain Number	Terrain Type	Zo m	Z min m
0	Coastal areas exposed to sea	0.003	1
I	Lakeshores and flat open areas with no obstacles	0.01	1
II	Area with low vegetation and isolated obstacles (average obstacle separations is more than 20 times obstacle height)	0.05	2
III	Villages and suburbs	0.3	5
IV	City centers and similar areas	1.0	10
	(more than 15% of the areas are covered with structures taller than 15 m)		

Basic wind velocity

The basic wind velocity is the value of the 10-minute average wind velocity measured at 10 m height in an open field that has a mean recurrence interval of 50 years. This is a probability 0.02 of exceeding the value.

The basic wind velocities for various regions are published by national and international authorities/organizations.

30.14.3 SURFACE ROUGHNESS COEFFICIENT

The surface roughness coefficient gives the effect of the terrain surface and height on wind velocity:

$$\text{For } Z > Z_{\min}, C_e = K_r \, \log_e \left(\frac{Z}{Z_o} \right) \tag{30.35}$$

$$K_r = 0.23 \left(Z_o \right)^{0.07} \quad \text{For } Z < Z_{\min}, C_e = C_e \left(Z_{\min} \right) \tag{30.36}$$

where

Z_o = surface friction length, m

$Z_{\min}$ = minimum friction length where the surface friction is constant, m

Z_o and $Z_{\min}$ values of the different terrain types are given in Table 30.22.

30.14.4 TOPOGRAPHY COEFFICIENT

This is calculated by the following formula:

$$C_t = 1 + 0.001 \Delta$$

where Δ = the height of the location above mean sea level. m

30.14.5 WIND VELOCITY CALCULATION

The wind velocity is defined as

$$V(Z, t) = U_m (Z) + W(Z, t) \tag{30.37}$$

where
$V(Z, t)$ = total wind velocity at height Z at time t
$V_m(Z)$ = average wind velocity at height Z
$W(Z, t)$ = turbulence of wind at height Z at time t
Notations
A = surface area, m²
$C_e(Z)$ = height-dependent friction coefficient
C_p = surface pressure coefficient
$C_p, 1$ = surface pressure coefficient for 1.0 m² area or less
$C_p, 10$ = surface pressure coefficient for 10.0 m² area or more
$C_q(Z)$ = height-dependent wind pressure coefficient
C_t = topography coefficient
d = width/length of the exposed surface perpendicular to wind direction
h = height of the exposed surface
$I_w(Z)$ = height-dependent turbulence factor
$q_p(Z)$ = wind pressure for unit area at height Z
V_b = basic wind speed
$V(Z, t)$ = total wind speed
$V(Z, t)$max = maximum total wind speed at height Z
W_{max} = maximum turbulence velocity
Z_o = surface friction coefficient
Z_{min} = minimum height in which surface friction is constant
ρ = mass density of air (= 12.5 kg/m³)
σ_w = standard deviation of turbulence

30.14.6 MAXIMUM WIND LOAD ON A SURFACE

$$Q(Z) = q_p(Z) \, C_p \, A \tag{30.38}$$

$$q_p(Z) = \frac{1}{2}\rho \, V_m{}^2(Z) + \rho \, V_m(Z) \, W_{max} \tag{30.39}$$

ρ = density of air ~ 12.5 kg/m³
A = surface area exposed to wind, m²
$C_p = C_p, 1$ for surface area of 1 m² or smaller
$C_p = C_p, 10$ for surface area of 10 m² or more
For surface area between 1 and 10 m²

$$C_p = C_p, 1 - (C_p, 1 - C_p, 10)\log_{10} A \tag{30.40}$$

The values of $C_p, 1$ and $C_p, 10$ are as per Table 30.23 for various values of h/d where h is the total height of the exposed surface and d is the length of the exposed surface perpendicular to the wind (Table 30.23).

For h/d > 5, use values of h/d = 5.
For other intermediate values, use interpolation.

30.14.7 EXAMPLE OF WIND LOAD CALCULATION

Maximum wind velocity V_b = 160 km/h
    Height of top of transformer from ground level = 6.5 m

TABLE 30.23

C_p, 1 and C_p, 10 values

h/d	C_p, 1	C_p, 10
5	1.0	0.8
1	1.0	0.8
<0.25 p	1.0	0.7

Center of gravity of transformer from ground level = 3.0 m
Exposure B, C_z at top = 0.878
Exposure B, C_z at bottom = 0.801
Velocity to pressure conversion coefficient

$$C_c = 4.72 \times 10^{-5}$$

Structure importance factor = 1.25

$$q_z = C_c\ C_T C_Z V_b^2 = 1.326\ \text{kN/m}^2\ \text{ at top}$$

= 1.210 kN/m² at bottom
L/B = 1.2, H/B = 1.11
Pressure coefficient C_p = 1.3
Gust coefficient C_g = 1.29
Design wind pressure Pz = $C_g\ C_p\ q_z$
= 2.244 kN /m² at top
= 2.029 kN/m² at bottom
Average wind pressure P_z = 2.12640 kN/m²
Force resulting from wind = 2.126 × 6.85 × 6.3
= 91.76 kN
Acting at a height 6.5–3 = 3.5 m from the top

The height of the transformer is 6.5 m and the wind force acts at a level 3.5 m below the top level. The seismic force applicable at the location also will have to be considered for the design of anchor bolt, which is explained in Section 26.9. The calculation of anchor bolt design is explained in Section 26.9.

A.1 Design of 1000 kVA, 11/0.4 kV, ONAN Transformer

A.1.1 SPECIFICATIONS

• kVA	-	1000
• Phases	-	3
• Frequency	-	50 Hz
• Cooling system	-	ONAN
• Primary volts	-	11000 (Delta)
• Secondary volts	-	400 (Star)
• Vector group	-	Dyn11
• Off circuit taps on HV	-	±7.5% variation in 2.5% steps
• Ambient temperature	-	50°C
• Top oil rise	-	45°C
• Average winding temperature rise	-	50°C
• HV BIL	-	75 kV
• AC test voltage	-	28 kV
• No load loss	-	850 W (Max)
• Load loss	-	6800 W (Max)
• Percentage impedance	-	5%
• Hermetically sealed with gas cushion		

A.1.2 LINE VOLTS, PHASE VOLTS AND CURRENTS

Description	LV	HV		
		Max. TAP	R TAP	L TAP
Line volts	400	11825	11000	10175
Phase volts	230.9	11825	11000	10175
Line current	1443.4	48.82	52.48	56.24
Phase current	1443.4	28.19	30.30	32.76

A.1.3 CALCULATION OF CORE CIRCLE, VOLT PER TURN, AND HV AND LV TURNS

Volt/turn $\sim 0.4\sqrt{kVA} \cong 0.4\sqrt{1000} = 12.649$

From the emf equation, the initial core circle diameter can be calculated.

The no-load loss required is comparatively low, and therefore, a flux density near to 1.6 T is assumed initially.

12.649 = 4.44 f B A

$f = 50\,\text{Hz}$

B = Flux density, initial value is taken as 1.6 T

$$\therefore A = \frac{12.649}{4.44 \times 50 \times 1.6} = 0.0356\;\text{m}^2$$

The core circle to get the net area of $0.0356\,\text{m}^2$ depends on the following factors.

- Core stacking factor of 0.97
- Number of steps of core laminations
- Requirement of standard widths of the core laminations

Assuming that 11 core steps are required and the widths need to be in multiples of 10, the first approximation of core circle is calculated as follows:

From Table 4.2, the maximum theoretical gross area feasible with 11 steps is 95.8% of the area of the circle, provided the optimum widths as per the table are used. Since the widths in multiples of 10 are required, the effective area is reduced. Therefore, 94% effective area is assumed for calculation,

$$A = \frac{\pi}{4}\,d^2 \times 0.97 \times 0.94.$$

Substituting the value of A,

$$d = \sqrt{\frac{4 \times 0.0356}{\pi \times .97 \times 0.94}} = 0.2229\;\text{m}.$$

∴ Core circle selected = 223 mm.

The widths of laminations are calculated using the factors from Table 4.2 and the sizes are rounded off to the nearest 10. The calculations of the widths, stack thicknesses and areas of cross section are shown in Table A1.1.

The net core cross section = $36135.3\,\text{mm}^2 = 0.03613\,\text{m}^2$

The actual working flux density depends on the number of turns of the winding:

LV Line Voltage = 400

LV phase voltage = $\dfrac{400}{\sqrt{3}} = 230.94$

Applying the initial volt / turn value of 12.649, the number of turns of LV $= \dfrac{230.94}{12.649} = 18.25$.

The number of turns on the LV will have to be an integer and 18 turns are selected.

$$\therefore \text{Volt/turn} = \frac{230.94}{18} = 12.83$$

TABLE A1.1
Core Step Width and Core Area

Step Number	Factor from Table 4.2	Core Circle × Factor	Lamination Width, mm (Nearest 10)	Stack Thickness, mm	Gross Area, mm²	Net Area, mm²
1	0.982	218.9	220	36.45	8020	7779.6
2	0.943	210.2	210	38.57	8101	7858.1
3	0.893	199.1	200	23.62	4723	4581.4
4	0.832	185.5	190	18.10	3439	3335.8
5	0.762	169.9	170	27.58	4689	4548.4
6	0.707	157.6	160	11.02	1762	1709.5
7	0.648	144.5	150	9.69	1454	1410.2
8	0.555	123.7	120	22.95	2754	2671.4
9	0.450	100.3	100	11.36	1136	1102.1
10	0.33	73.6	70	12.40	868	841.9
11	0.19	42.4	40	7.65	306	296.9
Total	-	-	-	219.39	37252	36135.3

TABLE A1.2
Number of Turns and Ratio Error

Tap Number	Volts	Turns	Ratio Error %
1	11825	922	0.072
2	11550	900	0.025
3	11275	879	0.091
4	11000	857	0.042
5	10725	836	0.111
6	10450	814	0.061
7	10175	793	0.008

The working flux density is now calculated using the actual volt/turn and actual net core area.

$$\text{Working flux density} = \frac{12.83}{4.44 \times 50 \times 0.03613} = 1.599 \text{ Tesla}$$

Calculation of the number of turns of HV
The number of HV turns and ratio errors are given in Table A1.2.

A.1.4 WINDING DESIGN

The load loss specified is comparatively low and therefore a low current density is selected. The low voltage winding is designed using copper foil and the high voltage winding is designed with rectangular paper-covered copper conductor. (All dimensions are in millimetres.)

Particulars	LV Winding	HV Winding
Number of turns per phase	18	922 (Highest) – 857 (Rated)
Turns/coil	18	922
Number of layers	18	17
Turns per layer	1	56
Bare width of conductor	560	9.6
Bare thickness of conductor	1.4	2.4
Insulated size of conductor	-	10×2.8
Area of cross section of conductor, mm²	784	22.49
Current density, A/mm²	1.84	1.35
Inner diameter of the winding	232	312
Interlayer insulation	0.25	0.125
Number of ducts	1	1
Size of the ducts	3	3
Radial thickness of the winding	30	53
Outer diameter of the winding	292	419
Winding axial height	560 (electrical)	560 (electrical)
	590 (with edge strip)	590 (with edge strip)
Gradient, °C	11	12
Resistance/phase Ω	4.03×10^{-4}	0.9246 (rated tap)
I²R loss (3 phase) at 75°C, watts	2518	2547
Eddy current loss, watts	97	535

Reactance Calculation

$$\%X = 8\pi^2 f \frac{I}{V} T^2 \, K\Delta \times 10^{-8}$$

$$= 8 \times \pi^2 \times 50 \times \frac{1443.4}{230.94} \times 18^2 \times 0.94 \times 69.15 \times 10^{-8}$$

$\%X = 5.19$

Load Loss Calculation (in watts)
LV I^2R Loss = 2518
LV Eddy current loss = 97
HV I^2R loss = 2547
HV Eddy current loss = 535
Stray loss on tank due to LV bushing = Not applicable (monoblock is used)
Tank loss = 540
Stray loss on tank due to HV Bushing = 32
HV and LV lead wire loss = 430
Total load loss = 6699 W
Guaranteed is 6800 W (about 1.5% Margin)

Calculation of Impedance

$$\% \, R = \frac{6699}{1000 \times 10} = 0.67$$

$\%X = 5.19$

$$\therefore \% Z = \sqrt{5.19^2 + 0.67^2} = 5.23\%$$

Guarantee is 5% with a+10% tolerance.

A.1.5 CORE WEIGHT AND CORE LOSS CALCULATION

Core circle diameter = 223 mm
Core window height = 620 mm
Core Leg centre = 431 mm
Core weight = 1140 kg
23 ZDKH 80 grade core material is selected

The specific loss at 1.6 T is 0.6 W/kg Building factor of 1.22 is considered for this grade with step lap construction
$\therefore$ Core loss = $1140 \times 0.6 \times 1.22 = 835$ W
The guarantee is 850 W (about 1.8% margin)

A.1.6 CALCULATION OF SIZE OF TANK

Tank length = $3 \times$ HV winding outer diameter +

$2 \times$ phase to phase clearance +

$2 \times$ winding to tank short side clearance

$= 3 \times 419 + 2 \times 12 + 2 \times 30$

$= 1341 = 1345$ mm

Tank width = HV winding outer diameter +

Clearance to tank on HV side +

Clearance to tank on LV side

$= 419 + 40 + 40 = 499$ mm

$= 500$ mm

Tank height = Core window height + $2\,W_{\max} + E + F + G$

$W_{\max}$ = Maximum step width of core
D = Minimum oil level above top yoke at the lowest ambient temperature and no load
$= 25$ mm for 11 kV
E = Oil level above top yoke at the time of oil filling at 30°C
G = Distance from the core bottom surface to tank bottom (60 mm for 11 kV)
F = Gas cushion above E

Note: If tap switch is mounted on top of the core, the minimum oil level (D) shall be calculated from the top level of the tap switch lead.

Assuming an initial value of gas cushion as 190 mm, the tank height and oil level inside the transformer are calculated.

Oil height (inside the tank)

$= G + 2 \times W_{max} + LL + E$

$= 60 + 2 \times 220 + 620 + 160 = 1280$ mm

Tank Height $= 1280 + 190 = 1470$ mm

Tank inner dimensions $= 1345 \times 500 \times 1470$ (L×W×H)

A.1.7   CALCULATION OF WINDING GRADIENT

By the specified top oil temperature rise of 45°C and average winding temperature rise of 50°C, the oil to winding gradient is $(50 - 0.8 \times 45) = 14$°C

The calculated gradients of HV and LV are HV gradient $= 0.075\ W_{HV}{}^{0.75} g_{HV}$

where $W_{HV} = $ W/m² of HV winding

$$= \frac{I^2 R \text{ loss} + \text{Eddy current loss of HV}}{\text{Cooling area of HV}}$$

$= 553$ W/m²

$\therefore g_{HV} = 0.075 \times 553^{0.75} = 8.55$°C

$$\text{LV gradient} = w_{LV} = \frac{I^2 R \text{ loss of LV} + \text{Eddy current loss of LV}}{\text{Cooling area of LV}}$$

$= 451.5$ W/m²

$\therefore g_{LV} = 0.095 \times 451.5^{0.75} = 9.3$°C

The calculated gradients are less than 14°C, and therefore, the cooling calculations can be proceeded based on 45°C top oil temperature rise.

A.1.8   CALCULATION OF COOLING AREA AND NUMBER OF RADIATORS

The total heat to be dissipated by the radiator and the tank surface (excluding the tank bottom plate area) $= 6699 + 835 = 7534$ W.

Only the wet area of the tank (i.e. the surface area of the tank up to the oil level) is considered. The top oil temperature rise allowed is 45°C, and assuming 2°C as a margin, the heat dissipation corresponding to 43° oil temperature rise is estimated.

Heat dissipation from the plane vertical surface with light grey paint is considered as 12 W/m²°C.

$\therefore$ Heat dissipation from the tank $=$ Tank area $\times 12 \times 43$

Tank area of vertical plate up to oil level

$= (1.3345 + 0.5)\ 2 \times 1.28 = 4.723$ m²

Heat dissipation from tank $= 4.723 \times 12 \times 43 = 2437$ W

Total loss to be dissipated $=$ No load loss $+$ load Loss

$= 6699 + 835 = 7534$

$\therefore$ Heat dissipation from the radiator $= 7534 - 2437 = 5097$ W

A.1.9   SELECTION OF THE SIZE AND NUMBER OF RADIATORS

The radiator top header pipe shall be located so that the centre line is at least 75 mm below the oil level at the lowest ambient temperature and there is no load. It is better to position the bottom radiator pipe at the bottom level of the winding. The centre-to-centre distance of the radiator is selected on this basis.

Centre-to-centre distance of radiator

$= 1280 - 75 - (60 + 220 + 15) = 910.$

The radiator centre is selected to be 900 mm and the radiator top level is $1280 - 75 = 1205$ mm (which is the height from the tank inner bottom and the height from the tank bottom is calculated by adding the thickness of the tank bottom plate).

A standard radiator panel width of 520 mm is selected. The heat dissipation of the radiator panel of 900 mm at 43°C is 358 W/section.

∴ Number of radiator sections required $= \dfrac{5097}{358} = 14.23$

A correction factor is to be applied to the number of radiator panels

$$\text{Correction factor} = K_V K_H K_N$$

K_V is the correction factor for thermal head. For this design thermal head is 185 mm and the corresponding K_V is 0.85.

∴ Number of radiator panels $= \dfrac{14.23}{0.85} = 16.74.$

It is proposed to use two radiators with nine panels each to achieve symmetry.

(K_H and K_N are unity.)

Oil quantity in one radiator panel of 900 mm $= 3.12$ L

∴ Oil quantity in 18 radiator panels $= 3.12 \times 18 = 56.16$ L

Oil quantity in the tank is calculated by subtracting the displacement volume of the core and coil assembly from the tank volume up to the oil level.

Tank volume up to oil level $= 1.28 \times .5 \times 1.345$

$= 0.8608 \, m^3$

$= 860.8$ L

The displacement volume is to be subtracted from this and is calculated in Table A1.3

∴ Total oil volume of transformer $= 860.8 - 391.9 + 56.16 = 525.06$ L

which is rounded off to 525 L.

A.1.10   CALCULATION OF MAXIMUM PRESSURE IN THE TANK

The transformer is sealed type with nitrogen gas cushion and therefore, the pressure on the gas cushion will go up and down with the increase or decrease of the oil temperature consequent to the load and ambient temperature.

TABLE A1.3
Displacement Volume of Oil

Material	Quantity in kg	Density g/cc	Displacement Litres
Core	1140	7.65	149
Copper	910	8.90	102
Insulation	110	1.10	100
Bottom channel (steel)	45	7.85	5.7
Top wooden beam	38	1.2	31.6
Other steel materials	28	7.85	3.6
Total displacement			391.9

Note: The oil quantity displaced by the insulation material includes the oil absorbed by the paper also.

The pressure in the gas cushion is calculated as follows:

$$\frac{P_1\,V_1}{T_1} = \frac{P_2\,V_2}{T_2}$$

where

P_1 = pressure at the time of filling = $1.0\,\text{kg}/\text{cm}^2$
V_1 = volume of gas cushion at the time of filling
T_1 = temperature at the time of oil filling
    = $373 + 30 = 403$
V_2 = volume of gas cushion at temperature T_2
P_2 = pressure of gas cushion at temperature T_2
T_2 = temperature at maximum ambient and full load
V_1 is the volume corresponding to 190 mm of gas cushion
V_1 = $0.19 \times 0.5 \times 1.345 = 0.127.7\,\text{m}^3 = 127.7\,\text{L}$
V_2 = V_1 – expansion of oil from temperature T_1 to T_2

Oil expansion is calculated based on the average oil temperature rise and by ignoring the effect of solubility of the nitrogen gas in oil with change in pressure and temperature
$V_2 = V_1 - 0.000718 \times 525\,[50 + 45 \times 0.8 - 30]$
   = $127.7 - 21.9 = 106.6\,\text{L}$
Substituting in the equation

$$\frac{1 \times 127.7}{(273 + 30)} = \frac{P_2 \times 106.6}{(273 + 50 + 45)}$$

$\therefore P_2 = 1.45\,\text{kg/cm}^2$
    This is $0.45\,\text{kg/cm}^2$ above the atmospheric pressure.
    The pressure relief device of $0.5\,\text{kg/cm}^2$ set pressure is selected. Corresponding to this, the tank test pressure of $0.7\,\text{kg/cm}^2$ is selected. Based on the above, the 190 mm gas cushion height assumed initially is sufficient.

A.1.11 CALCULATION OF TANK WEIGHT

a. Weight of side walls
     Select tank wall thickness = 5 mm
     Weight of tank side walls = $2\,[0.5 + 1.345]\,\dfrac{1.47 \times 7850 \times 5}{1000}$
     = 212.9 kg

b. Weight of top cover
     Length of the top cover = Tank length + 2 × tank curb
     = $1345 + 2 \times 50 = 1445$
     Width of the top cover = Tank width + 2 × Tank curb
     = $500 + 2 \times 50 = 600$
     Selecting 6 mm thickness for top cover,
     Weight of top cover = $1.445 \times 0.6 \times \dfrac{7850 \times 6}{1000} = 40.8\,\text{kg}$

c. Weight of tank curb
     Selecting 50×12 mm flat for the tank curb,
     Total length of the curb = $2[500 + 1345] + 4 \times 50 = 3890$
     Weight of the tank curb = $3.89 \times 0.05 \times \dfrac{12}{1000} \times 7850 = 18.3\,\text{kg}$

d. Tank bottom plate

Selecting 6 mm thickness for the bottom plate of the tank,

Length of bottom plate = inner length of the tank + 2×tank wall thickness

= 1345 + 2×5 = 1355

Width of the bottom plate = 500 + 2×5 = 510

Weight of the bottom plate of the tank = $1.345 \times 0.510 \times \dfrac{6}{1000} \times 7850 = 32.5\,\text{kg}$

e. Weight of tank stiffeners

Selecting 65×10 mm flat as stiffeners and providing four vertical stiffeners and two horizontal stiffeners for a safe withstanding of 0.7 kg/cm² test pressure, the stiffener dimensions and weights are calculated:

Length of Horizontal stiffener = (1345 + 50) 2×2 = 7380 mm

Weight of Horizontal stiffener = $7.38 \times 0.01 \dfrac{65}{1000} \times 7850 = 37.6\,\text{kg}$

Length of vertical stiffener = 1470×4 = 5880 mm

Weight of vertical stiffener = $5.88 \times 0.01 \times \dfrac{65}{1000} \times 7850 = 30\,\text{kg}$

f. Weight of base channel

Using two 150×75 mm base channels

Length of base channel = 2×(1345 + 250) = 3190 mm

Weight of base channel = 16.4×3.190 = 52.3 kg

(weight / m of 150×75 channel = 16.4 kg)

g. Weight of miscellaneous welded parts

This includes lifting lugs, small welded brackets and flanges for fixing radiators (radiator fixing pipes). For this design, the approximate weight of these items is 50 kg. The weights of cable boxes are calculated separately as these are not welded to the tank.

Total weight of the tank = a + b + c + d + e + f + g

= 212.9 + 40.8 + 18.3 + 32.5 + 30 + 52.3 + 50

= 436.8 = 440 kg

h. Weight of radiators

Number of radiators = 2

Number of radiator panels/radiator = 9

Total weight of radiator panels = 2×9×9.55 = 171.9 kg

A.1.12   CALCULATION OF WINDING TIME CONSTANT AND TRANSFORMER TIME CONSTANT

$$T_w = \frac{2.75\ g}{\left(1 + P_e\right) \delta^2}$$

where

T_w = winding time constant in minutes (for copper)

g = winding gradient

P_e = proportion of eddy current loss to I²R loss

For HV winding

$$T_w = \frac{2.75 \times 8.3}{1.35^2 \left(1 + \dfrac{535}{2547}\right)} = 10.3\ \text{minutes}$$

For LV winding

$$T_w = \frac{2.75 \times 9.3}{\left(1 + \dfrac{97}{2518}\right)1.84^2} = 7.3 \text{ minutes}$$

Calculation of Transformer Time Constant

$$T_0 = \frac{C \times \text{average oil temperature rise} \times 60}{P}$$

where

T_0 = Time constant of the transformer in minutes
C = 0.13 [weight of core and coil assembly] +
= 0.0882 [tank and radiator weight] + 0.40×weight of oil
P = No load loss+load loss

Solving for T_0
$\therefore T_0 = 200$ minutes

A.1.13 CALCULATION OF SHORT CIRCUIT STRENGTH

Thermal ability to withstand short circuit for 2 seconds. The system impedance is neglected as 1000 kVA is a category 1 transformer

$$\frac{\text{Rated current} \times 100}{\% \text{Impedance}}$$

$$= \frac{1443.4 \times 100}{5.23} = 27,597.9 \text{ A}$$

J = Short circuit current density for LV

$$= \frac{27,597.9}{784} = 35.2 \ A/mm^2$$

The winding temperature after 2 seconds of short circuit

$$\theta_{LV} = \theta_o + \frac{2\left(\theta_o + 235\right)}{\left(\dfrac{106000}{J^2\,t} - 1\right)} = 116°C$$

$$J \text{ for HV} = \frac{579.55}{22.54} = 25.7 \ A/mm^2$$

$$\theta_{HV} = 108.5°C$$

The short circuit temperatures of both HV and LV windings after 2 seconds are within the limits of 250°C.

Calculation of Winding Hoop Stress during Short Circuit

$$\text{Hoop stress} = \frac{0.031\, I^2 R \times 10^5}{Z^2 H}$$

where

Hoop stress is in kg/cm²

Z = Percent impedance

I = Phase current

R = Resistance of winding in ohms

H = Average winding height in mm

Hoop stress for LV = 174 kg/cm²

Hoop stress for HV = 173 kg/cm²

The Hoop stresses are safe values for copper winding

A.2 Design of 20 MVA, 33/11.5 kV, ONAN Transformer

A.2.1 SPECIFICATIONS

• kVA	-	20000
• Phases	-	3
• Frequency	-	50 Hz
• Cooling	-	ONAN
• Primary volts	-	33000 V (Delta)
• Secondary volts	-	11500 V (Star)
• Vector group	-	Dyn11
• On circuit tap changer on HV	-	+10% to −15% variation in 1.25% steps
• Ambient temperature	-	50°C
• Top oil rise	-	45°C
• Average winding temperature rise	-	50°C
• HV BIL	-	200 kV
• HV AC test voltage	-	70 kV
• LV BIL	-	75 kV
• LV AC test voltage	-	28 kV
• No load loss	-	13 kW (max)
• Load loss	-	80 kW (max)
• Percentage impedance	-	12.0%
• Maximum flux density	-	1.6 T

A.2.2 CORE DIAMETER AND VOLTS/TURN

Core diameter selected = 525 mm

Core net area with 13 core steps = 0.19478 m²

Initial volt per turn = 4.44 × 5. × 1.6 × 0.19478 = 69.1858

$$\text{Turns in LV} = \frac{11500}{\sqrt{3} \times 69.1858} = 95.966 = 96$$

$$\text{Actual volt/turn} = \frac{11500}{\sqrt{3} \times 96} = 69.1617$$

$$\text{Working flux density} = \frac{69.1617}{4.44 \times 50 \times 0.19478} = 1.599 \text{ T (which is below 1.6 T)}$$

A.2.3   DESIGN OF CORE AND COIL

The coarse–fine winding arrangement is selected for the onload tap changer. Rectangular paper-covered copper conductor is selected for the winding.

The LV winding is designed as a helical winding. The HV and tap windings are wound in two axially connected parallel windings.

Particulars	LV	HV	Tap
Phase voltage/coil	6639.5	28,010	4150
Phase current	1004	202	202
Number of turns	96	405	60
Winding type	Helical	Disc	Disc
Bare conductor size	7.5×1.75	12.2×1.7	12×2.25
Covered conductor size	7.9×2.15	12.7×2.2	12.5×2.75
Number in axial direction	2	1	1
Number in radial direction	19	3	2
Area of cross section of the conductor	472.3	122.28	105.8
Current density	2.13	1.65	1.91
Number of layers	2	-	-
Turns / disc	-	14 (13.5)	6
Number of discs	-	30	10
Insulation between discs	2	3	3
Extra blocks in winding	18	12	12
Block between coils	-	7.5	-
Shrinkage factor of insulation on conductor	0.9	0.9	0.9
Shrinkage factor for block	0.933	0.933	0.933
Electrical height of winding	949	951	322
Radial depth of winding	98	94.5	34
Length of conductor/phase (meter)	202.63	1130.9	207.7
Weight of bare conductor	2555	3692	586.8
Weight of covered conductor	2632	3816	602.8
Resistance/phase at 75°C	0.009052	0.19514	0.04143 / 0.04971
I²R loss (3 phase)	27,381	23,892	6087
Eddy current loss	2349	3758	20.3
Gradient	11.3	9.4	-
Number of blocks/row	20	20	20
Width of block	40	50	50
Clearance at top	112	120	132
Clearance at bottom	59	69	77

A.2.4   CORE WEIGHT AND CORE LOSS

Core diameter = 525
   Core window height = 1140
   Core leg centre = 1100
   Maximum step width of core = 520
   Total core weight = 13310 kg
   Grade of core = 30 MOH
   W/kg at 1.6 T = 0.85

Core loss = $13310 \times 0.85 \times 1.12 = 12671$ W

(For step lap construction of 30 MOH, a building factor of 1.12 is considered.)

Guaranteed maximum loss is 13000 W

A.2.5   LOAD LOSS

I^2R Loss = 57360

Eddy current loss = 6127

Lead wire loss = 689

Stray loss in tank = 11120

(unshielded tank)

Other stray losses = 3710

Total load loss = 79006 W (at rated top)

Guaranteed load loss = 80000 W

A.2.6   WINDING DIMENSIONS

Core diameter = 525 mm

$2 \times$ Core to LV = 26 mm

LV inner diameter = 551 mm

$2 \times$ LV radial depth = 196 mm

LV outer diameter = 747 mm

$2 \times$ LV–HV gap = 40 mm

HV inner diameter = 787 mm

$2 \times$ HV radial depth = 189 mm

HV outer diameter = 976 mm

$2 \times$ HV–tap gap = 36 mm

ID of HV Tap coil = 1012

$2 \times$ Radial depth of tap coil 68 mm

Outer diameter of HV tap coil = 1080 mm

Phase–phase clearance = 20

Core leg centre 1100

A.2.7   IMPEDANCE AT RATED TAP

Percent reactance = 11.665

Percent resistance = 0.40

Percent Impedance = 11.67

Guaranteed is 12% with IEC tolerance

A.3 Design of 72/90 MVA, 132/34.5 kV, ONAN/ONAF Transformer

A.3.1 SPECIFICATIONS

• kVA	-	72000/90000
• Phases	-	3
• Frequency	-	50 Hz
• Cooling	-	ONAN/ONAF
• Primary volts	-	132000 V (Star)
• Secondary volts	-	34500 V (Delta)
• Vector group	-	YNd1
• Off circuit taps on HV	-	±10% variation in 1.25% steps
• Ambient temperature	-	50°C
• Top oil rise	-	45°C
• Average winding temperature rise	-	50°C
• HV BIL	-	550 kV
• HV AC test voltage	-	230 kV
• LV BIL	-	170 kV
• LV AC test voltage	-	70 kV
• No load loss	-	40 kW (Max)
• Load loss	-	310 kW (Max)
• Percentage impedance	-	12.5%

A.3.2 CORE DIAMETER AND VOLTS/TURN

Select a core diameter of 795 mm and a core net area of 0.4372 m^2.

Selecting 1.65 T as flux density, the volt/turn is calculated as

volt/turn = $4.44 \times 50 \times 1.65 \times 0.4372 = 160.146$

In order to get low ratio error, it is a good practice to select integer number of turns for a tap step.

$$\text{Tap step voltage} = \frac{132000 \times 1.25}{\sqrt{3} \ 100} = 952.6279$$

$$\therefore \text{turns/step} = \frac{952.6279}{160.146} = 5.948$$

The turns per step is selected as 6.

Actual volt/turn = $952.6274 \div 6 = 158.771$

$$\text{Actual flux density} = \frac{158.771}{4.44 \times 50 \times 0.4372} = 1.635 \text{ T}$$

A.3.3 CALCULATION OF LINE VOLTS, PHASE VOLTS, LINE CURRENTS AND PHASE CURRENTS

Particulars	LV	HV
Line voltage	34500	132000
Line current ONAN rating	1204.9	314.92
Line current ONAF rating	1506.1	393.64
Phase voltage	34500	76210
Phase current ONAN rating	695.65	314.92
Phase current ONAF rating	869.56	393.64

A.3.4 ONLOAD TAP CHANGER AND CALCULATIONS OF TAP TURNS

Reversing switch type onload tap changer is selected for this design.

The tap winding is placed as the outermost winding. The number of turns per phase for LV and HV windings are shown in the following table.

Description	LV	HV Main + Tap		
		Max TAP	R TAP	MIN TAP
Volts	34500	83833	76210	69589
Turns	217	528	480	432
Current in the winding (ONAF rating)	869.56	357.85	393.64	431.1

The HV main winding has 480 turns and the tap winding has 48 turns. The main winding and tap winding conductor sizes are selected considering the maximum current in operation.

A.3.5 CONDUCTOR SELECTION

LV winding – continuously transposed conductor (CTC)
HV main winding – bunched conductor (one bunched conductor consists of $10 \times 1.8 \times 2$ radial)
HV tap winding – paper-covered conductor

Particulars	LV	HV Main	Tap
Turns per phase	217	480	48
Number of coils	1	2-axial parallel	2-axial parallel
Type of winding	Disc	Disc	Disc
Type of conductor	CTC	Bunched	PICC
Conductor size	6.8×1.7	10×1.8	16×2.2
Insulation covering	0.15+0.8 overall	0.4+0.6 overall	1.0
Number in axial direction	2	2 Bunched	1
Number in radial direction	14	2 Bunched	2
	(27 conductors)		
Area of cross section	302.4	141.17	139.36
Current density	2.87	2.79	2.82
Insulated conductor size	14.9×26.9	11×5	17×3.2
Number of discs	74	42	16
Turns per disc	2.93	11.42	3

(Continued)

Particulars	LV	HV Main	Tap
Block between discs	3	4	4
Extra block	6	40	20
Block between coils	-	6	300
Shrinkage factor of conductor insulation	0.9	0.9	0.9
Shrinkage factor for the block	0.933	0.933	0.933
Electrical height	1307	1302	970
Radial depth	82.5	122.5	20
Conductor length per phase (metre)	630.6	1827.5	226.6
Weight of bare conductor	5092	6886	843
Weight of covered conductor (kg)	5243	7236	886
Resistance per phase at 75°C	0.04372	0.2726	0.0343
I^2R loss (3 phase)	99184	126745	-
Eddy current loss	8892	13482	-
Size of the radial spacer	24 number×40	24 number×50	24 number×60
Clearance at top	163	163	329
Clearance at bottom	73	75	241

A.3.6 CORE WEIGHT AND CORE LOSS CALCULATION

Core diameter = 795 mm
    Window height = 1540 mm
    Core leg c-c = 1500 mm
    Maximum step width = 780 mm
    Total core weight = 41102 kg
    Core grade = 23 ZH 90
    W/kg at 1.635 T = 0.78
    Core loss = Core weight × w/kg × building factor
    = 41102 × 0.78 × 1.2 = 38471 W
    Guaranteed value is 40000 W (maximum)

A.3.7 LOAD LOSS AT ONAF RATING

I^2R Loss = 225929 W
    Eddy current loss = 22374 W
    Lead wire loss = 1458 W
    Stray loss on tank = 18,250 W
    (Shield is not provided on tank)
    Stray loss on core clamp, flitch plates etc. = 33910 W
    Total load loss = 302191 W
    Guaranteed load loss = 310000 W (maximum)

Note: This design is without shield on tank. If shield is provided the stray loss on tank is reduced.

A.3.8 WINDING DIMENSIONS AND CLEARANCES

Core diameter = 795 mm
    2 × Core to LV = 40 mm
    LV inner diameter = 835 mm
    2 × LV radial depth = 165 mm

LV outer diameter = 1000 mm 2×LV-HV main gap = 90 mm HV Main inner diameter = 1090 mm 2×HV Main radial depth = 245 mm HV main outer diameter = 1335 mm 2×HV main to HV tap = 86 mm

HV tap inner diameter = 1421 mm
2×HV tap radial depth = 40 mm
HV tap outside diameter = 1461 mm
Phase–phase clearance = 39 mm
Core leg centre = 1500 mm

A.3.9   REACTANCE AND IMPEDANCE

% Reactance at rated tap = 12.64

% Resistance $\dfrac{30,2191}{90,000 \times 10} = 0.34$

% Impedance 12.65

A.3.10   CALCULATION OF WEIGHT OF THE ACTIVE PART, TANK ETC.

Tank length = 4820 mm
Tank width = 2010 mm
Tank height = 3310 mm
Thickness of top cover = 20 mm
Thickness of side plate = 12 mm
Thickness of bottom plate = 16 mm

Weight of Core and Winding
Core weight = 41102 kg
Insulated conductor = 13366 kg
Core Frames = 4110 kg
Insulation = 2005 kg
Connections = 401 kg
Wooden parts = 802 kg
Hardware = 205 kg
Onload tap changer = 250 kg
Absorbed oil = 301 kg
Miscellaneous = 100 kg
Total core and winding weight = 62650 kg

Weight of Tank and Fittings
Tank and parts = 21260 kg
Radiator (without oil) = 12790 kg
HV line bushings = 369 kg
HV neutral bushings = 15 kg
LV bushings = 45 kg
Components = 991 kg
Total weight of tank and fittings= 35470 kg

Total Weight
Core and windings = 62650 kg
Tank and fittings = 35470 kg
Transformer oil = 28780 kg
Total weight = 126900 kg

A.4 Finite Element Methods for Transformer Design

A.4.1 INTRODUCTION

A transformer is a complex non-homogenous system consisting of coupled or interrelated entities like magnetic, electric, insulation, thermal components etc. The accurate estimate of any of the parameters of the transformer such as reactance, eddy current loss, stray loss, temperature rise, surge response, and seismic capability require the analysis of a very large number of nonlinear variables. The finite element method reduces the problem to one of a finite number of unknowns by dividing the problem into very small elements and then solving it by using assumed approximate functions. These functions are defined in terms of the small elements that divide the solution region. The corners of the elements are called nodes, and the nodes are connected to the adjacent elements. The behaviour of the variable within the element is defined by the nodal values of the variables and≈the interpolation function.

The finite element method formulates the solution for the individual element and puts them together for solving the entire problem. An example is the solution of the stiffness of a complex steel bar, where the force–displacement function in terms of the modulus of elasticity of small elements are calculated first and then the total elements are assembled as a matrix to solve the stiffness problem of the bar.

Although there are several procedures available for the solution of the finite element problem, the method of weighted residuals is commonly used for solutions of problems in electromagnetics, heat transfer, fluid dynamics etc.

A.4.1 STEPS INVOLVED IN SOLVING USING THE FINITE ELEMENT METHOD

A.4.2.1 DISCRETIZATION

A.4.2.2 Selection of Interpolation Function

The solution region is divided into a number of discrete elements. The shapes of the elements can be triangle, rectangle, tetrahedron etc. The number of elements, the shape of the element and the interlinking of the elements are decided at this stage.

A.4.2.3 Determination of the Properties of the Element

The nodes are assigned (numbered) and the interpolation functions are selected. The variation in the variable over the element is represented by this function. Normally polynomials are selected as interpolation functions.

A.4.2.4 Form System Equation

At this stage, the matrix equations of the individual elements are combined to form the solution matrix.

A.4.2.5 Initial Conditions and Boundary Conditions

The known values of the nodes and loads are substituted. The solution will have to be consistent with these values.

A.4.2.6   Solution

The set of linear or nonlinear equations are solved to get the unknown nodal values for a steady state problem. Same instances my require additional calculations to get the unknown values or functions based on the results.

A.4.3   APPLICATION TO TRANSFORMER DESIGN

The transformers are designed as per the customer requirements and the design process is complex as the parameters are linked and many being nonlinear. The outcome will have to comply with the customer specifications, standards and limitations of losses, impedance, temperature rise etc. The geometry of the core, winding and the associated systems are nonuniform. Accurate analytical calculations of the reactance, losses, leakage, fluxes, stresses etc. are not possible because of the complexities, and the finite element method can be used for better analysis of the parameters to derive an accurate and more reliable design.

This section gives the governing equations for the calculation of few key design parameters of the transformer using Finite Element Method.

A.4.3.1   Calculation of the Electromagnetic Field inside the Core Window

The axial and radial components of the flux distribution in the core window is necessary for the calculation of the eddy current losses, stray losses, forces etc. Also the magnetic field distribution inside the winding gap and that linking the winding are required for reactance calculation.

The electromagnetic field, vector potential current density and voltage are related by the following relationship:

$$\nabla \times \left(\frac{1}{\mu} \nabla \times A \right) = J - \gamma \frac{\delta A}{\delta t} - \gamma \nabla V$$

where
  A = Vector potential
  μ = Permeability
  J = Current density
  γ = Electrical conductivity
  V = Voltage

A.4.3.2   Calculation of Core Loss

The magnetic vector potential in the core considering a 2-dimensional system is given by

$$\frac{\delta}{\delta x} \left(V_y \frac{\delta A}{\delta x} \right) + \frac{\delta}{\delta y} \left(V_x \frac{\delta A}{\delta y} \right) = 0$$

where
  V_x, V_y = Reluctance of the core material in the x and y directions respectively

When eddy currents are present, the equation becomes

$$\frac{\delta}{\delta x} \left(V_y \frac{\delta A}{\delta x} \right) + \frac{\delta}{\delta y} \left(V_x \frac{\delta A}{\delta y} \right) = -J_o - J_e$$

where

$$J_e = -\sigma \frac{\delta A}{\delta t} + \frac{\delta}{\delta y} \iint_s s \frac{\delta A}{\delta t} d_s$$

J_e = eddy current density
J_o = magnetizing current density
σ = conductivity of lamination
S_t = cross sectional area of lamination
V_x = reluctance in the oriented direction
V_y = reluctance in the non-oriented direction

The effect of core joints can be evaluated by the following method:

$$b = B_1 \sin wt + \sum B_n \sin n \left(w_t + \theta_n\right)$$

where
b = instantaneous value of the disturbed flux at the corner
B = magnitude of the fundamental harmonic
B_n = amplitude of the nth harmonic
θ_n = phase angle of the nth harmonic

$$W = W_h + W_e$$

$$W_e = w_{ei}\left(B_e\right)$$

$$W_h = W_h(B_m) + 2\sum_1^m W_{ni}\left(B_{ki}\right)$$

W_h = hysteresis loss, w/kg
W_e = eddy current loss, w/kg
$W_{hi}\,(B_m)$ = function of hysteresis loss
$W_{ei}\,(B_e)$ = function of eddy current loss

$$B_e = \sqrt{\sum(n\,B_m)^2}$$

B_{ki} = amplitude of the minor loop and m is the number of minor loops.

A.4.3.3 Calculation of Heat Transfer of Oil Filled Transformer

Oil is considered as a non-compressible homogenous fluid with
$\nabla.\vec{U} = 0$

$$\left(\rho.\vec{U}.\nabla\right)\vec{u} + \nabla P - \mu\,\nabla^2\,\vec{U} = f$$

$$\nabla.\left(\rho C_P\,\vec{U}\,T\right) - \nabla.\gamma\,\nabla T = S_T$$

where
U = oil velocity vector
ρ = oil density
P = pressure
μ = dynamic viscosity of oil
γ = heat conductivity of oil
C_P = specific heat
f = external force vector
S_T = heat source

A.4.3.4 Temperature Rise of a Dry-Type Cast Resin Transformer

The heat transfer of a naturally cooled, dry-type cast resin transformer without enclosure (IPOO) is by conduction, convection and radiation. The conduction mode contributes the least of the total heat transferred. The total heat generated include the core losses and the winding losses including eddy current loss.

a. Heat transfer by conduction

$$\frac{\delta}{\delta x}\left(Kx\,\frac{\delta T}{\delta x}\right)+\frac{\delta}{\delta y}\left(Ky\,\frac{\delta T}{\delta y}\right)+\frac{\delta}{\delta z}\left(Kz\,\frac{\delta T}{\delta z}\right)+q^{1}=PC\,\frac{\delta T}{\delta t}$$

where

Kx, Ky, Kz = heat conductivity in the x, y, and z direction
T = temperature
ρ = density of the material
c = specific heat of the material
q^{1} = heat generated from unit volume

The thermal conductivities of the core material in the oriented direction and non-oriented direction are different.

$$q^{1}=\left(\frac{I^{2}\,R+e}{V}\right)\text{ for winding}$$

e = eddy current loss
V = volume of conductor in winding
I = rated current
R = resistance of winding

$$q^{1}=\frac{W_{core}}{V_{core}}$$

W_{core} = core loss
V_{core} = volume of core

b. Heat transfer by convection

$$-K\,\frac{\delta T}{\delta x}-K\,\frac{\delta T}{\delta y}=h\,(T_{2}-T_{1})$$

T_{1} = Ambient temperature or adjacent bulk temperature
  h = convection heat transfer coefficient
c = heat transfer by radiation

$$-K\,\frac{\delta T}{\delta x}-K\,\frac{\delta T}{\delta y}=\varepsilon\,\left(T_{2}^{4}-T_{1}^{4}\right)$$

ε = emissivity of the surface

A.5 Total Owning Cost (TOC) of a Transformer

A.5.1 INTRODUCTION

The transformer is expected to operate for more than 25 years under normal service conditions and it is one of the expensive assets in the power system. When a user buys a unit, comparing the initial cost and selecting the lowest priced asset is not a logical step. A comparison of the life time operating cost together with the initial buying cost is the optimum approach. The TOC method allows the buyer to evaluate the values of the losses of the transformer and consider the tax rate, borrowing rate, different energy rates etc.

A.5.2 BASIC TOC FORMULA

The simple version of the TOC formula includes the life time cost of no load loss and load losses discounted to the present value.

$$TOC = \text{Initial cost of transformer} + A \times \text{No load loss} + B \times \text{Load loss}$$

where

A = present value of the life time cost of one unit of no-load loss, \$/watt
B = present value of the life time cost of one unit of load loss, \$/watt
In the simple version of the TOC formula, the A and B factors are calculated based on the
    following variables:
P = cost of one unit of energy, \$/unit
If the unit is taken as 1 kwh, it is \$/kwh
n = number of years considered for loss evaluation
r = discounting rate (per unit value)

This can be taken as the interest rate for the simple version. Alternatively opportunity cost of the capital can be considered.
LF = load factor of the transformer.

Calculation of the A Factor
The no-load loss is fixed and it does not change with the load.

One watt of no-load loss $= \dfrac{1 \times 365 \times 24}{1000}$ kwh

= 8.76 kwh/year
The cost of one watt of no-load loss/year = 8.76 P
The same cost continues to occur for every year for the life time.
    Since the comparison is on present day cost, the future costs needs to be discounted to the present value.
    Assuming r as the discounting rate, the present cost of no-load loss for n years can be calculated to get the A factor in \$/W

$$\therefore A = \frac{8.76\,P}{r}\left[1 - \frac{1}{(1+r)^n}\right]$$

Calculation of the B Factor

The load loss changes in proportion to the square of the load current. Load factor gives the average load, and the load loss changes with the square of the load factor. From this, the cost of 1 watt of load loss discounted to the present value can be calculated.

$$B = (\text{LF})^2 \times A$$

An example of calculation of A and B factors is given here:

$P = 0.08$ \$/kwh
$r = 7\% = 0.07$ PU
$\text{LF} = 0.45$
$n = 20$ years

$$\therefore A = \frac{8.76 \times 0.08}{0.07} \left[1 - \frac{1}{(1.07)^{20}} \right]$$

$= 7.42$ \$/W
$B = 7.42 \times .45^2 = 1.50$ \$/W

A.5.3   TOC CALCULATION CONSIDERING SYSTEM AND COST PARAMETERS

The basic calculation in Section A.5.2 does not include the levelized energy cost, efficiency of transmission, fixed charge rate etc.

IEEE C.57.120 "IEEE Loss evaluation guide for Power Transformers and reactors" can be used as a guideline for determining the A factor and B factor where the basis of calculation of the cost parameters is explained.

Simplified version of the calculation of A and B factors which considers a system's capacity cost, fixed charges rate, efficiency of transmission etc. is as follows:

$$A = \left[\frac{C_{SC} + P \times 8760}{ET \times RFC \times IF} \right] \times \frac{1}{1000}$$

where
C_{SC} = System capacity cost, \$/kw/year
P = Levelized annual cost of energy, \$/kwh
R_{FC} = Fixed charge rate
R_F = Responsibility factor
L_F = Load factor
P_L = Uniform annual peak load
ET = Efficiency of transmission = 1 – transmission loss in PU
IF = Increasing factor
LLF = Loss load factor = $0.2 \, LF + 0.8 \, LF^2$

$$B = \left[\left\{ \frac{C_{SC} \times RF^2 + (P \times 8760 \, LF)}{ET \times R_{FC} \times IF \times 1000} \right\} \times PL^2 \right] \tag{A.5.1}$$

The following is how A and B Factors are calculated. The system data are also given.

C_{SC} = 180 \$/kw/year
P = 0.1 US \$/kwh
R_{FC} = 0.1571
ET = 0.892
IF = 1

$$R_F = 0.91 = \frac{\text{Transformer load at system peak load}}{\text{Transformer peak load}}$$

L_F = Load Factor = 0.48

LLF = 0.2 LF+0.8 LF^2
     = 0.2×0.48+0.8×0.48^2 = 0.28

$$P_L = \left[\frac{(LF)^2}{(LLF)}\right]^{0.5} = \left[\frac{0.48^2}{0.28}\right]^{0.5} = 0.91$$

Substituting the values,

$$A = \left[\frac{180+0.1\times8760}{0.892\times0.157\times1}\right] \times \frac{1}{1000} = 7.54 \text{ \$/W}$$

$$B = \left[\frac{180 \times 0.91^2 + (0.1 \times8760 \times0.28)}{0.89 \times0.1571 \times1 \times1000}\right] 0.91^2$$

= 2.34 \$/W

A.6 Comparison of IEC 60076 and ANSI/IEEE C.57.12 Standards

Though the IEC 60076 and ANSI / IEEE C.57.12 standards are similar, some provisions and clauses have differences that affects the design. This section brings out these differences. The ongoing attempts to bring both standards to full compatibility levels will bring down the differences in the days to come.

Sl. No.	Subject	IEC 60076- Standard	ANSI/IEEE C.57.12 Standard
1	Rated kVA	Defined as the apparent power input to the transformer	Defined as the output that can be delivered at the rated secondary voltage and frequency
2	Reference temperature for the calculation of load losses and impedance	75°C for liquid-filled transformer	85°C for liquid-filled transformer
3	Ambient temperature	• Ambient air temperature between −25°C and +40°C • Water temperature at inlet not exceeding 25°C for water-cooled design	• Ambient air temperature not exceeding 40°C and average air temperature for a 24-hour period not exceeding 30°C • Minimum water temperature shall not be less than 1°C
4	Tolerance for losses	Component losses +15% subject to the condition that total losses are within +10%	No load loss +10% Load loss +6%
5	Tolerance for impedance	For 2-winding transformers If the declared impedance is < 10%, tolerance allowed is ± 10% If the declared impedance is > 10%, tolerance of impedance is ± 7.5% (the above are applicable for the principal tapping) For 2-winding transformers with any other tapping (other than the principal tapping) • If impedance declared is < 10%, tolerance is ± 15% • If impedance value declared is > 10%, tolerance is ± 10% For multi winding transformers tolerance is ±10% at principal tap. For other taps or pair of windings tolerance ≥ ±15% (to be agreed between manufacturer and purchaser)	For 2-winding transformers • For impedance < 2.5%, tolerance is ± 10% • If impedance is > 2.5%, tolerance is ± 7.5% • Tolerance is ± 10% for 3 winding, zig zag winding and auto transformer
6	Temperature rises	Top oil temperature rise is 60°C Average winding temperature rise is 65°C Hot spot temperature rise is 78°C	Top oil temperature rise is 65°C Average winding temperature rise is 65°C Hot spot temperature rise is 80°C

(Continued)

Sl. No.	Subject	IEC 60076- Standard	ANSI / IEEE C.57.12 Standard
7	Tolerance ratio	a. At no-load and principal tapping, ±0.5% of specified ratio or ±1/10 of the impedance at principal tapping whichever is less b. For other tappings, ±0.5% of the design value of the turns ratio	a. 0.5% of the name plate voltage at no load b. When volts / turn of the winding exceeds 0.5% of the name plate voltage, turns ratio of the winding is taken as the nearest turn
8	Tolerance on no-load current	+ 30%	Not specified
9	Transport of transformer	During transportation, the transformer shall be capable of withstanding 1 g acceleration in all directions	Not specified
10	Climate conditions at place of installation	• Normal pollution rate • Ground acceleration level below 2 ms^2	Not specified
11	Transformer rating at on-load tap changer taps	All taps shall be full-rated taps unless otherwise specified	Transformers can have reduced capacity onload tap changer taps unless otherwise specified. This implies that if there is no specification, reduced kVA can be offered when OLTC is provided
12	Leak, vacuum and pressure test	Details of tightness test (leak test), vacuum test and pressure test with limits for deflection of tank are specified	• These test requirements are not covered • For sealed transformers the tank pressure under normal operating conditions shall not exceed 1 kg/cm^2
13	Load loss and temperature rise requirements	a. For tapping range within ±5% or rated power up to 2500 kVA, the guaranteed loss and temperature rise correspond to those of principal tapping b. For tapping above ±5% or rated power above 2500 kVA, the temperature rise is to be guaranteed at maximum current tapping	
14	Short circuit requirements	a. Transformers are categorized into three: Category I – 25 to 2500 kVA Category II – 2501–100,000 kVA Category III – Above 100,000 kVA b. Thermal ability to withstand is calculated for a 2 second duration of short circuit c. For category I transformers, the system impedance is ignored d. The short circuit test duration for dynamic withstand test is 0.5 second for category I and 0.25 second for category II and category III	a. Transformers are categorized into four: Category / Single-phase, kVA / 3-phase, kVA I / 5–500 / 15–500 II / 501 to 1667 / 501 to 5000 III / 1668 to 10000 / 5001 to 30000 IV / Above 10000 / Above 30000 b. Thermal ability to withstand short circuit is demonstrated by calculation c. The short circuit test duration for each shot is 0.25 second, except for the long duration test. For category I, the long duration test is for

(Continued)

Sl. No.	Subject	IEC 60076- Standard	ANSI / IEEE C.57.12 Standard
		e. For single-phase transformers, the number of shots are three, one each on the highest, rated and lowest tapping. For 3-phase transformers, the number of shots are 9 (3 each on the highest, rated and lowest tappings)	$t=\dfrac{1250}{I^2}$ seconds, where $I=$The symmetrical short circuit current as a multiple of normal base current. For category II, the long duration test is for 1 second and for category III the duration is 0.5 second. There is no long duration test for category IV transformers
15	Temperature correction factor for no-load loss	Not specified	Specified
16	Categories of voltage variation	Specified (CFVV, VFVV and combined voltage variation)	Not specified
17	Voltage harmonic level	Total harmonic content not to exceed 5% and even harmonic content not to exceed 1%	Not specified
18	Name plate information	a. Impedance values of extreme tappings to be included in the name plate in addition to the impedance at rated tapping for tapping ranges above+5% b. The information required on the name plate is specified by IEC	a. No specific requirement b. Three categories of name plates are specified c. Information required on the name plate include the following additional details (not a requirement in IEC) • Conductor material for each winding • Suitability for step up operations • Reference to installation and operating instructions • Statement that no PCB is used
19	Load rejection of generator transformer	During load rejection, the transformer shall withstand 1.4 times the rated voltage for 5 seconds	Not specified
20	Induced over voltage test	a. Test duration $$=\frac{120 \times \text{Rated Frequency}}{\text{Test frequency}}$$ Minimum test time is 15 second	a. Test duration=7200 cycles b. Minimum test time is not specified c. Maximum test time is 60 seconds
21	Partial discharge measurement	PD acceptance limit 250 pC	PD acceptance limit is 500 pC

Bibliography

1. RELEVANT IEC STANDARDS FOR TRANSFORMERS

IEC 60076-1-2011 Power Transformers Part 1 – General.
IEC 60076-2-2011 Power Transformers Part 2 – Temperature Rise for Liquid Immersed Transformers.
IEC 60076-3-2013 Power Transformers Part 3 – Insulation levels, dielectric tests and external clearances in air.
IEC 60076-4-2012 Power Transformers Part 4 – Guide to the lightning Impulse and switching Impulse testing – Power Transformers & Reactors.
IEC 60076-5-2006 Power Transformers Part 5 – Ability to withstand short circuit.
IEC 60076-6-2007 Power Transformers Part 6 – Reactors.
IEC 60076-7-2005 Power Transformers Part 7 – Guide for Loading Oil Immersed Power Transformers.
IEC 60076-8-1997 Power Transformers Part 8 – Application Guide.
IEC 60076-10-2016 Power Transformers Part 10 – Determination of Sound Levels.
IEC 60076-10-1-2016 Power Transformers Part 10–1 – Determination of Sound Levels application guide.
IEC 60076-11-2004 Power Transformers Part 11 – Dry type transformers.
IEC 60076-12-2008 Power Transformers Part 12 – Loading guide for Dry Type Power Transformers.
IEC 60076-13-2006 Power Transformers Part 13 – Self protected Liquid filled Transformers.
IEC 60076-14-2013 Power Transformers Part 14 – Liquid immersed Power Transformers using high temperature insulation materials.
IEC 60076-15-2015 Power Transformers Part 15 – Gas filled Power Transformers.
IEC 60076-16-2011 Power Transformers Part 16 – Transformers for wind Turbine applications.
IEC 60076-18-2012 Power Transformers Part 18 – Measurement of frequency response.
IEC 60076-19-2013 Power Transformers Part 19 – Rules for the determination of uncertainties in the measurement of the losses on Power Transformers and reactors.
IEC TS 60076-20-2017 Power Transformers Part 20 – Energy efficiency.
IEC 60076-21-2011 Power Transformers Part 21 – Standard requirements, terminology and test code for step voltage regulators.
IEC 60137-2008 Insulated Bushings for alternating voltage above 1000 volts.
IEC 60214-1-2003 Tap changers – Part -1.
Performance requirements and test methods.
IEC 60296-2003 Fluids for electro technical applications – unused mineral insulating oils for transformer and switchgear.
IEC 60721-34-1995 Classification of Environmental conditions.
Part 3: Classification of groups of environmental parameters and their severities. Section – 4- Stationary use at non-weather protected locations.
IEC 60310-2016 Traction transformers and traction reactors.
IEC 61378-3:2015 Converter transformers – Part 3: Application guide.
IEC 61378:2001 Converter Transformers for HVDC applications.
IEC 60270: 2000 High Voltage Test techniques – partial Discharge Measurement.
IEC 60027 Letter symbols to be used in Electrical Technology.
IEC 60028 International standard of resistance for copper.
IEC 60038 IEC Standard Voltages.
IEC 60044 Instrument Transformer.
IEC 60050 International Electro Technical Vocabulary.
IEC 60059 IEC Standard current ratings.
IEC 60060 High Voltage test techniques.
IEC 60071 Insulation Coordination.
IEC 60085 Electrical Insulation.
IEC 60099 Surge Arresters.
IEC 60233 Tests on Hollow insulators for use in Electrical Equipment.
IEC 60269 Low Voltage Fuses.
IEC 60298 High Voltage switchgear in Metallic Enclosures.
IEC 60317 Specifications for particular types of winding wires.

2. RELEVANT IEEE / ANSI STANDARD FOR TRANSFORMERS

IEEE Std 4-2012, IEEE Standard for High-Voltage Testing Techniques.

IEEE Std 259™-1999 (R2010), IEEE Standard Test Procedure for Evaluation of Systems of Insulation for Dry-Type Specialty and General – Purpose Transformers.

IEEE Std 638™-2013, IEEE Standard for Qualification of Class IE Transformers for Nuclear Power Generating Stations.

IEEE Std 1276™-1997 (R2006), IEEE Guide for the Application of High-Temperature Insulation Materials in Liquid-Immersed Power Transformers.

IEEE Std 1277™-2010, IEEE Standard General Requirements and Test Code for Dry-Type and Oil-Immersed Smoothing Reactors for DC Power Transmissions.

IEEE Std 1538™-2000 (R2005), IEEE Guide for Determination of Maximum Winding Temperature Rise in Liquid-Filled Transformers.

IEEE Std 1538™-2015, IEEE Guide for Determination of Maximum Winding – Temperature Rise in Liquid Filled Transformers.

IEC/IEEE 65700-19-03:2014, Bushings for DC application.

IEEE Std C57.12.00™-2015, IEEE Standard for General Requirements for Liquid-Immersed Distribution, Power and Regulating Transformers.

IEEE Std C57.12.01™-2015, IEEE Standard for General Requirements for Dry-Type Distribution and Power Transformers.

IEEE Std C57.12.10™-2010, IEEE Standard Requirements for Liquid- Immersed Power Transformers.

IEEE Std C57.12.10™-2010, Errata to IEEE Standard Requirements for Liquid-Immersed Power Transformers.

IEEE Std C57.12.10™-2010/Cor 1–2012, IEEE Standard Requirements for Liquid-Immersed Power Transformers Corrigendum 1: Correction of 5.1.9 Sudden Pressure Relay.

IEEE Std C57.12.10™-2010/Cor 2–2013, IEEE Standard Requirements for Liquid-Immersed Power Transformers Corrigendum 2: Correction of A.3.2.13 Autotransformer LTC application considerations.

IEEE Std C57.12.20™-2011, IEEE Standard for Overhead-Type Distribution Transformers 500 kVA and Smaller: High Voltage, 34,500 V and Below: Low Voltage, 7970/13 800Y V and Below.

IEEE Std C57.12.23™-2009, IEEE Standard for Submersible Single-Phase Transformers: 167 kVA and Smaller: High Voltage 25 000 V and Below: Low Voltage 600 V and Below.

IEEE Std C57.12.24™-2009, IEEE Standard for Submersible, Three-Phase Transformers, 3750 kVA and Smaller: High Voltage, 34,500 GrdY/19 920 Volts and Below: Low Voltage, 600 Volts and Below.

IEEE Std C57.12.28™-2014, IEEE Standard for Pad-Mounted Equipment-Enclosure Integrity.

IEEE Std C57.12.29™-2014, IEEE Standard for Pad-Mounted Equipment-Enclosure Integrity for Coastal Environments.

IEEE Std C57.12.30™-2010, IEEE Standard for Pole-Mounted Equipment Enclosure Integrity for Coastal Environments.

IEEE Std C57.12.31™-2010, IEEE Standard for Pole-Mounted Equipment-Enclosure Integrity.

IEEE Std C57.12.31™-2010/Cor 1–2014, IEEE Standard for Pole –Mounted Equipment-Enclosure Integrity Corrigendum 1: Correction to the SCAB Corrosion Test in 4.5.6.

IEEE Std C57.12.32™-2002(R2008), IEEE Standard for Submersible Equipment-Enclosure Integrity.

IEEE Std C57.12.34™-2015, IEEE Standard Requirements for Pad-Mounted, Compartmental-Type, Self-Cooled, Three-Phase Distribution Transformers, 10 MVA and Smaller; High-Voltage, 34.5 kV Nominal System Voltage and Below; Low-Voltage, 15 k V.

IEEE Std C57.12.35™-2013, IEEE Standard Bar Coding for Distribution Transformers and Step-Voltage Regulators.

IEEE Std C57.12.36™-2007, IEEE Standard Requirements for Liquid-Immersed Distribution Substation Transformers.

IEEE Std C57.12.37™-2015, IEEE Standard for the Electronic Reporting of Distribution Transformer Test Data.

IEEE Std C57.12.38™-2014, IEEE Standard for Pad-Mounted-Type, Self-Cooled, Single-Phase Distribution Transformers 250 kVA and Smaller: High Voltage, 34 500 GrdY/19 920 V and Below: Low Voltage, 480/240 V and Below.

IEEE Std C57.12.40™-2011, IEEE Standard for Network, Three-Phase Transformers, 2500 kVA and Smaller: High Voltage, 34 500 GrdY/19 920 and Below; Subway and Vault Types (Liquid Immersed).

IEEE Std C57.12.44™-2014, IEEE Standard Requirements for Secondary Network Protectors.

IEEE Std C57.12.51™-2008, IEEE Standard for Ventilated Dry-Type Power Transformers, 501 kVA and Larger, Three-Phase, with High-Voltage 601 V to 34 500 V; Low-Voltage 208Y/120 V to 4160 V-General Requirements.

IEEE Std C57.12.52™-2012, IEEE Standard for Sealed Dry-Type Power Transformer, 501 kVA and Higher, Three-Phase with High-Voltage 601 to 34500 Volts, Low-Voltage 208Y/120 to 4160 Volts-General Requirements.

IEEE Std C57.12.58™-1991(R2008), IEEE Guide for Conducting a Transient Voltage Analysis of a Dry-Type Transformer Coil.

IEEE Std C57.12.59™-2015, IEEE Guide for Dry-Type Transformer Through-Fault Current Duration.

IEEE Std C57.12.60™-2009, IEEE Standard Test Procedure for Thermal Evaluation of Insulation Systems for Dry-Type Power and Distribution Transformers, Including Open-Wound, Solid-Cast, and Resin-Encapsulated Transformers.

IEEE Std C57.12.60™-2009/Cor 1–2013, IEEE Standard Test Procedure for Thermal Evaluation of Insulation Systems for Dry-Type Power and Distribution Transformers, Including Open-Wound, Solid-Cast, and Resin-Encapsulated Transformers Corrigendum 1.

IEEE Std C57.12.70™-2011, IEEE Standard for Standard Terminal Markings and Connections for Distribution and Power Transformers.

IEEE Std C57.12.80™-2010, IEEE Standard Terminology for Power and Distribution Transformers.

IEEE Std C57.12.90™-2015, IEEE Standard Test Code for Liquid-Immersed Distribution, Power and Regulating Transformers.

IEEE Std C57.12.91™-2011, IEEE Standard Test Code for Dry-Type Distribution and Power Transformers.

IEEE Std C57.13™-2016, IEEE Standard Requirements for Instrument Transformers.

IEEE Std C57.13.1™-2006, IEEE Guide for Field Testing of Relaying Current Transformers.

IEEE Std C57.13.2™-2005(R2010), IEEE Standard Conformance Test Procedure for Instrument Transformers.

IEEE Std C57.13.3™-2014, IEEE Guide for Grounding of Instrument Transformer Secondary Circuits and Cases.

IEEE Std C57.13.5™-2009, IEEE Standard for Performance and Test Requirements for Instrument Transformers of a Nominal System Voltage of 115 kV and Above.

IEEE Std C57.16™-2011, IEEE Standard for Requirements, Terminology and Test Code for Dry-Type Air-Core Series Connected Reactors.

IEEE Std C57.17™-2012, IEEE Standard Requirements for Arc Furnace Transformers.

IEEE Std C57.18.10™-1998(R2003), IEEE Standard Practices and Requirements for Semiconductor Power Rectifier Transformers.

IEEE Std C57.18.10™-1998(R2003), Errata to IEEE Standard Practice and Requirement for Semiconductor Power Rectifier Transformer.

IEEE Std C57.18.10a™-2008, IEEE Standard Practices and Requirements for Semiconductor Power Rectifier Transformers Amendment 1: Added Technical and Editorial Corrections.

IEEE Std C57.19.00™-2004(R2010), IEEE Standard General Requirements and Test Procedure for Power Apparatus Bushings.

IEEE Std C57.19.00™-2004, Errata to IEEE Standard General Requirements and Test Procedure for Power Apparatus Bushings.

IEEE Std C57.19.01™-2000(R2005), IEEE Standard Performance Characteristics and Dimensions for Outdoor Apparatus Bushings.

IEEE Std C57.19.01™-2000(R2005), Interpretation to IEEE Standard Performance Characteristics and Dimensions for Outdoor Apparatus Bushings.

IEEE Std C57.19.100™-2012, IEEE Guide for Application of Power Apparatus Bushings.

IEEE Std C57.21™-2008, IEEE Standard Requirements, Terminology, and Test Code for Shunt Reactors Rated over 500 kVA.

IEEE Std C57.32–2015, IEEE Standard for Requirements, Terminology and Test Procedures for Neutral Grounding Devices.

IEEE Std C57.19™-2011, IEEE Guide for Loading Mineral-Oil-Immersed Transformers and Step-Voltage Regulators.

IEEE Std C57.93™-2007, IEEE Guide for Installation and Maintenance of Liquid-Immersed Power Transformers.

IEEE Std C57.94™-2015, IEEE Recommended Practice for Installation, Application, Operation and Maintenance of Dry-Type Distribution and Power Transformers.

IEEE Std C57.96™-2013, IEEE Guide for Loading Dry-Type Distribution and Power Transformers.

IEEE Std C57.98™-2011, IEEE Guide for Transformer Impulse Tests.

IEEE Std C57.100™-2011, IEEE Standard Test Procedure for Thermal Evaluation of Insulation Systems for Liquid Immersed Distribution and Power Transformers.

IEEE Std C57.104™-2008, IEEE Guide for the Interpretation of Gases Generated in Oil-Immersed Transformers.

IEEE Std C57.105™-1978(R2008), IEEE Guide for Application of Transformer Connections in Three-Phase Distribution Systems.

IEEE Std C57.106–2015, IEEE Guide for Acceptance and Maintenance of Insulating Mineral Oil in Electrical Equipment.

IEEE Std C57.109™-1993(R2008), IEEE Guide for Liquid-Immersed Transformer Through-Fault-Current Duration.

IEEE Std C57.110™-2008, IEEE Recommended Practice for Establishing Liquid-Filled and Dry Type Power and Distribution Transformer Capability When Supplying Nonsinusoidal Load Currents.

IEEE Std C57.111™-1989(R2009), IEEE Guide for Acceptance of Silicone Insulating Fluid and Its Maintenance in Transformers.

IEEE Std C57.111™-1989, Interpretation to IEEE Guide for Acceptance of Silicone Insulating Fluid and Its Maintenance in Transformers.

IEEE Std C57.113™-2010, IEEE Recommended Practice for Partial Discharge Measurement in Liquid-Filled Power Transformers and Shunt Reactors.

IEEE Std C57.116™-2014, IEEE Guide for Transformers Directly Connected to Generators.

IEEE Std C57.119™-2001, IEEE Recommended Practice for Performing Temperature Rise Tests on Oil-Immersed Power Transformers at Loads Beyond Nameplate Ratings.

IEEE Std C57.120™-1991(R2006), IEEE Loss Evaluation Guide for Power Transformers and Reactors.

IEEE Std C57.121™-1998, IEEE Guide for Acceptance and Maintenance of Less Flammable Hydrocarbon Fluid in Transformers.

IEEE Std C57.123™-2010, IEEE Guide for Transformer Loss Measurement.

IEEE Std C57.124™-1991(R2002), IEEE Recommended Practice for the Detection of Partial Discharge and the Measurement of Apparent Charge in Dry-Type Transformers.

IEEE Std C57.125™-2015, IEEE Guide for Failure Investigation, Documentation, Analysis and Reporting for Power Transformers and Shunt Reactors.

IEEE Std C57.127™-2007, IEEE Guide for the Detection and Location of Acoustic Emissions from Partial Discharges in Oil-Immersed Power Transformers and Reactors.

IEEE Std C57.129™-2007, IEEE Standard for General Requirements and Test Code for Oil-Immersed HVDC Converter Transformers.

IEEE Std C57.130™-2015, IEEE Guide for the Use of Dissolved Gas Analysis Applied to Factory Temperature Rise Tests for the Evaluation of Mineral Oil-Immersed Transformers and Reactors.

IEEE Std C57.131™-2012, IEEE Standard Requirements for Tap Changers.

IEEE Std C57.134™-2013, IEEE Guide for Determination of Hottest-Spot Temperature in Dry-Type Transformers.

IEEE Std C57.136™-2000(R2005), IEEE Guide for Sound Level Abatement and Determination for Liquid-Immersed Power Transformer and Shunt Reactors Rated Over 500 kVA.

IEEE Std C57.138™-1998(R2005), IEEE Recommend Practice for Routine Inpulse Test for Distribution Transformers.

IEEE Std C57.139–2015, IEEE Guide for Dissolved Gas Analysis in Transformer Load Tap Changers.

IEEE Std C57.140™-2006, IEEE Guide for the Evaluation and Reconditioning of Liquid Immersed Power Transformers.

IEEE Std C57.142™-2010, IEEE Guide to Describe the Occurrence and Mitigation of Switching Transients Induced by Transformers, Switching Device, and System Interaction.

IEEE Std C57.143™-2012, IEEE Guide for Application for Monitoring Equipment to Liquid-Immersed Transformers and Components.

IEEE Std C57.144™-2004, IEEE Guide for Metric Conversion of Transformer Standards.

IEEE Std C57.146™-2005, IEEE Guide for the Interpretation of Gases Generated in Silicone-Immersed Transformers.

IEEE Std C57.147™-2008, IEEE Guide for Acceptance and Maintenance of Natural Ester Fluids in Transformers.

IEEE Std C57.148™-2011, IEEE Standard for Control Cabinets for Power Transformers.

IEEE Std C57.149™-2012, IEEE Guide for the Application and Interpretation of Frequency Response Analysis for Oil-Immersed Transformers.

IEEE Std C57.150™-2012, IEEE Guide for the Transportation of Transformers and Reactors Rated 10000 kVA or Higher.

IEEE Std C57.152™-2013, IEEE Guide for Diagnostic Field Testing of Fluid – Filled Power Transformers, Regulators and Reactors.

IEEE Std C57.153™-2015, IEEE Guide for Paralleling Regulating Transformers.

IEEE Std C57.154™-2012, IEEE Standard for the Design, Testing and Application of Liquid – Immersed Distribution, Power, and Regulating Transformers Using High-Temperature Insulation Systems and Operating at Elevated Temperatures.

IEEE Std C57.155™-2014, IEEE Guide for Interpretation of Gases Generated in Natural Ester and Synthetic Ester-Immersed Transformers.

IEEE Std C57.157™-2015, IEEE Guide for Conducting Functional Life Tests on Switch Contacts Used in Insulating Liquid-Immersed Transformers.

IEEE Std C57.163™-2015, IEEE Guide for Establishing Power Transformer Capability while under Geomagnetic Disturbances.

IEEE Std C57.637™-2015, IEEE Guide for the Reclamation of Mineral Insulating Oil and Criteria for Its Use.

3. BOOKS & OTHER PUBLICATIONS

ABB Transformer and Engineering Services North America, *Service Handbook for Power Transformers*, 3rd Edition, ABB, New York, USA, 2010.

H Akagi, H Fujita, S Yonetani, and Y Kondo, "A 6.6 kV transformerless STATCOM based on a five-level diode-clamped PWM converter: system design and experimentation of a 200 V 10-kVA laboratory model", *IEEE Transactions on Industry Applications*, Vol. 44, No. 2, pp. 672–680, 2008.

S Akishita, S Kawanaka, K Moritsu, and H Shinagawa, "A new cooler with a low-noise fan for forced-oil forced air cooled large power transformers", *Mitsubishi Denki Giho* (Japan), Vol. 34, No. 3, pp. 37–40, March 1980. 3 refs (In Japanese) (1980–47136).

AM Alshehavy, DEA Mansoar, and M Ghali, "Condition assessment of aged transformer oil using optical spectroscopy techniques", 2016 *IEEE Conference on Electrical Insulation and Dielectric Phenomenon*.

EI Amoiralis, PS Georgilakis, TD Kefalas, MA Tsili, and AG Kladas, "Artificial intelligence combined with hybrid FEM-BE techniques for global transformer optimization", *IEEE Transactions on Magnetics*, Vol. 43, pp. 1633–1636, 2007.

C Bartoletti, M Desiderio, D Di Carlo, G Fazio, F Muzi, G Sacerdoti, and F Salvatori, "Vibro-acoustic techniques to diagnose power transformers", *IEEE Transactions on Power Delivery*, Vol. 19, No. 1, pp. 221–229, 2004.

G Bertagnolli, *Short – Circuit Duty of Power Transformers*, 3rd Revised Edition, ABB Management Services Ltd Transformers, Zurich, Switzerland, 2006.

Bharat Heavy Electricals Limited, *Transformers*, McGraw-Hill, New York, 2005.

LF Blume, A Boyajian, G Camili, TC Lennox, S Minneci, and VM Montsinger, *Transformer Engineering* – 2nd Edition, John Wiley & Sons, Inc., New York, 1951.

R Boehr, "Transformer technology state of the art and trends of future development", *CIGRE Electra*, No. 198, pp. 13–19, October 2001. 11 refs.

JL Brooks, "Solid state transformer concept development", Final Report, October 1978–September 1979, Naval Construction Battalion Center, Port Hueneme, CA.

JH Brunke, and KJ Frohlich, "Elimination of transformer inrush currents by controlled switching – part I: theoretical considerations", *IEEE Transactions on Power Delivery*, Vol. 16, No. 2, pp. 276–280, 2001.

"Cast resin insulated (GEAFOL) dry type transformers", *Electrical India*, Vol. 29, No. 4, 28 Feb. 1989, pp. 9–12, no refs (1989–79225).

LS Coelho, VC Mariani, FA Guerra, MVF Luz, and JV Leite, "Multi objective optimization of transformer design using a chaotic evolutionary approach", *IEEE Transactions on Magnetics*, Vol. 50, No. 2, pp. 669–672, 2014

"Color affects temperature", *Electrical World* (USA), Vol. 180, No. 5, September 1, 1973, p. 34. refs.

RM Del Vecchio, "Eddy-current losses in a conducting plate due to a collection of bus bars carrying currents of different magnitudes and phases", *IEEE Transactions on Power Delivery*, Vol. 39 No. 1, pp. 549–552, January 2003. 4 refs.

M Del Vecchio, B Poulin, PT Feghali, DM Shah, R Ahuja, *Transformer Design Principles*, 2nd Edition – With Application to Core-Form Power Transformers, CRC Press, Taylor & Francis Group, Boca Raton, FL, 2010.

JP Donlon, ER Moto, et al., The Next generation 6.5kV IGBT, Proceedings of 2014 IEEE Energy Conversion Congress and Exposition (ECCE) Pittsburg, Pennsylvania, USA, 14–18 September 2014.

L Fagarasan, S Costinas, and IS Stelian, "Monitoring and diagnostic methods for high voltage power transformer", *University Politechnica of Bucharest Bulleting Series* C, Vol. 70, No. 3, 2008.

WM Flanagan, *Handbook of Transformer Design & Applications*, 2nd Edition, McGraw-Hill Book Company, New York, 1993.

JAC Forrest, and B Allard, "Thermal problems caused by harmonic frequency leakage fluxes in three-phase, three-winding converter transformers", *IEEE Transactions on Power Delivery*, Vol. 19, No. 1, pp. 208–213, January 2004. 3 refs.

SL Foster, "How to get emergency transformer capacity", *Allis Chalmers Electrical Review* (USA), Vol. 21, No. 2, Second Quarter, pp. 16–23, 1956.

K Funaki, M Iwakuma, K Kajikawa, et al., "Development of a 22 kV/6.9 kV single-phase model for a 3 MVA HTS power transformer", *IEEE Transactions on Applied Superconductivity*, Vol. 11, No. 1, 2001, pp. 1578–1581.

J Fyvie, *Design Aspects of Power Transformers*, Arima Publishing, Bury S T Edmunds, Suffolk, UK, 2009.

GDC Industries, Catalogue (www.gdc-industries.com) Dayton, OH, USA, The covetics specialists.

PS Georgilakis, "Recursive genetic algorithm – finite element method technique for the solution of transformer manufacturing cost minimization problem", *IET Electric Power Applications*, Vol. 3, No. 6, pp. 514–519, 2009.

PS Georgilakis, *Spotlight on Modern Transformer Design*, Springer, New York, 2009.

HE Greene, "Transformers in hot climates", *Electrical Review* (GB), Vol. 208, no. 7, pp. 20–21, 20 February 1981. No. refs. (1981–22384).

R Haller and A Welsch, "Thermal modeling of enclosed cast resin transformers", *Elektrizitatswirtschaft* (Germany), Vol. 97, No. 10, pp. 39–45, 4 May 1998, 5 refs (In German) (1999–32388).

JH Harlow, *Electric Power Transformer Engineering*, 3rd Edition, CRC Press, Taylor & Francis Group, Boca Raton, FL, 2012.

RG Hatch and RCH Johnson, "The application of dry type transformers and their special features", *Electric Energy Conference*, Melbourne, Australia, 406, October 1981, Institution of Engineers, Australia, pp. 124–129. No. refs (1982–33655).

M Heathcote, *J & P Transformer Book*, 13th Edition, Elsevier Science, Oxford, Great Britain, 2011.

C Hernandez, MA Arjona, and JP Sturgess, "Optimal placement of a wall-tank magnetic shunt in a transformer using FE models and a stochastic – deterministic approach", *Proceedings of 12th Biennial IEEE Conference on Electromagn. Field Comput.*, Miami, FL., USA, 2006, pp. 206 ff.

B Hochart, editor, *Power Transformer Handbook*, Butterworths & Co. Ltd., London, 1987.

FL Holloway, and WH Mears, "A review of gaseous dielectric development", *Electrical Engineering* (AIEE, USA), Vol. 78, No. 2, pp. 137–140, February 1959. 16 refs.

JR Holmquist and JA Rooks, "Temperature and loading considerations for cast coil transformers a user's perspective", *1995 IEEE Pulp and Paper Conference*, Vancouver, B C, Canada, June 12–16, 1995, IEEE Pub. 95CH3572-5, pp. 142–147. 11refs (1995–72528).

M Horii, N Takahashi, and J Takehara, "3-D optimization of design variables in x-, y-, and z-directions of transformer tank shield model", *IEEE Transactions on Magnetics*, Vol. 17, No. 5, pp. 3631–3634, September 2001. 5 refs.

JE Huber, and JW Kolar, "Applicability of solid-state transformers in today's and future distribution grids", *IEEE Transactions on Smart Grid*; August 2017.

M Iwakuma, K Funaki, K Kajikawa, H Tanaka, T Bohno, A Tomioka, H Yamada, S Nose, M Konno, Y Yagi, H Maruyama, T Ogata, S Yoshida, K Ohashi, K Tsutsumi, and K Honda, "AC loss properties of a 1 MVA single phase HTS power transformer", *IEEE Transactions on Applied Superconductivity*, Vol. 11, No. 1, pp. 1482–1485, 2001.

P Jaspal, and P Sharma, "Performance comparison of carbon Nanotubes with Copper and Aluminium as winding materials", *Indian Journal of Science and Technology*, Vol. 9, 2016.

E Jawaorski, "Cast resin transformers come of age", *Electrical Review* (GB), Vol. 216, No. 9, pp. 30, 32, March 1985.

S Jeszenzsky, "History of transformers", *IEEE Power Engineering Review*, Vol. 16, No. 12, pp. 9–12, December 1996. No. refs. Also, letter by G Zingales, Vol. 17, No. 6, June 1997, p. 33.

JFE Catagolue – "Super Core", Electrical Steel sheets for high Frequency Application (JFE Steel Corporation, Tokyo, Japan – www.jfe-steel.co.jp).

H-L Jou, J-C Wu, K-D Wu, W-J Chiang, and Y-H Chen, "Analysis of Zig-zag transformer applying in the three-phase four wire distribution power system", *IEEE Transactions on Power Delivery*, Vol. 20, No. 2, pp. 1168–1173, 2005.

Y Junyou, T Renyuan, and L Yan, "Eddy current fields and overheating problems due to heavy current carrying conductors", *IEEE Transactions on Magnetics*, Vol. 30, No. 5, pp. 3064–3067, September 1994. 11 refs

K Karsai, D Kerenyi, and L Kiss, *Large Power Transformers*, (Studies in Electrical and Electronic Engineering, Vol. 25), Elsevier Company, New York, 1987.

LO Kaser, "Dry type transformers in the United States", *4th BEAMA International Electrical Insulation Conference*, Brighton, England, May 11, 1982. 12 refs.

M Kazmierski, M Kozlowski, J Lasocinski, I Pinkiewicz, and J Turowski, "Hot spot identification and over heating hazard preventing when designing a large transformer", CIGRE, 30th Session, pp. 12–12, 1984. 4 refs. (1994–90065).

D Kerenyi, and L Szabo, "Hot spots caused by the stray field of transformer windings", *International Conference on Electrical Machines*, Budapest, Hungary2, Vol. 2, pp. 707–710, 5–9 September 198. 1 ref. (1983–16487).

JB Kessler, "Substations in hot zones", *Specher News* (Switzerland), pp. 9–11, March 1985. No. refs. (1986–51694).

A Khatri, and OP Rahi, "Optional design of transformer: a compressive bibliographical survey", *International Journal of Scientific Engineering and Technology*, Vol. 1, No. 2, pp. 159–167, 2012.

DH Kim, S-Y. Hahn, and S-Y Kim, "Improved design of cover plates of power transformers for lower eddy current losses", *IEEE Transactions on Magnetics*, Vol. 15, No. 5, pp. 3529-3511, September 1999. 4 refs.

J-W Kim, BK Park, SC Jeong, SW Kim, and PG Park, "Fault diagnosis of a power transformer using an improved frequency-response analysis", *IEEE Transactions on Power Delivery*, Vol. 20, No. 1, pp. 169–178, 2005.

F Knapp, "Epoxy casting resins", *The Brown Boveri Review* (Switzerland), Vol. 52, No. 8, pp. 575–589. 42 refs. (In English), August 1965.

A Kolling, "Improved cooling with air channels", *Elektrotechnik* (Germany), Vol. 66, pp. 28–29, 10 October 1984. No. refs. (In German) (1985–9959).

DA Koppikar, SV Kulkarni, PN Srinivas, SA Khaparde, and R Jain, "Evaluation of flitch plate losses in power transformers", *IEEE Transactions on Power Delivery*, Vol. 14, No. 3, pp. 996–1001, July 1999. 7 refs.

SV Kulkarni, and SA Khaparde, *Transformer Engineering Design & Practice*, Marcel Dekker, Inc., New York, 2004.

SV Kulkarni, JC Olivares, R Escarela-Perez, VK Lakhiani, and J Turoowski, "Evaluation of eddy losses in cover plates of distribution transformers", *IEE Proceedings – Science, Measurement and Technology*, Vol. 151, No. 5, pp. 313–318, September 2004.

J Kreuzer, "Brown Boveri's new cast-resin transformer", *The Brown Boveri Review* (London), Vol. 72, No. 6, pp. 298–303, June 1985.

V Kumar, and UC Savardekar, "Super conducting transformers", *International Journal of New Innovations in Engineering and Technology* Issue 3, March 2016.

GB Kumbhar, SM Mahajan, and WL Collett, "Reduction of loss and local overheating in the tank of a current transformer", *IEEE Transactions on Power Delivery*, Vol. 25, No. 4, pp. 2519–2525. 25 refs.

T Kunimoto, M Fujita, and HM Shafey, "The effect of paint coating on heat transfer", *Heat Transfer-Japanese Research* (USA), Vol. 15, No. 6, pp. 17–31, 1986. 5 refs.

KD Lin, and BB Ellis, "Progress of core technology in transformer manufacturing". *Insulation / Circuits* (USA), Vol. 23, No. 13, pp. 25–30.31 refs, December 1977.

XM Lopez Fernandez, P Penabad Duran, J Turowski, and PM Rribeiro, "Non-linear heating hazard assessment on transformer covers and tank walls", *Przeglad Elektrotechn* (Elect. Rev.) Vol. 88, No. 7b, 2012, pp. 28–31.

XM Lopez Fernandez, P Penabad-Dunn, and J Turowwski, "Three-dimensional methodology for the overheating hazard assessment on transformer covers", *IEEE Transactions on Industry Applications*, Vol. 48, No. 5, pp. 1549–1555, September/October 2012.

LE Lundgaard, W Hansen, D Linhjell, TJ Painter, "Aging of oil –impregnated paper in power transformers", *IEEE Transactions on Power Delivery*, Vol. 19, No. 1, 2004, pp. 230–239.

B Ma, and R Koritala, *David Forrest Nanocarbon – Infused Metals: A New Class of Covetic Materials for Energy Applications Conference Paper at Materials*. Science and Technology Technical Meeting, Salt Lake City, UT, USA, Oct 23–27, 2016.

S Magdaleno-Adame, R Escarela-Perez, JC Olivares-Galvan, E Campero-Littlewood, and R Ocon-Valdez, "Temperature reduction in the clamping bolt zone of shunt reactors: Design enhancements", *IEEE Transactions on Power Delivery*, Vol. 29, No. 6, pp. 2648–2655, December 2014. 42 refs.

M Majrekar, R Kieferndorf, and G Venkataraman, Power Electronic Transformer for utility, *Applications Conference Record of IEEE Industry Application Conference* – 2000 (PP2496-2502).

ML Manning, "The electrical insulation challenge for dry type transformers", *Insulation / Circuits* (USA), Vol. 19, No. 10, pp. 87–92, September 1973. No. refs (1974–1713).

S Markalous, S Tenbohlen, and K Feser, "Detection and location of partial discharges in power transformers using acoustic and electromagnetic signals", *Transactions on Dielectrics and Electrical Insulation*, Vol. 15, No. 6, pp. 1576–1583, 2008.

D Martin, T Sha, O Krause, Y Cui, et al., "Effect of rooftop-PV on power transformer insulation and on-load tap changer operation", *Power and Energy Engineering Conference* (APPEEC), 2015 IEEE PES Asia-Pacific.

S Maximov, JC Olivares-Galvan, S Magdaleno-Adame, R Escarela-Perez, and E Campero-Littlewood, "Calculation of nonlinear electromagnetic fields in the steel wall vicinity of transformer bushings", *IEEE Transactions on Magnetics*, Vol. 51, No. 6, pp. 1–11, June 2015.

S Maximov, R Escarela-Perez, JC Olivares-Galvan, J Guzman, and E Campero-Littlewood, "New analytical formula for temperature assessment on transformer tanks", *IEEE Transactions on Power Delivery*, Vol. 31, No. 3, pp. 1122–1131, June 2016. 14 refs.

Metglass Catalogue Magnetic Alloy 2605 SAI (Iron Based) metglas@metglas.com.

R Murray, and M Hlatshwayo "Transformers within photovoltaic generation plants: Challenges and possible solutions", *CIGRE 8th Southern Africa Regional Conference* 14–17-November 2017.

KRM Nair, and R Krishnan, "Transformer for Distributed Photovoltaic (DPV) Generation", *Conference Proceedings 2018 International Conference on Electrical*, Electronics, Computer and Optimization Techniques (ICEECCOT), 14–15 December 2018, Mysore, India.

KRM Nair, and R Krishnan, "Innovative materials for the transformers of the future", *Paper presented at the International Electrical Engineering Congress* (IEECON 2020), Chiang Mai, Thailand, March 4–6, 2020-Paper P1019.

KRM Nair, and R Krishnan, "Carbon foot print calculation of transformer and the potential for reducing CO2 emissions", *Paper Number 1570575832 presented at 4th International Conference on Technology, Informatics, Management, Engineering and Environment* (TIME-E 2019), 13–15 November at Bali, Indonesia.

KRM Nair, and R Krishnan, "Climatic, environmental and fire behaviour (C2 E2 F1) test of cast resin transformer", Published in American Journal of Electrical Power and Energy Systems; doi: 10.11648/j. epes.20180701.12; ISSN: 2326–912X (Print); ISSN: 2326–9200 (Online).

KRM Nair, and R Krishnan, "Fastener spacing and tightening torque of gasket joints of oil filled transformers", Vol. 11, No. 11, pp. 1–4, 2018. doi: 10.17485/ijst/2018/v11i11/119545.

KRM Nair, and R Krishnan, "A review of the monitoring & diagnostic methods of oil immersed transformers", DOI: http://dx.doi.org/10.17577/IJERTV7IS030156; Paper ID: IJERTV7IS030156; Volume & Issue: Volume 07, Issue 03 (March 2018).

JP Nelson, "High altitude considerations for electric power systems and components", *IEEE Trans. On Industry Applications*, Vol. IA-20, No. 2, March / April 1984, pp. 407–412. 12 refs.

T Ngneguey, M Maihot, A Munar, and M Sacotte, "Zero phase sequence impedance and tank heating model for three phase three leg core type power transformers coupling magnetic field and electric circuit equations in a finite element software", *IEEE Transactions on Magnetics*, Vol. 31, No. 3, pp. 2068–2071, May 1995. 8 refs. (1995–52875).

LPD Noia, D Lauria, F Mottola, and R Rizzo, "Design optimization of distribution transformers by minimizing the total owning cost", *International Transaction on Electrical Energy Systems*, June 2017.

R Nowicz, W Winiarski and R Partyka, "Determination of thermal short circuit strength of current transformers", *6th International Symposium on Short Circuit Currents in Power Systems*, Liege, Belgium, 6–8 September 1994, P 2/16/1–5. 3 refs (1995–19690).

FR den Outer, "Rating and protection of oil –cooled and cast-resin transformers", *Power Technology International* (UK), 1992, pp. 191–192, 194–195.

VI Panchenko, DV Tsyplenkov, et al., "Machine-transformer units for wind turbines", *Research Gate*, March 2016.

DF Peelo, "Cathedral Square- BC Hydro's urban underground station", *Transmission & Distribution* (USA), Vol. 38, No. 5, pp. 48, 51, 52, 56, 57, May 1986.

P Penabad-Duran, XM Lopez-Fernandez, J Turowski, and PM Ribeiro, "3D heating assessment on transformer covers arrangement decisions", *COMPEL – The International Journal for Computation and Mathematics in Electrical and Electronic Engineering*, Vol. 31, No. 2, 2012, pp. 703–715.

R Pfeiffer, "Behavior under electrical and thermal stress of insulation materials used in dry-type transformers", *Conference Record of the 1985 International Conference on Properties and Applications of Dielectric Materials*, Sian, China, 24–29 June 1985, IEEE Cat. No. 85CH2115-4, Vol. I, pp. 160–163. 11 refs.

LW Pierce, "An investigation of the temperature distribution in cast-resin transformer windings", *IEEE Transactions on Power Delivery*, Vol. 7, No. 2, pp. 920–926, April 1992. 19 refs (1992–53791).

LW Pierce, "Specifying and loading cast-resin transformers", *IEEE Transactions on Industry Applications*, Vol. 29, No. 3, pp. 590–599, May/June 1993. 17 refs (1993–64835).

Power Transformer Department, LF Blume, and A Boyajian, *Transformer Connections, General Electric*, Schnectady, New York, 1970.

H Pudelk, "Transformers under the tropical sun", *Elektrotechnik* (Germany), Vol. 63, No. 5, pp. 16–18, 19 March 1981. No. refs. (1981–35215).

Z Radakovic and K Feser, "A new method for the calculation of the hot-spot temperature in power transformers with ONAN cooling", *IEEE Transactions on Power Delivery*, Vol. 18, No. 4, pp. 1284–1292, 2003.

H Rahbari Magham, S Esmaili, and GB Gharehption, "Electromagnetic wave application for detection of mechanical defects in transformer windings", *IEEE Sensors Journal*, Vol. 17, No. 16, August 15 2017.

M Rizzo, A Savini, and J Turowski, "Influence of flux collectors on stray losses in transformers", *IEEE Transactions on Magnetics*, Vol. 36, No. 4, pp. 1915–1918, July 2000. 6 refs.

ER Ronan, SD Sudhoff, SF Glover, DL Galloway, "A power electronic-based distribution transformer", *IEEE Transactions on Power Delivery*, Vol. 17, No. 2, 2002, pp. 537–543.

TK Saha, "Review of modern diagnostic techniques for assessing insulation condition in aged transformers", *Transactions on Dielectrics and Electrical Insulation*, Vol. 10, No. 5, pp. 903–917, 2003.

TK Saha, and P Purkait, "Investigation of polarization an depolarization current measurements for the assessment of oil-paper insulation of aged transformers", *Transactions on Dielectrics and Electrical Insulation*, Vol. 11, No. 1, pp. 144–154, 2004.

KJ Sanders, "The corrugated tank transformer enters the UK fold", *Electrical Review* (GB), Vol. 210, No. 18, 7 May 1982, pp. 29, 32, 37. No refs (1982–44205).

KJ Sanders, "Corrugated tanks for distribution transformers", *Electronics and Power, Journal of the IEE*, Vol. 29, No. 10, pp. 729–732, October 1983 (1984–7988).

DG Say. "Trends in distribution transformer development", *Electrical Review* (GB), Vol. 205, no. 4, pp. 32–34, 3 August 1979.

J Schlabbach, "Improvement of permissible loading of transformers by solar shield", *EUROCON, Computer as a Tool, IEEE Region* 8, 2003, Vol. 2, pp. 305–309.

R Schlosser, H Schmidt, M Leghissa, and M Meinert, "Development of high temperature superconducting transformers for railway applications", *IEEE Transactions on Applied Superconductivity*, Vol. 13, No. 2, pp. 2325–2330, 2003.

E Schmidt, and P Hamberger, "Design optimization of power transformers Part 2 – Eddy current analysis for tank wall and core clamping parts", *Proceedings of International Conference on Power System Technology*, Singapore, pp. 21–24, 2004.

HR Sheppard, "A century of progress in electrical insulation. 1886–1986", *IEEE Electrical Insulation Magazine*, Vol. 2, No. 5, pp. 20–30, September 1986. 79 refs.

Y Sheng, and SM Rovnyak, "Decision trees and wavelet analysis for power transformer protection", *IEEE Transactions on Power Delivery*, Vol. 17, No. 2, pp. 429–433, 2002.

H Shertukde, *Transformers: Theory, Design and Practice with Practical Applications: New and Improved Look at Transformers with Emphasis on Partial Discharge, Core .... Finite Element Analysis and 18-Slot Designs*, VDM Verlag Dr. Muller, Saarbrüken, Germany, 2010.

J Smolka, AJ Nowak, and LC Wrobel, "Numerical modelling of thermal process in an electrical transformer dipped into polymerized resin by using commercial CFD package fluent", *Computers & Fluids*, Vol. 33, 2004, pp. 859–868.

N Soltan, D Eggers, K Hameyer, and RW De Doncker, "Iron losses in a medium frequency transformers operated in a High power DC-DC converter", *IEEE Transactions on Magnetics*, Vol. 50, No. 2, pp. 963–956, 2014.

CA Stitt, and CR Bost, "Past, present and potential of iron-silicon alloys", *Insulation / Circuits* (USA), Vol. 18, No. 11, pp. 25028. No. refs, November 1972.

CG Suits, "Seventy-five years of research in general electric", *Science* (USA), Vol. 118, pp. 451–456. No. refs, September 1953.

D Susa, M Lehtonen, and H Nordman, "Dynamic thermal modelling of power transformers", *IEEE Transactions on Power Delivery*, Vol. 20, No. 1, pp. 197–204, 2005.

N Takahashi, T Kitamura, M Horii, and J Takehara, "Optimal design of tank shield model of transformer", *IEEE Trans. On Magnetics*, Vol. 36 No. 4, pp. 1089–1093, July 2000. 4 ref.

WH Tang, QH Wu, and ZJ Richardson, "A simplified transformer thermal model based on thermal electric analogy", *IEEE Transactions on Power Delivery*, Vol. 19, No. 3, pp. 1112–1119, 2004.

"Tests with a large air-cooled oil cooler for transformers", *Bulletin Oerlikon* (London), No. 114, pp. 578–580, December 1930 (1931–679).

K Tekletsadik, and M Saravolac, "Calculation of losses in structural parts of transformers by FE method", *IEE Colloquium of Field Modelling: Application to High Voltage Power Apparatus*, London, UK, 17 January 1996, IEE Digest No. 1996/008, pp. 4/1–3. 4 refs. (1996–49407).

KF Thang, RK Aggarwal, AJ McGrail, and DG Esp. "Analysis of power transformer dissolved gas data using the self-organising map", *IEEE Transactions on Power Delivery*, Vol. 18, No. 4, pp. 1241–1248, 2003.

K Tomsovic, M Tapper, and T Ingvarsson, "Fussy information approach to integrating transformer diagnostic methods", *IEEE TPD*, Vol. 8, No. 3, July 1993.

"Transformer cooler retrofit assures capacity", *Transmission & Distribution World*, Vol. 51, No. 13, p. 8, December 1999.

"Transformer coolers", *Asea Journal* (Sweden), Vol. 49, No. 1, 1976, p. 23 (In English).

ED Treanor, and EL Raab, "Transformer Oil", *Transmission AIEE*, Vol. 69, Pt. II, pp. 1060–1070, 1950, disc. P. 1070, no refs.

J Turowski, and A Pelicant, "Eddy current losses and hot-spot evaluation in cover plates of power transformers", *IEE Proceedings – Electric Power Applications*, Vol. 144, No. 6, pp. 435–440, November 1997. 4 refs.

Ultrawire workshop 2017 "Workshop on Commercilisation of Nano-carbon wire Technology, Cambridge Nanomaterials Technology", Wolfson College Cambridge, UK, 12–13th July 2017.

E Van Brunt, L Cheng, et al., 27kV, 20A 4H-Sicn-IG-BTs Materials Science Forum June 2015 (pp. 821–823).

"Varnish protects 15kv dry type transformer", *Materials News from Dow Corning* (USA), may/June 1978, p. 3.

PA Venikar, MS Ballal, BS Umre, and HM Suryavanshi, "Search coil based on-line diagnostics of transformer internal faults", DOI 10.1109/TPWRD 2016–2682083 IEEE.

MAA Wahab, MM Hamada, AG Zeitoun, and G Ismail, "A newly modified forced oil cooling system and its impact on in-service transformer oil characteristics", *IEEE Transactions on Power Delivery*, Vol. 18, No. 3, pp. 827–834, July 2003.

Y Watanabe, T Takahashi, T Higashihara, and T Hasegawa, "Development of outdoor epoxy resin molded apparatus for distribution systems", *IEEE Transactions on Power Delivery*, Vol. 5, No. 1, pp. 204–211, January 1990.

M Waters, *The Short-Circuit Strength of Power Transformers*, Macdonald & Co., London, 1966.

G Wise, "Heat transfer research in general electric, 1910–1960; examples of the product-driven innovation cycle", *Heat Transfer Engineering* (GB), Vol. 12, No. 2, pp. 19–28, 1991. 34 refs.

A Kr. Yadav, A Singh, H Malik, and A Azeem, "Cost analysis of transformer's main material weight with artificial neural network (ANN)", *International Conference on Communication Systems and Network Technologies*, 2011.

XG Yao, AJ Moses, and F Anayi, "Normal flux distribution in a three-phase transformer core under sinusoidal and PWM excitation", *IEEE Transactions on Magnetics*, Vol. 43, No. 6, pp. 2660–2662, 2007.

L Yule, "Redesigned cooler regains transformer rated MVA", *Electrical World* (USA), Vol. 203, No. 10, pp. 41–42, October 1989.

J Zhao, F Yao, H Wang, Y Mi, and Y Wang. "Research on application of genetic algorithm in optimization design of transformer", *DRPT*, pp. 955–958, July 2011.

Y Zhu, F Zhuo, and H Shi, "Power management strategy research for a photovoltaic – hybrid energy storage system", ECCE Asia Downunder (ECCE Asia), 2013 IEEE.

Index

Printed in the United States
by Baker & Taylor Publisher Services